MATERIALS SCIENCE AND TECHNOLOGIES

X-RAY SCATTERING

MATERIALS SCIENCE AND TECHNOLOGIES

Additional books in this series can be found on Nova's website
under the Series tab.

Additional E-books in this series can be found on Nova's website
under the E-books tab.

Materials Science and Technologies

X-Ray Scattering

Christopher M. Bauwens
Editor

Nova Science Publishers, Inc.
New York

Library of Congress Cataloging-in-Publication Data

X-Ray scattering / editor, Christopher M. Bauwens.
p. cm.
Includes bibliographical references and index.
ISBN 978-1-61324-326-8 (hardcover)
1. Radiography, Industrial. 2. X-rays--Scattering. I. Bauwens, Christopher M.
TA417.25X74 2011
621.36'73--dc22

2011012584

Published by Nova Science Publishers, Inc. †New York

CONTENTS

PREFACE

X-ray scattering techniques are a family of non-destructive analytical techniques which reveal information about the crystallographic structure, chemical composition and physical properties of materials and thin films. These techniques are based on observing the scattered intensity of an X-ray beam hitting a sample as a function of incident and scattered angle, polarization, and wavelength or energy. In this book, the authors present current research in the study of X-ray scattering, including real-time synchrotron X-ray scattering, applications of X-ray scattering in edible lipid systems; X-ray scattering of bacterial cell wall compounds and their neutralization and small angle X-ray scattering analysis of nanomaterials for ultra large scale integrated circuits. (Imprint: Nova)

Chapter 1 - Morphology of interfaces in multi-layer structures influences many physical effects such as the giant and tunnel magentoresistances, ferromagnetic interlayer coupling and specular X-ray reflectivity. In order to inspect the interfaces, a non-destructive and non-local diagnostic technique statistically averaging over large areas is required. The grazing-incidence small-angle X-ray scattering (GISAXS) fulfills all these attributes. The GISAXS evaluation provides all basic quantities that appear in mathematical description of buried interfaces. The first section will introduce a brief theoretical background for the description of the X-ray scattering from rough interfaces. Based on real cases, two important classes of rough interfaces – self-affine (fractal) and mounded will be discussed in detail. The second section will present common experimental setups suitable for GISAXS from solid surfaces used at synchrotron GISAXS beamlines but also some recent table-top laboratory setups. The third section will demonstrate applications of GISAXS to multi-layer interference mirrors and multi-layers exhibiting giant magnetoresistance (GMR) effects. GISAXS results from various EUV and soft X-ray multi-layer mirrors (Mo/Si, W/B4C, Mo/B4C, La/B4C) deposited by various techniques (electron beam evaporation, magnetron sputtering, ion beam sputtering, deposition with in situ ion beam polishing) will be reviewed in detail to show differences in their interface properties and thermal stability. In particular, GISAXS allows detection of embedded cluster formation that is inherent to ultrathin multi-layer films. A NiFe/Au/Co/Au GMR multi-layer will be introduced as a typical representative of vertically correlated mounded interfaces. The authors will demonstrate how the surface versus bulk sensitivity can be tuned by changing the X-ray incident angle.

Chapter 2 - Advances in synchrotron X-ray scattering techniques and X-ray detector technology in recent years are allowing more and more possibilities for in situ, real-time experiments to be conducted. Many of these experiments require the construction of a specialised cell that can survive and/or contain the in situ environment (e.g. solution composition, temperature, gas pressure, etc.). In this chapter, cell design considerations will be discussed and examples given of two in situ cells that have been constructed and used in real-time experiments: nanoparticle synthesis, and electrochemical deposition of thin films.

Chapter 3 - The phase behavior as well as the thermal and structural behavior of many lipid systems such as chocolate, butter, margarine, milk fat and its fractions have been thoroughly studied by traditional X-ray diffraction techniques. The effect of the addition of emulsifiers or additives on crystallization kinetics of fat systems has also been investigated by X-ray. More recently, the polymorphic behavior of a new fat, cupuassu fat, with properties very similar to cocoa butter, has also been described. This behavior is very relevant since it is related to functional properties of final products. Studies with small angle X-ray scattering (SAXS) allowed describing the structural dynamics of several fats such as milk fat fractions in the early stage of crystallization. This early stage of crystallization is very important since it determines the later evolution of the system. A synchrotron source allows diffraction patterns to be acquired during real-time crystallization, and therefore, further and less speculative information about mechanisms of action can be obtained. SAXS was a valuable tool to explain some unexpected behavior in milk fat isothermal crystallization and also to clarify the effect of sucrose ester on fat crystallization. Several reports have dealt with the latter subject, both in bulk and in emulsion systems, but some of these results may be considered contradictory. With the aid of SAXS, it was possible to improve our understanding of the mechanism of action of a palmitic sucrose ester on low trans fat blends since it was strongly related to the effects on fat polymorphism.

Chapter 4 - Small angle X-ray scattering is a non-destructive versatile characterization technique that is becoming increasingly important in nanomaterial analysis. Its applications include nanolithography, quantum dots, quantum wires and nanoporous materials. In this chapter, the authors focus on the determination of the properties of nanoporous dielectrics by small angle X-ray scattering. After an overview and the historical background, the basic theory of small angle X-ray scattering is discussed. This theory is then applied to the characterization of two types of solution processed nanoporous dielectric films: (i) sol-gel derived silica and (ii) porous organic polymer films used for reducing interconnect parasitic capacitance in ultra large scale integrated circuits. For each type of sample, the small angle X-ray scattering pattern is collected in a transmission geometry and a parametric model is developed to fit to the measured data. By means of model fitting, the characteristic pore size and other critical parameters can be deduced. This information is useful to the development of nanomaterials for advanced integrated circuits.

Chapter 5 - Bacterial infections are still a major concern of human health. Despite the existence of antibiotics (AB), the increasing occurrence of resistant strains and the inability of the AB to neutralize bacterial pathogenicity factors (PF), which are released from the bacterial cells, are responsible for the high death rate on intensive care units. For the major pathogenicity factors of Gram-negatives, endotoxins (lipopolysaccharides, LPS), and of Gram-positives, lipoproteins (LP), which may cause in patients the life-threatening septic syndrome, there is still no effective therapy available. A new therapeutic approach to neutralize the PF is the use of suitable binding proteins from human or animal origin or peptides derived thereof (defense structures). In this way, the binding of the PF to receptors of the human immune system such as CD14 and the Toll-like receptors (TLR2 and 4), which represent the initial events of the inflammation reaction, may be inhibited competitively. The characterization of the binding process comprises, among others, the supramolecular structures of the PF. In the absence of the defense structures, cubic aggregates have been shown to be the active principle. In the presence of the defense structures, the PF aggregates are converted into a multi-lamellar organization which inhibit the binding to the cell receptors. The effect of a variety of human or animal proteins on the aggregate structures of the PF has been characterized. To this class belong albumin, hemoglobin, lactoferrin, Nk-lysin, and Limulus anti-LPS factor. It can be shown that the influence of these proteins on the aggregate structures of the PF is protein-specific: whereas albumin or hemoglobin do not change the aggregate structures or even convert them into a different type of cubic organization, the other proteins and the peptides derived from them lead to a multi-lamellar aggregate structure of the PF. These results directly correlate with the biological data, leading to no change or even to an increase in bioactivity for the former, but to a decrease for the latter. By constructing suitable peptides with high binding affinity to the PF, it is possible to obtain defense structures which are able to completely down-regulate the bioactivity of the PF, making them suitable effective compounds to combat bacterial infections in general and sepsis in particular.

Chapter 6 - A Monte–Carlo analysis technique is presented as a general tool for decomposing experimental WAXS diffractograms of semicrystalline polymers. The technique of simulated annealing used here is an adaptation of the Metropolis–Hastings algorithm for global optimization, and is intended as an analysis technique for polymers whose ideal crystalline and amorphous structures are well known. It is capable of simultaneously extracting from a single WAXS diffractogram (a) the degree of crystallinity (b) relative content of each phase, and (c) most probable lamellar thickness of the individual crystalline phases. It is applicable to specimens containing a single type of polymer capable of crystallizing in multiple phases (such as i-PP), or blends of different polymers. It is also capable of finding multiple solutions when more than one combination of the constituents is consistent with the experimentalWAXS diffractogram. The lamellar thickness predicted by the technique contains contributions from lattice strain, and thus, the deformation history of the specimen is required for a proper interpretation of the results. Prior knowledge of the different species that make up the specimen is assumed.

Chapter 7 - In this chapter, two recent analysis method of depth profile of surface layer structure using x-ray scattering at small glancing angles of incidence are presented. In the first, the analysis using x-ray diffraction is presented, and next, the analysis using x-ray reflectivity is presented. In the first, the analysis method for the depth profile of poly-crystalline surface using x-ray diffraction at small glancing angles of incidence is discussed. The peak profiles of the diffracted x-ray intensity are investigated as function of incidence angles at small glancing angle incidence of x-rays on a poly-crystalline surface. The intensity of x-ray propagation in surface layer materials characterized by a complex refractive index that changes continuously with depth is derived, and with use of the result, an analyzing method for evaluating the structure of the poly-crystalline layers and the distribution of crystal grain size in the surface layers is studied. The dependence of the diffracted x-ray intensities on the glancing angle lead to the depth profile of the poly-crystalline layers of oxidized iron with accuracy of the order of nanometers. The dependence of the peak width of diffracted x-rays on the glancing angle lead to the depth profile of the crystal grain size of iron in the surface layer. The dependence of the diffraction angle of diffracted x-rays on the glancing angle lead to the depth profile of the strain distribution in the surface layer. The derived analyzing method can be applied to the residual stress distribution analysis of the surface layer materials of which densities change continuously in depth as multi thin films, compound plating layers. Next, recent analysis of depth profile of surface layer using x-ray reflectivity is discussed. In the previous study, the x-ray reflectivity has been calculated based on the Parratt formalism, accounting for the effect of roughness by the theory of Nevot-Croce. However, the calculated result showed a strange phenomenon in that the amplitude of the oscillation due to interference effects increases in the case of a specific roughness of the surface. The strange result had its origin in a used equation due to a serious mistake in which the Fresnel transmission coefficient in the reflectivity equation is increased at a rough interface, and the increase in the transmission coefficient completely overpowers any decrease in the value of the reflection coefficient because of a lack of consideration of diffuse scattering. The mistake in Nevot and Croce's treatment originates in the fact that the modified Fresnel coefficients were calculated based on the theory which contains the x-ray energy conservation rule at surface and interface. In this chapter, a new accurate formalism that corrects this mistake is presented. The new accurate formalism derives an accurate analysis of the x-ray reflectivity from a multilayer surface, taking into account the effect of roughness-induced diffuse scattering. The calculated reflectivity by this accurate reflectivity equation should enable the structure of buried interfaces to be analyzed more accurately.

Chapter 8 - Small- and wide-angle X-ray scattering (SAXS and WAXS) are useful techniques for characterizing structures of disordered materials. An important branch of the disordered materials is sol-gel derived nanomaterials, which have many applications such as in catalysis, sorption, sensing, drug delivery, photovoltaics, and nanofabrication. In this chapter the authors first review the principles and methodology behind the SAXS and WAXS techniques, and then they focus on applications of these techniques to the characterization of sol-gel derived nanomaterials. Structural information such as nanoscale morphology, fractal dimension, surface area, defects, and short- and intermediate-range order can be obtained. Specific examples are provided with an emphasis on the characterization of the intermediate-range and nanoscale structures of the sol-gel derived nanomaterials. Finally the authors introduce synchrotron-based SAXS/WXAS techniques and address corresponding new opportunities for investigating nanostructured materials.

In: X-Ray Scattering
Editor: Christopher M. Bauwens

ISBN: 978-1-61324-326-8

Chapter 1

GISAXS - Probe of Buried Interfaces in Multi-Layered Thin Films

P. Siffalovic, M. Jergel, and E. Majkova
Institute of Physics, Slovak Academy of Sciences,
Bratislava, Slovakia

Abstract

Morphology of interfaces in multi-layer structures influences many physical effects such as the giant and tunnel magentoresistances, ferromagnetic interlayer coupling and specular X-ray reflectivity. In order to inspect the interfaces, a non-destructive and non-local diagnostic technique statistically averaging over large areas is required. The grazing-incidence small-angle X-ray scattering (GISAXS) fulfills all these attributes. The GISAXS evaluation provides all basic quantities that appear in mathematical description of buried interfaces. The first section will introduce a brief theoretical background for the description of the X-ray scattering from rough interfaces. Based on real cases, two important classes of rough interfaces – self-affine (fractal) and mounded will be discussed in detail. The second section will present common experimental setups suitable for GISAXS from solid surfaces used at synchrotron GISAXS beamlines but also some recent table-top laboratory setups. The third section will demonstrate applications of GISAXS to multi-layer interference mirrors and multi-layers exhibiting giant magnetoresistance (GMR) effects. GISAXS results from various EUV and soft X-ray multi-layer mirrors (Mo/Si, W/B_4C, Mo/B_4C, La/B_4C) deposited by various techniques (electron beam evaporation, magnetron sputtering, ion beam sputtering, deposition with in situ ion beam polishing) will be reviewed in detail to show differences in their interface properties and thermal stability. In particular, GISAXS allows detection of embedded cluster formation that is inherent to ultrathin multi-layer films. A NiFe/Au/Co/Au GMR multi-layer will be introduced as a typical representative of vertically correlated mounded interfaces. We will demonstrate how the surface versus bulk sensitivity can be tuned by changing the X-ray incident angle.

1. Introduction

The interfaces and their morphology play a fundamental role in all branches of physical sciences. Starting in the classical mechanics, the interfaces and their roughness define the friction coefficients. The contact angle of the liquids depends on the interface morphology as well. In classical optics, the visual appearance of an object is given by the wavelength dependent light scattering from the object interface. The rougher the object interfaces on the microscale, the more off-specular light is scattered from the object. Consequently, the polishing of the interfaces leads to the reduced diffuse scattering that was an important issue in the 17th century when the first mirror telescopes were constructed. The identical principles once defined in the classical optics are valid for the cutting-edge X-ray optics used for the exploration of the universe as well. Nowadays, a typical roughness of high quality X-ray mirrors is less than 0.3 nm. The atomic force microscopy (AFM) with appropriate sharp cantilevers is routinely used to locally image the surface morphology. The problem arises when the inspection of buried, unexposed, interfaces is required. This is mainly the case when the mirrors are realized as a stack of alternating layers with high refractive index contrast also called Bragg reflectors. The transmission electron microscopy of the mirror cross-sections brings qualitative information of the interfaces. However, the risk of the sample contamination and interface modification during the sample preparation is high and the procedure is time-consuming. The scattering of the X-ray photons and neutrons proved to be a very sensitive technique able to quantitatively describe the scattering by the buried interfaces. The X-ray scattering from the solid interfaces at small incident and exit angles is the main subject of this chapter. In the first part, we will summarize the theoretical background important for the description of rough interfaces. A brief introduction into the X-ray scattering from interfaces will be given as well. The second part will summarize experimental requirements for the grazing-incidence small-angle X-ray scattering (GISAXS). In addition to the synchrotron based experiments, we will focus on the laboratory systems as well. The third, final part will exemplify typical applications of GISAXS such as those on X-ray mirrors (Mo/Si, W/B_4C, Mo/B_4C, La/B_4C) and magnetic multi-layers (Ni-Fe/Au/Co/Au/).

2. Theoretical Description of Interfaces

Let us define the average of an arbitrary function $f(\tau)$ as follows

$$\langle f(\tau)\rangle = \frac{\int f(\tau)d\tau}{\int d\tau} \tag{1}$$

Assume a stack of two alternating layers as shown in Figure 1. Each interface designated with an index j is fully described by an interface height function $h_j(\vec{x})$, which gives the interface height fluctuation with respect to the mean height value at lateral position $\vec{x}$. It is reasonable to define $\langle h_j(\vec{x})\rangle = 0$.

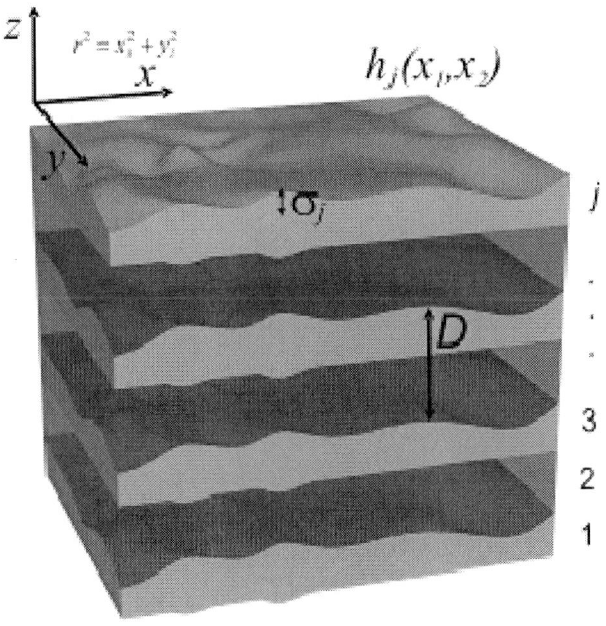

Figure 1. Sketch of a multilayer stack with rough interfaces.

The height function of the uppermost interface relates to the three-dimensional function which can be obtained experimentally by AFM technique. The height functions of buried interfaces are not accessible by a common scanning probe or imaging techniques without cutting the sample. In order to quantify various interface types, it is useful to define an autocorrelation function. The autocorrelation function $C_j(\vec{r})$ of the *j*-th interface is defined as [*1*]

$$C(\vec{r}) = \langle h(\vec{x})h(\vec{x}+\vec{r})\rangle \tag{2}$$

For isotropic interfaces, the autocorrelation function depends only on the modulus $|\vec{r}|$ and it is sufficient to define one-dimensional autocorrelation function $C(r)$ where $r = |\vec{r}|$. For most interfaces, the autocorrelation is a monotonous, rapidly decaying function that fulfills the boundary condition $\lim_{r\to\infty} C(r) = 0$. The root-mean-square (rms) roughness σ_j of the *j*-th interface can be determined from the autocorrelation function as follows:

$$\sigma_j = \sqrt{\langle h(\vec{x})\rangle} = \sqrt{C(0)} \tag{3}$$

It is reasonable to define the lateral correlation length ξ_j of the *j*-th interface as given by

$$C\left(\xi_j\right)=\frac{\sigma_j^2}{e} \tag{4}$$

The power spectral density (PSD) $P_j\left(\vec{q}\right)$ of *j*-th interface is given simply by the Fourier transform of the autocorrelation function $C_j\left(\vec{r}\right)$ [*1, 2*]

$$P_j\left(\vec{r}\right)=\iint C_j\left(\vec{r}\right).e^{i\vec{q}\vec{r}}\,d\vec{r} \tag{5}$$

For the isotropic interfaces, the equation (5) can be simplified to a one-dimensional Henkel transform

$$P_j\left(q\right)=2\pi\int C_j\left(r\right)J_0(qr)rdr \tag{6}$$

The integral of total area under the PSD function $P_j\left(\vec{q}\right)$ in *q* space (reciprocal space) is equal to the square of interface roughness, i.e. σ_j^2.

Based on the form of the autocorrelation function, we can generally define two classes of interfaces: self-affine and mounded ones. The self-affine interfaces on the scale smaller than the lateral correlation length ξ should satisfy the following scaling equation [*3*]:

$$h\left(x\right)\sim\varepsilon^{-H}h\left(\varepsilon x\right) \tag{7}$$

Here, $h\left(x\right)$ is the height profile of the interface in one dimension, *H* is the roughness exponent also called Hurst parameter and ε is the scaling parameter. The range of *H* parameter is between the 0 and 1. The smaller the *H* value, the rougher (jagged) is the local interface. Many autocorrelation functions fulfilling the scaling equation (7) have been proposed. The exponential correlation model proposed by Sinha [*4*] has the following form

$$C(r)=\sigma^2\exp\left[-\left(\frac{r}{\xi}\right)^{2H}\right] \tag{8}$$

The autocorrelation model for *H*=1 is reduced to a Gaussian function and for the *H*=0.5 to a simple exponential function. However for the values $H\rightarrow 0$ the function tends to a constant value and thus is not suitable for the description of self-affine interfaces. A more complex *K*-model autocorrelation has been proposed to overcome this problem [*5, 6*]

$$C(r)=\frac{\sigma^2 H}{2^{H-1}\Gamma(H+1)}(\frac{r}{\xi}\sqrt{2H})^H K_H(\frac{r}{\xi}\sqrt{2H}) \tag{9}$$

Here Γ is the Gamma function, K_H is the *H*-th order modified Bessel function of the second kind.

The PSD of a self-affine interface is a monotonously decreasing function without any maxima. Presence of a peak in the PSD function indicates a predominant waviness at some specific spatial frequency in the height interface function which renders the scaling equation (7) inappropriate. It can be shown that for $q \gg \xi^{-1}$ the PSD of self-affine interfaces can be approximated by the following function [*7*]

$$P(q) \propto q^{-2(H+1)} \tag{10}$$

This dependence allows for a simple estimation of the *H* parameter from the experimental data.

The mounded interfaces represent a class of interfaces with one or even more maxima in the PSD function [*1, 8*]. In that case, the autocorrelation function $C(r)$ is not monotonous but exhibits a damped oscillatory character at the length scales $r > \xi$. This feature can be incorporated by extending the *K*-model autocorrelation function (9) of self-affine interfaces by an oscillatory function as follows [*1*]

$$C(r)=\frac{\sigma^2 H}{2^{H-1}\Gamma(H+1)}(\frac{r}{\xi}\sqrt{2H})^H K_H(\frac{r}{\xi}\sqrt{2H})J_0\left(\frac{2\pi r}{\Lambda}\right) \tag{11}$$

The Λ is a typical wavelength of the mounded interface and J_0 is the zero-order Bessel function of the first kind. The PSD of the autocorrelation function (11) has a distinct peak in the reciprocal space at $q \approx 2\pi/\Lambda$. For the experimental PSD functions it allows a fast retrieval of the characteristic wavelength Λ of mounded interfaces.

In the case of multiple interfaces it is important to discuss the effect of inheritance of interface properties. For example, in the case of multilayers the upper interfaces have some relationship to the bottom ones which is defined by the growth model. The relationship or the similarity of two different interfaces is mathematically represented by a cross-correlation function defined as [*9*]

$$C_{jk}(\vec{r})=\langle h_j(\vec{x})h_k(\vec{x}+\vec{r})\rangle \tag{12}$$

Here h_j and h_k denote the height functions of *j*-th and *k*-th interfaces, respectively. Simplified models of the cross-correlation function independent of any specific growth model have been proposed. The cross-correlation function proposed by Ming et al. [*10*] has the form:

$$C_{jk}(r) = \sqrt{C_j(r)C_k(r)}\exp(-\frac{|z_j - z_k|}{L_v}) \quad (13)$$

The L_v is the vertical correlation length and $|z_j - z_k|$ is the vertical distance between the *j*-th and *k*-th interfaces. As the damping factor in equation (13) depends on the relative distance of the two interfaces only, the corresponding PSD of the cross-correlation function will be scaled only in terms of amplitude. The physical meaning is that the replication of the interface morphology between the two layers is identical for low and high spatial frequencies that is in contradiction with realistic growth models. Yet, the Ming model describes the experimental data in a limited range of spatial frequencies of PSD functions quite satisfactorily.

A more rigorous approach how to define the cross-correlation function (or PSD) is based on the thin films growth models. We introduce here the growth model based on the diffusion equation also known as Edwards-Wilkinson (EW) growth model [*11*].

$$\frac{\partial h(\vec{r},t)}{\partial t} = \nu.\nabla^2 h(\vec{r},t) + \eta(\vec{r},t) \quad (14)$$

The ν is the relaxation parameter describing the diffusion-like growth of the interface height function $h(\vec{r},t)$ and $\eta(\vec{r},t)$ is the noise function which takes into account random fluctuations in the flux of the deposited atoms. If we assume that the solution of EW equation (14) will be in the form of self-affine height function, the EW equation has to satisfy the scaling condition given by the equation (7). Consequently, the self-affine autocorrelation functions that are solutions of the EW equation must have the Hurst parameter equal to zero [*1, 3*].

The Stearns [*12*] and Spiller at al. [*13*] derived models of the propagation of the PSD functions of interfaces across the multi-layer stack based on the EW growth model. For the isotropic interfaces, they proposed that the frequency spectrum $w_j(q)$ of the height function $h_j(r)$ of the *j*-th interfaces depends on the (*j*-1)-th interface as follows:

$$w_j(q) = \gamma_j(q) + a_j(q)w_{j-1}(q) \quad (15)$$

where $\gamma_j(q)$ is the Fourier transform of the intrinsic height function of the growing *j*-th layer (produced by stochastic flux of deposited entities) and the $a_j(q)$ is the frequency dependent replication parameter. The replication parameter $a_j(q)$ is, in general, a complex number whose phase introduces the phase shifts in the interface replication process [*14*]. Therefore, the frequency spectra $w_j(q)$ rather than the PSD functions of height functions

$h_j(r)$ are used. Using the EW equation (14) and the Galileo transformation of the growth time to the film thickness we can find the replication parameter $a_j(q)$ in the form [*12, 14*]

$$a_j(q) = \exp(-\nu_j d_j q^2) \tag{16}$$

Here, the ν_j and d_j is the relaxation parameter and the thickness of the *j*-th layer, respectively. Finally, the PSD function of the *j*-th interface can be obtained as

$$P_j(q) = |w_j(q)|^2 \tag{17}$$

3. Grazing-Incidence Small-Angle X-Ray Scattering from Multiple Interfaces

A simplified scattering geometry is shown in Figure 2. The monochromatic incident X-ray beam with the wave vector $\vec{k}_i$ incident at the grazing angle α_i is elastically scattered into direction defined by the two angles α_f and $2\theta_f$ denoted by the final wave vector $\vec{k}_f$.

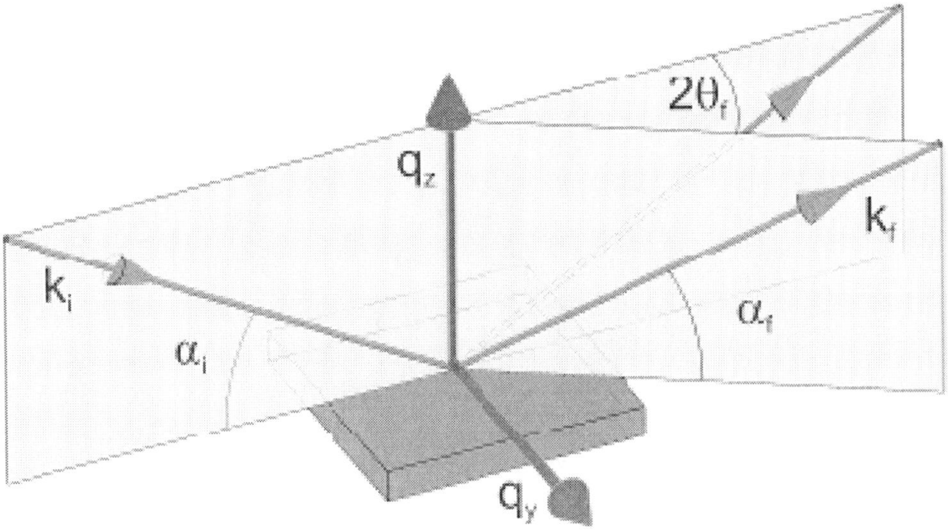

Figure 2. The simplified grazing-incidence small-angle X-ray scattering geometry.

The components of the wave vector transfer $\vec{q}$ can be written in the form [*15, 16*]

$$
\begin{aligned}
q_x &= \frac{2\pi}{\lambda}\left[\cos(2\theta_f)\cos(\alpha_f) - \cos(\alpha_i)\right] \\
q_y &= \frac{2\pi}{\lambda}\left[\sin(2\theta_f)\cos(\alpha_f)\right] \\
q_z &= \frac{2\pi}{\lambda}\left[\sin(\alpha_f) + \sin(\alpha_i)\right]
\end{aligned} \tag{18}
$$

where λ is the wavelength of the monochromatic X-ray beam. Figure 3 shows the accessible reciprocal space in the GISAXS geometry for the incident angle α_i=0.7°, λ=0.154 nm and the final scattering angles $\alpha_f \in \langle 0,5 \rangle$ and $2\theta_f \in \langle -5,5 \rangle$ degrees. Note completely different scales of the lateral wave vector transfers q_x and q_y, hence, the q_x scale axis is expanded by a factor of 10. In most experimental cases, the q_x scattering component is simply ignored as it has a negligible contribution. It is obvious from Figure 3 that the non-coplanar GISAXS geometry records identical length scales parallel and perpendicular to the inspected interface. The situation changes for the measurement in the coplanar geometry when $q_y = 0$.

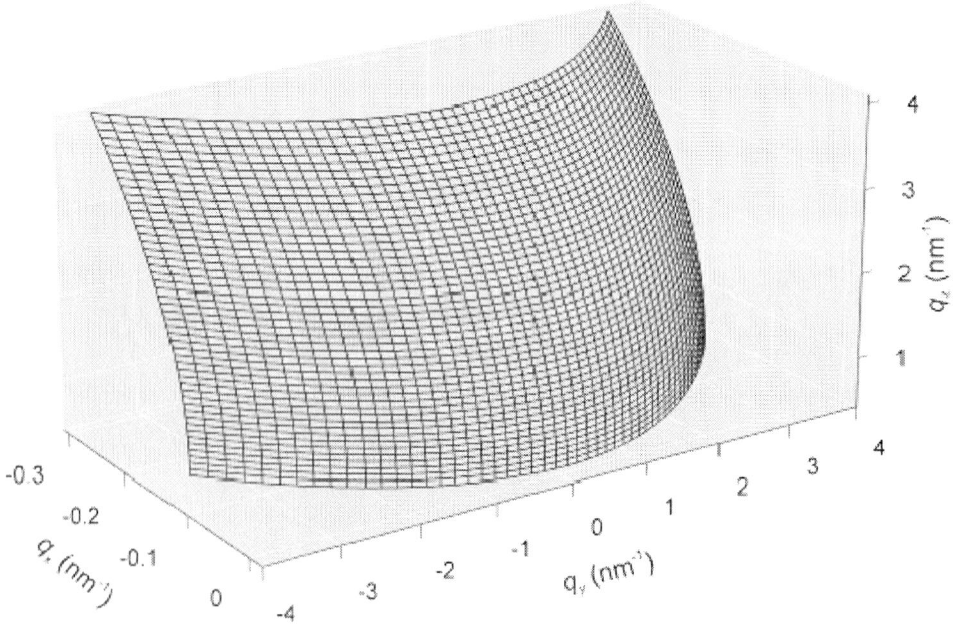

Figure 3. The accessible reciprocal space in GISAXS geometry.

The distribution of the scattered radiation intensity in reciprocal space at the point $\vec{q}$ is given by [*17-21*]

$$I(\vec{q}) = A(\vec{q}).A(\vec{q})^*$$
$$A(\vec{q}) = \iiint \rho(\vec{x}).e^{i\vec{q}\vec{x}}\,d\vec{x} \tag{19}$$

Here, the $\rho(\vec{x})$ is the electronic density of scattering volume at the position given by vector $\vec{x}$. For the case of scattering from a multi-layer structure with N isotropic interfaces, the equation (19) can be the reformulated as follows [*22, 23*]:

$$I(\vec{q}) \approx \sum_{j,k}^{N} Q_{jk} \left| T_j^i T_k^i T_j^f T_j^f \right| e^{i\left(q_{jz}z_j - q_{kz}z_k\right)} \tag{20}$$

$$Q_{jk} = \frac{1}{q_{jz}q_{kz}^*} e^{-\frac{1}{2}\left(\left(\sigma_j q_{jz}\right)^2 + \left(\sigma_k q_{kz}^*\right)^2\right)} \Delta\rho_j \Delta\rho_k \int_0^{\infty} rJ_0(q_{\|}r)\left(e^{q_{jz}q_{kz}^* C_{jk}(r)} - 1\right) dr \tag{21}$$

Here, the q_{jz} is the *z*-component of $\vec{q}$ at the *j*-th interface and $q_{\|}$ is the in-plane component of $\vec{q}$. The $\Delta\rho_j$ is the difference in the scattering density at the *j*-th interface. The $T_j^{i(f)}$ is the amplitude transmission coefficient [*24-27*] of the multi-layer stack from the top to the *j*-th interface for the incoming (scattered) wave. The equation (20) is the first term of a more general distorted wave Born approximation (DWBA) theory of scattering from multiple interfaces [*4, 9, 28-30*]. The DWBA theory takes into account multiple scattering effects occurring at small incident and exit angles. The generally four scattering channels that appear in the DWBA theory can be reduced to two channels with significant contribution by increasing the incident angle above the critical angle in the GISAXS geometry [*16*]. A further restriction to only the exit angles larger than the critical angle simplifies the interpretation and evaluation of the experimental data to the first term of the DWBA theory given by the equation (20) [*31, 32*]. A typical distribution of the scattered intensity in reciprocal space can be deduced by a careful analysis of the equation (20). Let us assume we have a multi-layer stack with spacing *D* and all the interfaces are identical i.e. $C_{jk}(r) = C(r)$. The integral in the equation (21) can be simplified by expanding the exponential function in Taylor series and taking only the linear part provided the interface roughness σ is small enough. In this case, the integral equals the Hankel transform of the interface autocorrelation function $C(r)$ that is the PSD function $P(q)$ of the interface. Consequently, the intensity distribution along $q_{\|} = \sqrt{q_x^2 + q_y^2}$ will be identical for all q_z values and reflects directly the PSD function of interfaces. The intensity along q_z will be modulated by the exponential term in equation (20) that has a maxima at the values $q_z = 2\pi/D.n$ where *n* is an integer. The interference fringes in the q_z direction are equivalent to the Kiessig fringes in conventional X-ray reflectivity [*15, 33*]. For a high number of bilayers *N* in multi-layer stack, the intensity distribution in reciprocal space is schematically shown in Figure. 4. The so-called Bragg sheets [*28, 34*] representing the radially symmetric PSD functions of interfaces will appear at the position of Bragg points given by $q_z = 2\pi/D.n$. The full width at half maxium (FWHM) of the Bragg sheets in the reciprocal space is given by

$$\delta q_z = 2\pi/(N.D) \tag{22}$$

where N is the number of bilayers in the multi-layer stack.

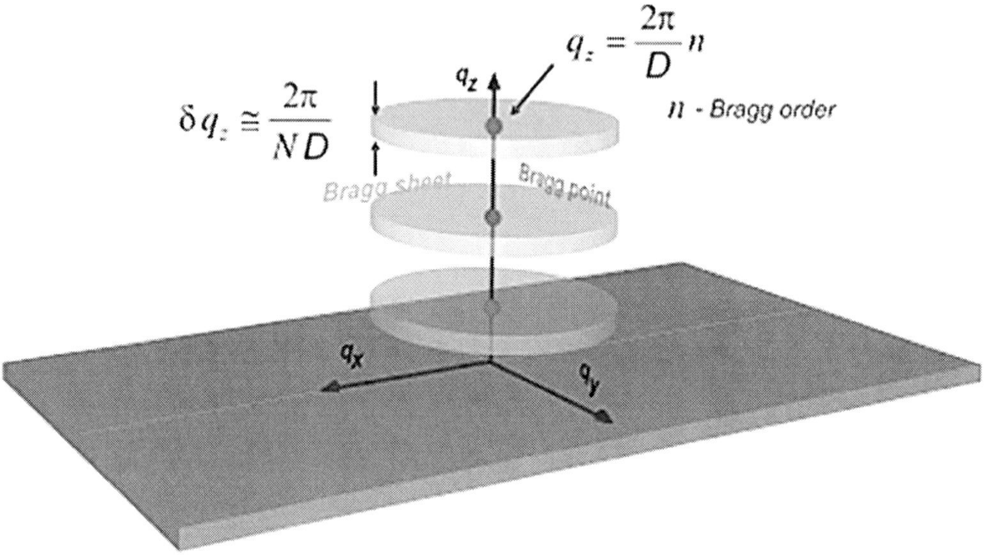

Figure 4. The graphical representation of the intensity distribution in the reciprocal space for a multi-layer stack.

For the real systems, it is reasonable to define the effective number of correlated bilayers $N_{eff}(q_{||})$ which is a function of lateral component $q_{||}$ of wave vector transfer $\vec{q}$ [*35*]

$$L_v(q_{||}) = D.N_{eff}(q_{||}) = \frac{2}{\delta q_z(q_{||})} \tag{23}$$

Here, the $L_v(q_{||})$ is the vertical correlation length and the $\delta q_z(q_{||})$ is the FWHM of the Bragg sheet at the $q_{||}$. In the framework of the Ming model [*10*], the vertical correlation length $L_v(q_{||})$ is constant and equal to L_v defined in equation (13). This fact renders the Ming model restricted primarily to the description of X-ray scattering in the coplanar geometry. The EW growth model introduced above accounts for the gradually decreasing vertical correlation length $L_v\left(q_{\|}\right)$ with the increasing spatial frequency $q_{\|}$. Starting from the equation (14) we can derive the effective number of bilayers $N_{eff}\left(q_{\|}\right)$ [*13*]

$$N_{eff}\left(q_{\|}\right)=\frac{1}{\ln(1+Dvq_{\|}^{2})} \tag{24}$$

The equation (24) allows experimental determination of the relaxation parameter □. On the other hand, frequency-dependent replication parameter defined in equations (15) and (16) aggravates a simple interpretation of the intensity decay within the Bragg sheet. In particular, the intensity within the Bragg sheet is not simply proportional to the PSD function of interfaces as the intensity decay is governed also by the relaxation parameter of the EW growth model. In order to extract the information on the PSD function one has to simulate experimental data with the relaxation parameter ν estimated according to equation (24). In special cases when the vertical correlation length is a slowly varying function with respect to the PSD function, a simple Ming model can be used to extract the parameters of the autocorrelation functions or at least to estimate their upper limits.

4. Experimental Aspects of Small-Angle Scattering from Multi-layers

Traditionally, the GISAXS experiments on solid surfaces have been performed at dedicated synchrotron beamlines [*16, 36-38*]. However, the synchrotron radiation is not a necessary prerequisite for a successful measurement and laboratory systems that can perform GISAXS measurements were realized [*39-41*]. The fundamental layout of each GISAXS beamline or laboratory system shares common features with the conventional small-angle X-ray scattering (SAXS) instrumentation [*18, 20*]. Typically, three GISAXS configurations are used: a) parallel beam configuration, b) configuration with a beam focused at the detection plane and c) configuration with a focused beam at the sample position. In the first configuration, the monochromatized X-ray beam is conditioned with a three pinhole system [*20*]. The first two pinholes define the divergence and size of the beam. The third one, also called a guarding or an anti-scatter pinhole, is reducing the scattering produced by the beam-conditioning pinholes. At the synchrotron beamlines, the pinholes are replaced by closely coupled horizontal and vertical variable size slit systems [*38*]. It has an advantage of fast adjustment of the beam divergence without breaking the vacuum of flight tubes. A strong directional emittance of the synchrotron radiation [*42*] is a big advantage over the conventional X-ray anode systems [*43*] where only a very small portion of radiation can be employed by a collimation system [*44*]. Due to this fact, the X-ray brilliance of the GISAXS synchrotron beamlines is by a few orders of magnitude higher than that of the laboratory systems. The second setup is based on imaging system focusing the X-ray beam at the detector plane. Here, the bent crystal monochromators [*20, 45*], focusing mirror optics [*46-48*] or the X-ray lenses are used [*49, 50*]. Selecting the appropriate focus size and the distance between the sample and detector one can obtain a very compact setup suitable for laboratory purposes. Table-top GISAXS systems having a maximum resolution of 200 nm and employing the focused X-ray sources were reported [*41, 51, 52*]. The third GISAXS setup relies on the X-ray beam focused at the inspected sample [*38*]. The advantage of this system is the possibility to perform spatially resolved GISAXS measurements where the focused X-ray beam can be scanned over the sample volume [*53, 54*]. Recent advances in X-ray focusing allow for a high resolution scanning at the sub-100 nm level [*55*]. In all of the above described GISAXS setups, the X-ray beam hits the sample surface under a small angle, typically 0.05°-1°. Tuning the angle of incidence around the critical angle of reflection, one can probe only the surface or near-surface regions which have only a weak contribution to the total scattering signal [*26*]. Such a suppression of the bulk scattering is especially important in multi-layered structures. The sample is typically located at a motorized goniometer supported by vertical and horizontal linear stages. The atmospheric air significantly influences the quality of GISAXS data due to relatively high scattering cross-section of air for the X-ray energies below 10 keV. Evacuated or He filled flight tubes are strongly recommended [*52*]. The best results are achieved when the sample is placed in an evacuated chamber that is directly incorporated in the synchrotron beamline. In the laboratory set-up, an evacuated tube with Be, kapton or mylar windows may be used along the X-ray beam path. Additional anti-scatter slits reduce the parasitic scattering originating from the windows. A knife-edge collimator located directly above the sample surface further reduces the parasitic scattering. Nevertheless, many valuable GISAXS measurements, especially those in-situ, have been performed at atmospheric pressure. As a rule, the distribution of the scattered X-ray radiation is recorded by a two-dimensional detector. Consequently, the GISAXS pattern is a projection of the reciprocal space map (RSM) onto the detector plane. There are four types of two-dimensional detectors used for GISAXS. The imaging plate detectors provide the best ratio between the sensitive area and price [*52, 56*]. They are suitable for large-scale RSM like simultaneous recording of GISAXS and WAXS (Wide Angle X-ray Scattering). The disadvantage is a slow readout procedure and inability of single photon counting. The second class of detectors suitable for the GISAXS are the X-ray sensitive CCDs (Charge-Coupled Devices). Here, X-ray fluorescence screen is de-magnified via an optical fiber taper to a CCD chip usually held at low temperature to avoid noise integration [*52, 57*]. The CCD detectors have a good linear response, moderate maximum counting rate, single photon counting capability, fast readout and low noise parameters. Large detection planes are tiled from multiple CCD devices. The third class of the two-dimensional detectors are gas wire ones [*52, 58*]. These detectors are based on the multiple cell architecture which detects the locally amplified noble gas discharge when ionized by incoming X-ray photons. They excel in linearity of response, high counting range, single photon counting capabilities and fast readout. They do not require cooling, have large sensitive areas and practically no background. The disadvantage is the cross-talk effect between the adjacent pixels at the high X-ray intensities resulting into broadening the detector point-spread function [*59*]. This problem is overcome in the X-ray CMOS detectors [*60, 61*]. A pixelated Si sensor is directly bonded to a CMOS chip which guarantees an individual control over each pixel in the detector array. The high linear counting range, zero background, zero cross-talk between the neighboring pixels and radiation hardness rank the X-ray CMOS detectors among the best candidates for the GISAXS detectors. A large detection area can be arranged by forming arrays from single detection modules [*62, 63*]. Apart from the dedicated small-angle scattering beamlines, the GISAXS measurements can be performed with commercial instrumentation available from the companies like Bruker AXS, Hecus XRS, Anton Paar, Rigaku and Forvis Technologies. Recently, a compact GISAXS system with low-energy consumption which is suitable for surface as well as solid/solid and liquid/air interface studies was developed at the Institute of Physics SAS (Bratislava, Slovakia) [*51*] (Figure 5).

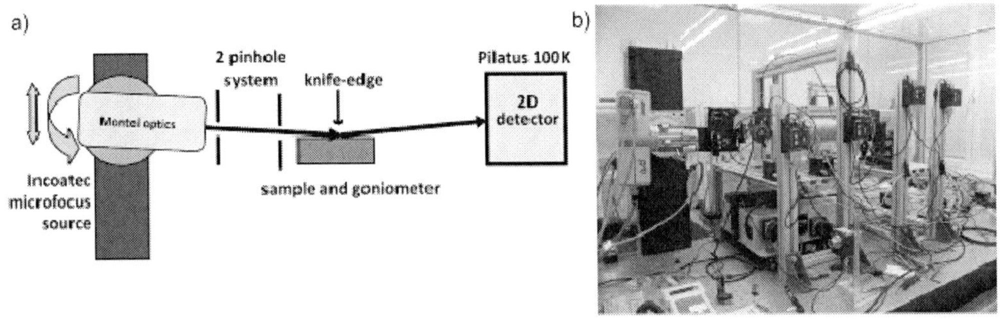

Figure 5. (a) Scheme of the table-top GISAXS system working at 1 atm. (b) The photograph of the system.

The Cu-K_α radiation emitted from a 50 μm spot at the anode is collected and refocused by a motorized multi-layer Montel optics [*64, 65*] into a 250 μm (FWHM) spot at the 560 mm distance downstream the X-ray source. The X-ray source is attached to a rotation stage supported by a vertical linear stage. The collimation system attached to the X-ray source is based on a set of two pinholes separated by 100 mm. The pinholes are precision laser drilled holes into a 0.25 mm tungsten foil. Exchanging the pinholes, we choose between the two standard operating modes – high flux and high resolution. The high flux mode has a moderate resolution of 0.1 nm^{-1} and a photon flux of 10^8 photons/s at the detection plane in the primary beam. The high resolution mode has a resolution of 0.04 nm^{-1} and 10^6 photons/s in the primary beam. In many circumstances, the high flux mode is suitable for the GISAXS studies of multi-layered samples. When analyzing the subtle changes mainly concerning the Kiessig fringes along the detector scan direction, a high resolution mode is an option. The inspected sample is placed at a goniometer coupled to motorized horizontal and vertical linear stages. The grazing incidence angle is set with a precision better than 0.02 deg. The scattered X-ray radiation is recorded by a CMOS coupled Si X-ray detector with an active area of 83.8x33.5 mm^2 [*60*]. The distance between the sample and detector is adjusted according to the required size of reciprocal space. An optional evacuated flight tube is placed between the sample and detector to reduce the air absorption and scattering. The tungsten beamstop for the primary and specularly reflected beam is located directly in front of the detector. In many applications, beamstop for the specularly reflected beam can be removed and the X-ray reflectivity of the sample can be recorded by the same experimental apparatus [*15*]. The presented table-top GISAXS system is versatile and can be converted to an in-situ monitoring/control technique for arbitrary deposition/etching process or to inline inspection device.

5. Interface Studies of Multilayered System

5.1. Interface Properties of Mo/Si Multilayers

The Molybdenum/Silicon (Mo/Si) is the most promising material combination for the EUV (Extreme UltraViolet) mirrors of the next generation lithography steppers working at 13.5 nm wavelength [*66-69*]. The high quality Mo/Si mirrors are also important for control of high brilliance pulses at third-generation synchrotrons and FEL (Free Electron Laser) sources [*70*]. Dedicated Mo/Si mirrors with a tailored phase response to the incident radiation were able to support the ultrashort EUV pulses in the attosecond temporal domain [*71-73*]. In order to achieve the maximum reflectivity of the Mo/Si mirrors, the interfaces have to exhibit a low roughness and a high density contrast. Perfectly flat interfaces with an abrupt step in the electronic density across the interfaces are impossible to fabricate. The interface roughness and material diffusion across the interface always occur. The goal in achievement of the highest possible mirror reflectivity is to tightly control the interface morphology and composition. The GISAXS measurement technique addresses the most problems concerning the quality judgement of the mirror interfaces. Numerous deposition techniques such as e-beam evaporation [*74*], ion beam assisted e-beam evaporation [*75*], single and dual ion beam sputtering [*67, 76*], DC and RF magnetron sputtering [*77*] and pulsed laser deposition [*78*] were adopted for the Mo/Si mirror fabrication. We show the impact of the above deposition techniques on the quality of buried Mo/Si interfaces. Using GISAXS technique, we determined mean values of the following parameters: multi-layer period, gamma value ($\Gamma = d_H / D$ where d_H is the thickness of absorber layer), interface roughness, lateral and vertical correlation length and Hurst parameter. We show the GISAXS reciprocal space maps obtained at the synchrotron beamline and laboratory equipment as well. In the following, we use the approximation of the stationery multi-layer growth where we assume that a single mean autocorrelation function can properly describe all of the multi-layer interfaces. This is based on the experience from multiple TEM (Transmission Electron Microscopy) analysis of the multi-layers cross-sections that all of the interfaces are very similar and within the statistical uncertainties can be described by a mean autocorrelation function. The mean roughness and thickness of the Mo and Si layers were determined by fitting the experimental X-ray reflectivity data by a modified genetic algorithm [*79*]. The GISAXS measurements of Mo/Si mirrors were performed at the BW4 Beamline at HASYLAB, Hamburg [*38*]. The monochromatic X-ray beam (65x35 μm^2, λ=0.138 nm) was focused at the sample and the scattered X-ray radiation was recorded at a distance 225 cm with an X-ray CCD detector (marCCD165). The incident angle was set to 0.7°. The laboratory GISAXS measurements were done at the commercial SAXS/GISAXS device – Nanostar (Bruker AXS) equipped by a microfocus X-ray source (IμS source, Incoatec). The instrument uses a collimated beam with a size 0.5x0.5 mm^2. The distance between the sample and detector (Vantec-2000) was 108 cm. The X-ray refelectivity curves were measured on an X-ray diffractometer (Bruker D8 Discover) equipped with Cu X-ray rotating anode. The following deposition techniques were employed for the fabrication of the studied Mo/Si multi-layers: the e-beam evaporation and the ion beam assisted e-beam evaporation were done at the University Bielefeld, Germany [*74, 80, 81*], the RF sputtering and ion beam sputtering were done at the Tohoku University, Japan [*82, 83*].

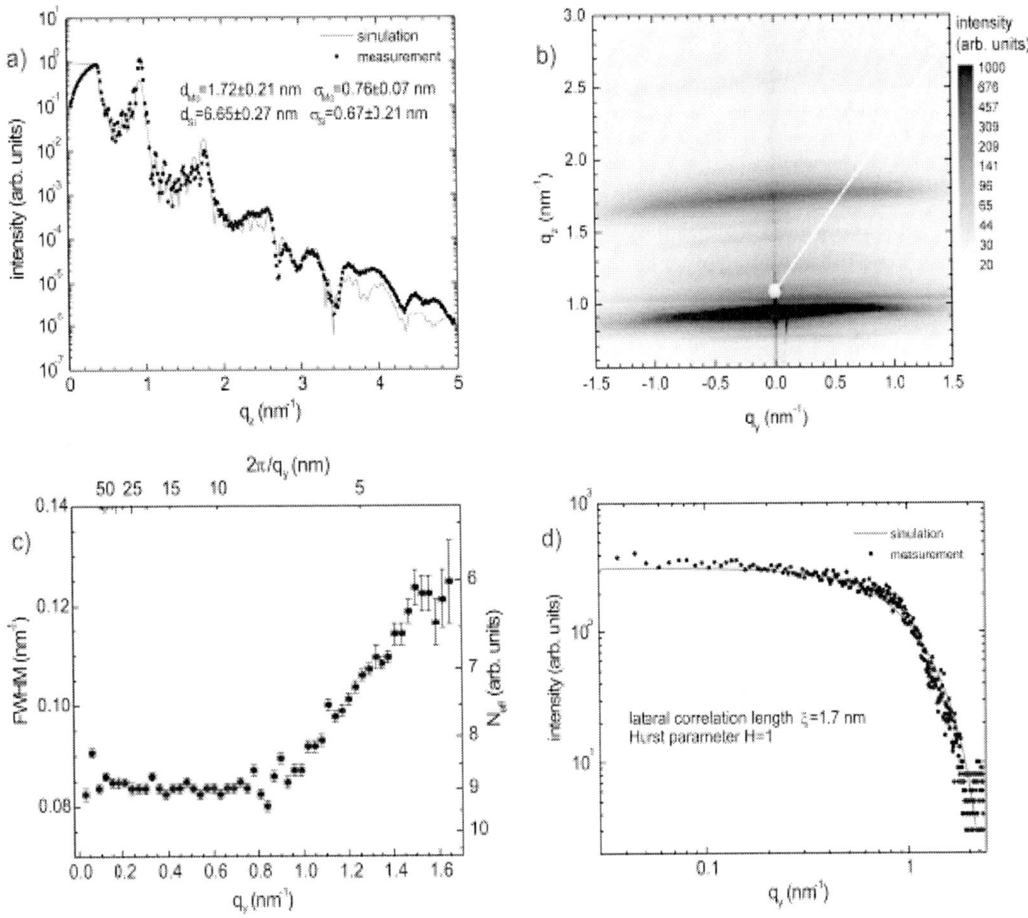

Figure 6. a) The X-ray reflectivity and b) the GISAXS reciprocal space map of the Mo/Si mirror prepared by e-beam evaporation. c) The extracted width of the 2nd Bragg sheet and d) the measured and simulated intensity along the q_y direction in the 2nd Bragg sheet.

Figure 6a shows the measured and simulated X-ray reflectivity of the Mo/Si multi-layer mirror fabricated by e-beam evaporation with a total number of 10 periods. The thermal kinetic energy of the atoms deposited on the substrate is few hundreds of meV that results in small atoms mobility at the surface. The substrate was held at room temperature. The low mobility is responsible for relatively high intrinsic layer roughness. The X-ray simulation gives the average roughness value of 0.76 nm and 0.67 nm for Mo and Si layers, respectively. The mean multi-layer period is D=8.37 nm. The GISAXS pattern measured at synchrotron is shown in Figure 6b. The three visible Bragg sheets indicate vertically correlated roughness in a wide range of spatial frequencies. Figure 6c shows the width (FWHM) of the second Bragg sheet and the effective number of correlated periods N_{eff} (q_z=1.75 nm^{-1}). It follows from Figure 6c that N_{eff} is nearly constant up to the lateral frequencies $q_y \approx 1$ nm^{-1} and equal to 9 bilayers that practically corresponds to the complete multi-layer stack. Within the q_y range $\langle 1\ \mathrm{nm}^{-1}, 1.6\ \mathrm{nm}^{-1}\rangle$ of spatial frequencies, the vertical correlation length decreases to the half of the initial value. This drop in the vertical correlation length is attributed to the loss of the vertical roughness replication across the multi-layer stack at these lateral frequencies. This behavior is in agreement with the EW growth model. Using the equation (24) that describes the number of effectively correlated bilayers, one can estimate the relaxation parameter governing the EW growth model. However, the numerical simulation of the scattered intensity that takes into account not only the PSD functions of interfaces but also the EW growth model would be rather slow and tedious. The integrated intensity of the 2nd Bragg sheet along q_y direction is shown in Figure 6d. The q_z was integrated within the interval $\Delta q_z = \pm 0.05$ nm^{-1}. The intensity along the Bragg sheet is rather constant up to approximately 1 nm^{-1} followed by a strong intensity drop by almost two orders magnitude. If we assume that the number of effectively correlated bilayers decreased by 50% in this interval (see above), the intensity drop related to the vertical replication effects would account for a fourfold drop in the observed intensity at maximum. Therefore, we can attribute the intensity decay of nearly two orders of magnitude mostly to the nature of the PSD function of interface roughness. Ignoring the vertical correlation effects at this stage in the Mo/Si multi-layer mirrors under study simplifies significantly the numerical modeling of scattered intensity. We will return to the problem of the vertical correlation phenomena when evaluating the GISAXS data of the W/B_4C multi-layer mirrors in the next paragraph. The simulation of the intensity along the Bragg sheet within the Ming correlation model gives the mean lateral correlation length ξ = 1.7 nm and mean Hurst parameter H = 1. This is one of the rare examples of the interface roughness autocorrelations described by the Gaussian function.

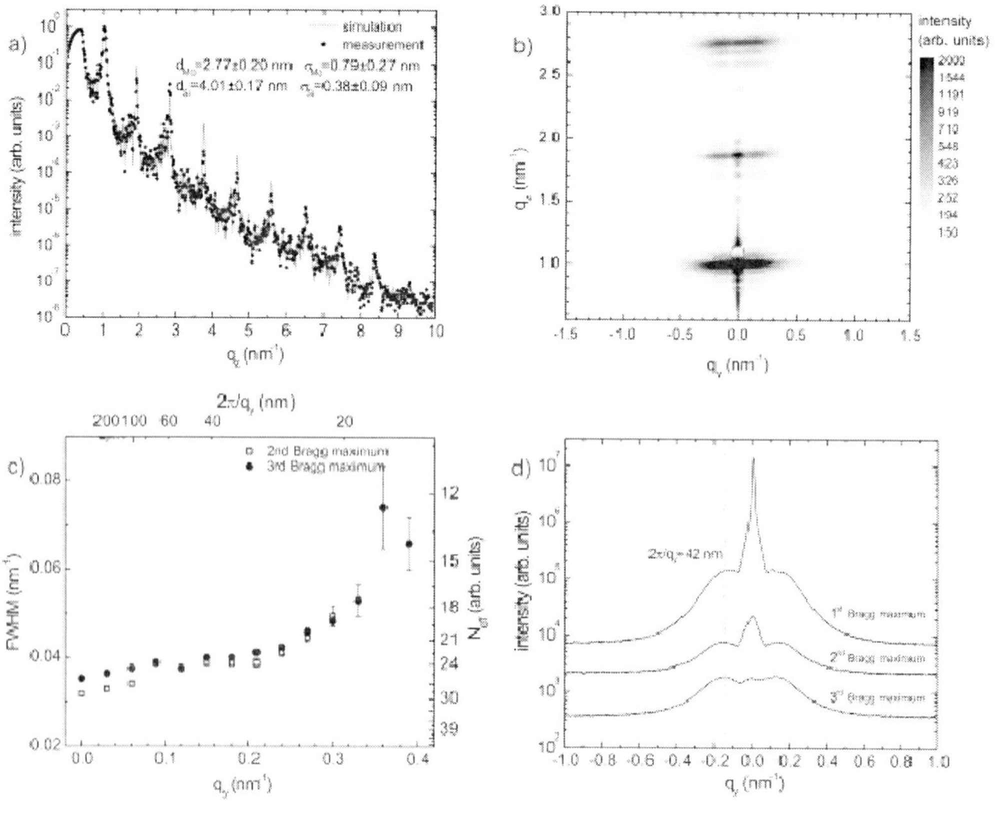

Figure7. a) The X-ray reflectivity and b) the GISAXS reciprocal space map of the Mo/Si mirror prepared by e-beam evaporation on a heated substrate. c) The extracted width of the 2nd and 3rd Bragg sheet and d) the measured intensity along the q_y direction in the 1st, 2nd and 3rd Bragg sheet.

In order to increase the adatom mobility on the substrate, the multi-layer structure was evaporated on the substrate held at the elevated temperature of T=170°C. The total number of periods was set to 30 with a period of D=6.78 nm. The simulation of the X-ray reflectivity curve (Fig. 7a) indicates a reduced roughness of 0.38 nm for the Si layer as expected. The mean roughness of the Mo layers remained at a comparable value of 0.79 nm as in the case of unheated substrate. The GISAXS reciprocal space map measured at synchtrotron is shown in Figure 7b. The three Bragg sheets with a significantly lower integral intensity are visible when compared to the GISAXS map of the unheated substrate. The extension of the Bragg sheets along the q_y direction was considerably reduced indicating a changed form of the interface morphology. The number of the effectively correlated bilayers N_{eff} is shown in Figure 7c. The N_{eff} evaluated form the 2nd and 3rd Bragg sheets shows comparable values within the measurement uncertainties. This confirms quite low vertical replication disorder as the second-type disorder [*19, 84-86*] would result in a linear broadening of the width of Bragg sheets with increasing order number. Figure 7d shows the intensity distribution in the first three Bragg sheets along the q_y direction. All three intensity profiles show a common peak located at $q_y \approx 0.15$ nm^{-1}. This is an indication of the vertically correlated mounded interfaces with a typical wavelength of 42 nm. A low vertical replication disorder allows an observation of the correlated mounded interfaces up to the third Bragg sheet. The mounded interfaces along with a higher roughness of the Mo layers suggest polycrystalline Mo layers [*77, 87-91*].

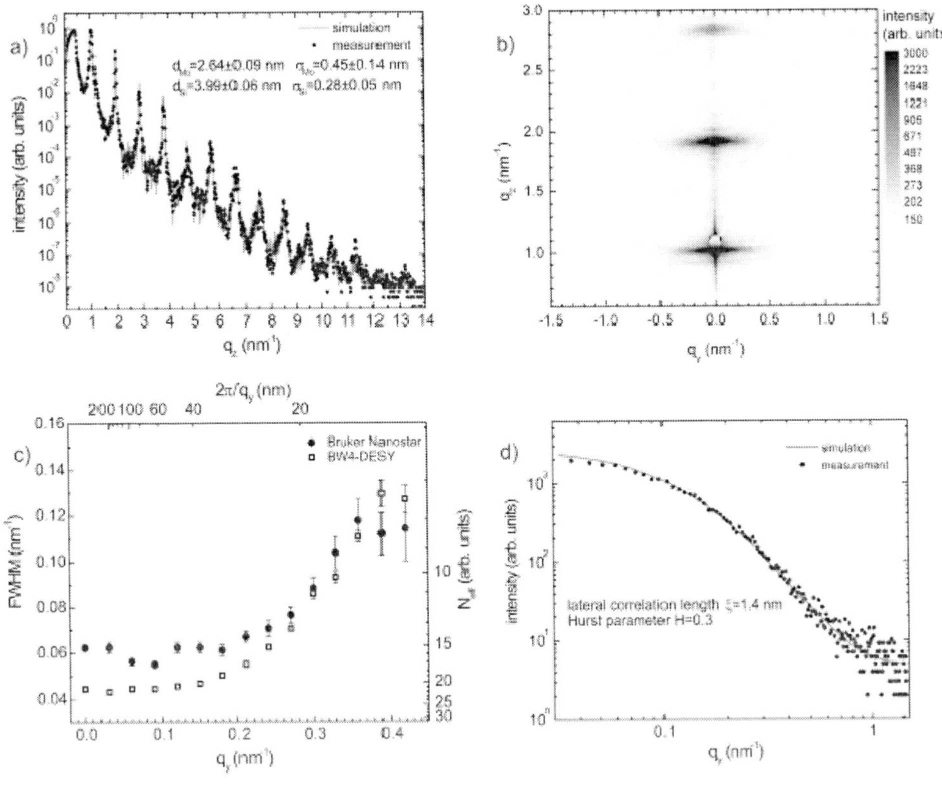

Figure 8. a) The X-ray reflectivity and b) the GISAXS reciprocal space map of the Mo/Si mirror prepared by ion beam assisted e-beam evaporation. c) The extracted width of the 2nd Bragg sheet measured at the synchrotron and laboratory GISAXS system (Nanostar). d) The measured and simulated intensity profile along the q_y direction of the 2nd Bragg sheet.

To improve the interface quality, we employed further the ion-beam assisted e-beam evaporation [75]. The number of the deposited bilayers and their period was 50 and 6.63 nm, respectively. After the deposition of each Si layer, the noble gas ion-beam was used to polish the deposited layer. The ion-beam sputtering of the Si layer leads to the thinning and simultaneous smoothening and densification of the layer. The simulation of the measured X-ray reflectivity (Figure 8a) yields the mean roughness of 0.45 nm and 0.28 nm for Mo and Si layers, respectively. The GISAXS pattern measured at the synchrotron is shown in Figure 8b. The recorded reciprocal space map shows the three Bragg sheets extending up to $q_y \approx 1$ nm^{-1} and exhibiting no signs of mounded interfaces. The widths of the 2nd Bragg sheet extracted from the synchrotron and laboratory measurements are shown in Figure 8c. The total number of the vertically correlated bilayers saturates for the low lateral frequencies at a value of N_{eff}=20. The extracted widths of the 2nd Bragg sheet measured by the laboratory GISAXS equipment reaches the instrumental resolution of some 0.05 nm^{-1}. This drawback can be further improved by switching to the high resolution mode at the expense of the lower intensity in the primary beam. The instrumental resolution in the high resolution mode is approximately 0.03 nm^{-1} that would correspond to the maximum value of N_{eff}=30 vertically correlated layers. The N_{eff} was evaluated up to the spatial frequency $q_y \approx 0.4$ nm^{-1} where the number of effectively correlated bilayers dropped roughly to half of the initial value.

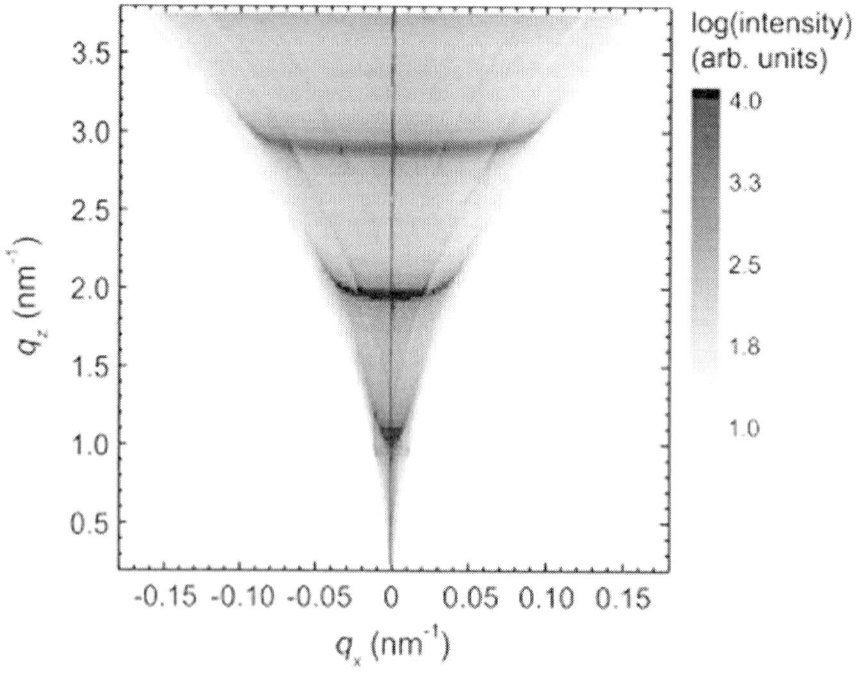

Figure 9. The reciprocal space map of the Mo/Si mirror deposited by the ion-beam assisted e-beam evaporation measured in the coplanar geometry. The reciprocal space map of the identical multi-layer mirror measured in the GISAXS(non-coplanar) geometry is shown inFigure 8b.

Figure 8d shows the scattered intensity profile across the 2nd Bragg sheet. The intensity decay is slower when compared to the intensity profile shown in Figure 6d corresponding to the Hurst parameter H=1. In accordance with the equation (10), that relates the slope of the PSD function to the scaling Hurst parameter, the interfaces fabricated by the ion-beam assisted e-beam evaporation should posses a lower Hurst parameter when compared to the interfaces deposited without the ion-beam polishing. The simulation of the scattered intensity using the Ming replication model gives mean lateral correlation length ξ=1.4 nm and mean Hurst parameter H=0.3. It is important to notice that in order to make an appropriate estimation of the Hurst parameter using equation (10) from the slope of the intensity decay in reciprocal space, the intensity profile in q_y range <0.1 nm^{-1}, 1 nm^{-1}> is substantial. To illustrate the difficulties of obtaining the Hurst parameter from the coplanar geometry, a reciprocal map of the identical mirror was measured and is shown in Figure 9. The reciprocal space map measured in coplanar geometry shows the first three Bragg sheets that are identical with the Bragg sheets shown in Figure 8b. The lateral q space is limited to the range ±0.05 nm^{-1} and ±0.1 nm^{-1} for the 2nd and 3rd Bragg sheet, respectively. This serious limitation restricts a simple evaluation of the Hurst parameter and lateral correlation length based on the equation (10) as the measured range does not cover the relevant q space for estimation. Moreover, dynamical effects [*9, 92*] occurring at the low incident and exit angles hinders the use of a simple Born approximation and the full DWBA has to be applied. The further limitation of the coplanar geometry is the inability to distinguish between the self-affine and mounded interfaces as the typical wavelength of the mounded interfaces are around 50 nm that corresponds to some 0.13 nm^{-1} in reciprocal space that are inaccessible for high intense low order Bragg sheets in coplanar geometry.

The ion beam deposition technique achieves one of the highest kinetic energies of the atoms reaching the substrate.Using this technique, we deposited a Mo/Si multi-layer mirror consisting of 40 bilayers with a period D=6.61 nm. The simulation of the X-ray reflectivity (Fig. 10a) gives similar values for the interface roughness as in the case of ion-beam assisted e-beam evaporation. The mean roughness of 0.52 nm and 0.21 nm for Mo and Si layers were obtained from the simulation, respectively. The GISAXS reciprocal space map measured at the laboratory system is shown in Figure 10b. The bright circular spot below the first Bragg sheet is the specularly reflected beam from the sample surface that was stopped by the beamstop in case of synchrotron measurements. The lateral extent of the Bragg sheets to merely 0.3 nm^{-1} indicates the PSD function with relatively larger lateral correlation length as in the previous case. The number of the effectively correlated bilayers was larger than N_{eff}>18 for all the evaluated lateral frequencies (Fig. 10c) and was limited by the instrumental resolution of the Nanostar system running in the high flux mode. The measured intensity decay along the 2nd Bragg sheet together with the simulated profile are shown in Fig. 10d.

The simulation gives the mean lateral correlation length ξ=7 nm and mean Hurst parameter H=0.5. The interface evaluation shows that the ion beam sputtering belongs to the promising deposition techniques for the fabrication of multi-layers with a low intrinsic interface roughness. Using the dual ion beam deposition technique, the second ion beam can provide extra smothering and densification of the deposited layers [*67*].

Another widely used deposition technique for the Mo/Si multi-layer fabrication is the RF sputtering. The kinetic energy of atoms reaching the substrate is lower than in case of the ion beam sputtering but is still significantly higher than the thermal kinetic energy of e-beam evaporation. The Mo/Si multi-layer mirror prepared by RF sputtering had 30 bilayers with a multi-layer period D=10.7 nm. The measured X-ray reflectivity along with the simulated profile is shown in Figure 11a. The fitted mean layer roughness of 0.51 nm and 0.22 nm for Mo and Si layers, respectively, are similar to the values of multi-layer fabricated by ion beam sputtering. The reciprocal space map measured at Nanostar system in shown in Figure 11b. The first five Bragg sheets are visible whereas the first Bragg sheet is overlapped with the intense specularly reflected beam. The lateral extension of the Bragg sheet is similar to the reciprocal space map shown in the Figure 10b that indicates a similar morphology of interfaces.

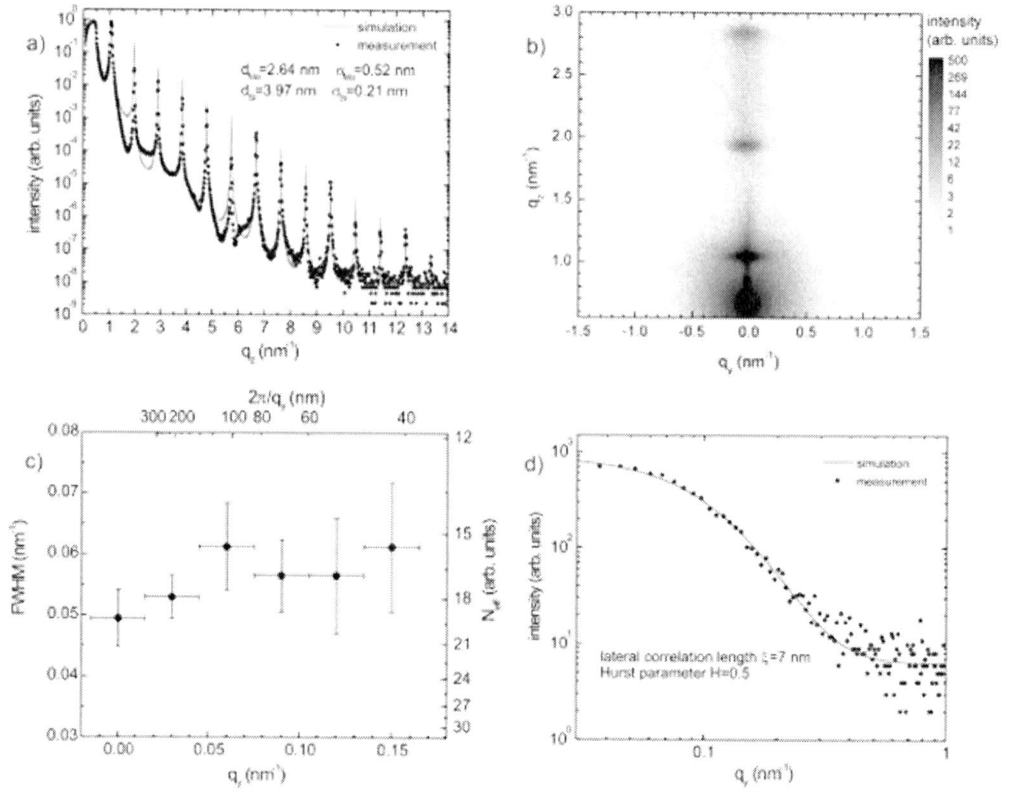

Figure 10. a) The X-ray reflectivity and b) the GISAXS reciprocal space map of the Mo/Si mirror prepared by ion beam sputtering. c) The extracted width of the 2nd Bragg sheet measured at laboratory GISAXS system (Nanostar). d) The measured and simulated intensity profile along the q_y direction of the 2nd Bragg sheet.

The number of vertically correlated bilayers $N_{eff}>12$ is limited by the instrumental resolution of the laboratory setup up to the lateral frequency $q_y\approx0.15$ nm^{-1} and then drops to the value of approximately $N_{eff}=6$ at the $q_y\approx0.25$ nm^{-1} as shown in Figure 11c. The integrated intensity along the 3rd Bragg sheet is shown in Figure 11d. In order to show the unambiguity in determining of the Hurst parameters, three simulations with the fixed Hurst parameter at the values $H=1$, $H=0.5$ and $H=0.3$ were performed and are shown in Figure 11d. It is common for the simulations of diffuse scattering in the coplanar geometry that the lateral correlation length and the Hurst parameters are coupled with a significant uncertainty due to a limited lateral q space available. As can be seen from Figure 11d, only one simulation can properly fit the experimental data in the non-coplanar (GISAXS) geometry. The simulation provides the following values, $\xi=1.9$ nm and $H=0.3$ for the mean lateral correlation length and mean Hurst parameter, respectively.

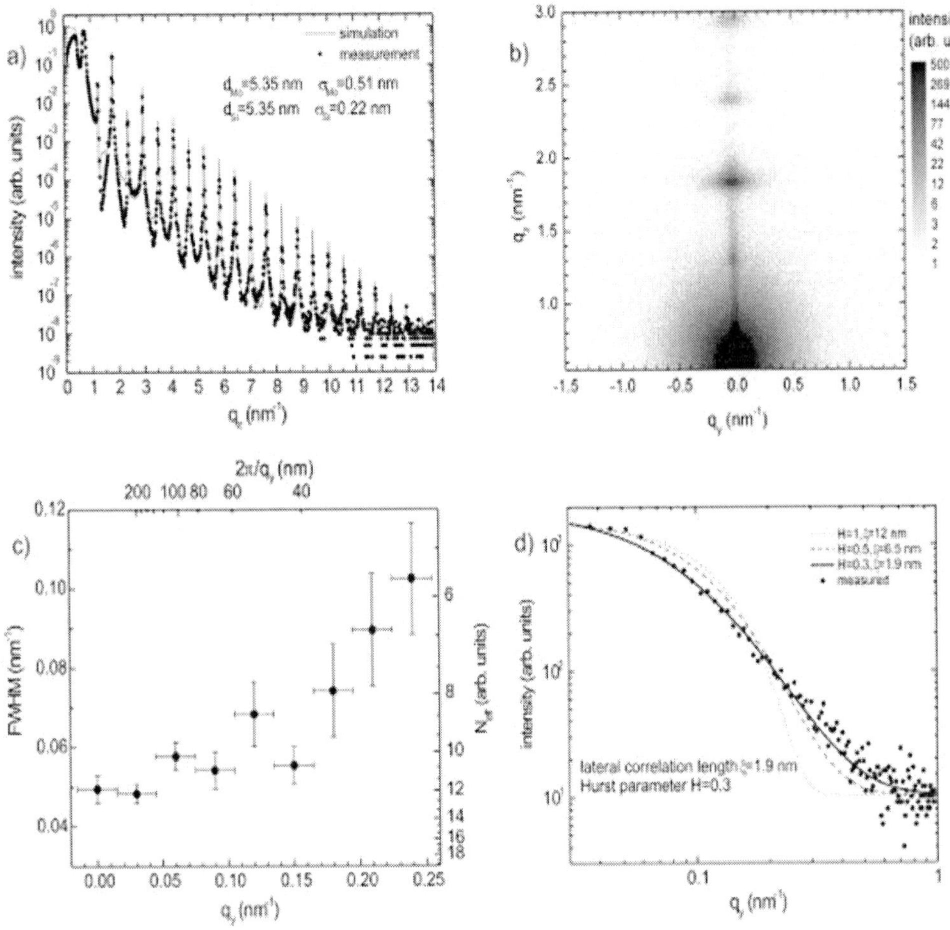

Figure 11. a) The X-ray reflectivity and b) the GISAXS reciprocal space map of the Mo/Si mirror prepared by RF sputtering. c) The extracted width of the 3rd Bragg sheet measured at laboratory GISAXS system (Nanostar). d) The measured and simulated intensity profile along the q_y direction of the 3rd Bragg sheet for different H values.

In many cases, qualitative information on the interface morphology is required, especially when two or more deposition techniques have to be compared. Figure 12 shows the cuts through the Bragg sheets extracted from the reciprocal space maps of the Mo/Si multi-layer mirrors deposited by various techniques. Based on the integral intensity under the curves, one can easily figure out the best deposition technique that produces the lowest scattering in the reciprocal space. The deposition by the ion beam and RF sputtering produces the flattest interfaces among all of the deposition techniques inspected.

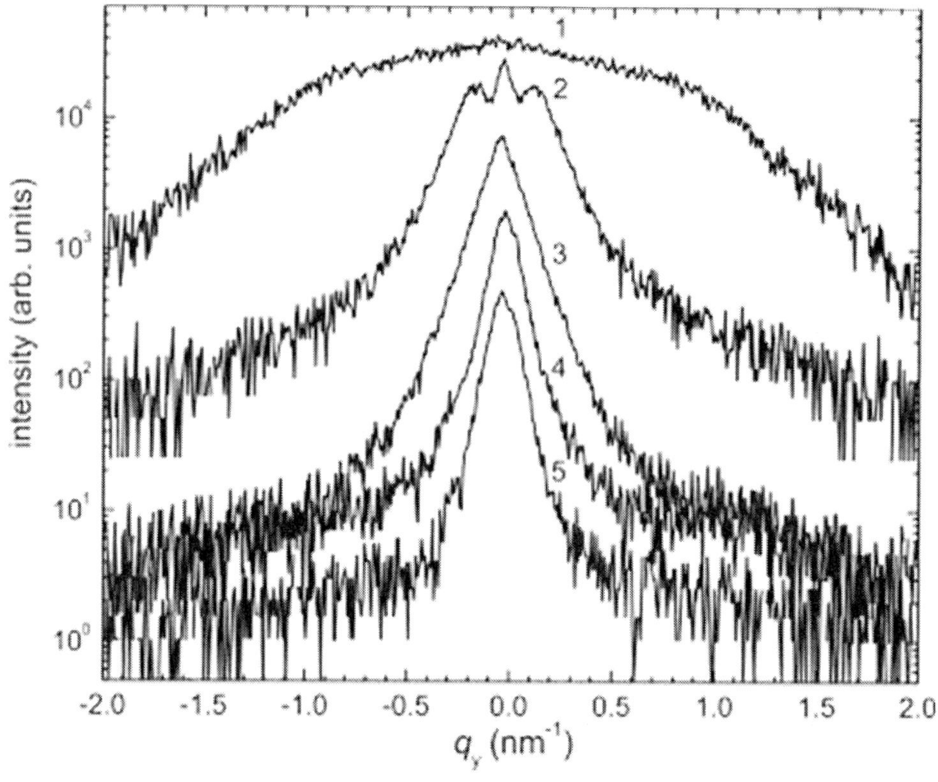

Figure 12. The intensity profiles of the Bragg sheets extracted from the reciprocal space maps of the Mo/Si mirrors deposited by different techniques: 1) e-beam evaporation, 2) e-beam evaporation on heated substrate, 3) ion-beam assisted e-beam evaporation, 4) ion beam sputtering and 5) RF sputtering.

5.2. Interface Properties and Thermal Stability W/B4C Multi-layers

The multi-layer mirrors based on the W/B_4C material combination are suitable for reflecting the soft as well as hard X-rays [*93-95*]. In recent years, we are witnessing a gradual progress in the laboratory X-ray instrumentation where especially multi-layer mirrors play a key role [*64, 96*]. Highly efficient Goebel mirrors [*97*] for collection, monochromatization and collimation of the emitted X-ray radiation from laboratory X-ray sources with line focus are now standard.

The Montel optics [*46*] based on the two orthogonally coupled multi-layer mirrors are used to collimate or focus the X-ray radiation in two dimensions. All of the recent advances in the multi-layer optics rely on the deposition of high quality multi-layer systems with sharp and flat interfaces. The minimization of the aberration errors in the multi-layer optics calls for the systems with a larger deflection angle and consequently, with the periods close to 1 nm. Such a delicate multi-layer system with the layer thicknesses close to the roughness of the layers themselves requires a non-destructive and rapid diagnostic technique of buried interfaces. In this paragraph, we show a complete characterization of a W/B_4C multi-layer mirror based on the X-ray reflectivity and GISAXS reciprocal space mapping [*98*].

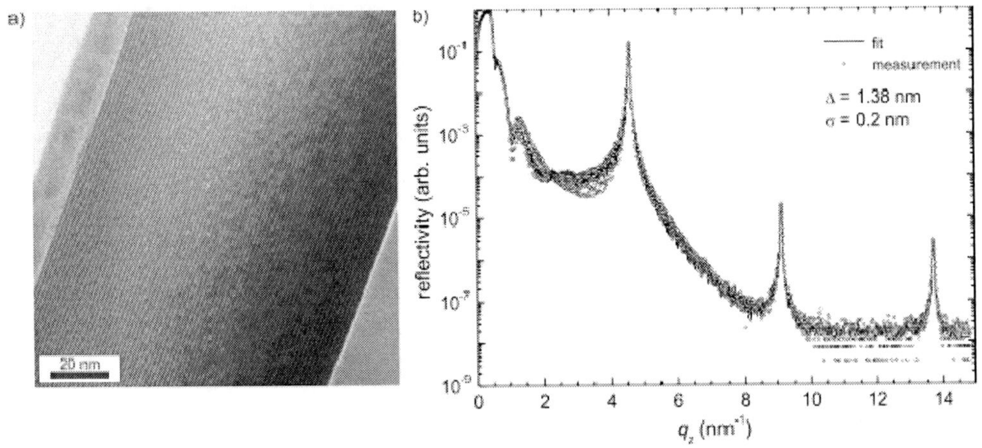

Figure 13. a) TEM micrograph of W/B_4C mirror. b) The measured and simulated X-ray reflectivity.

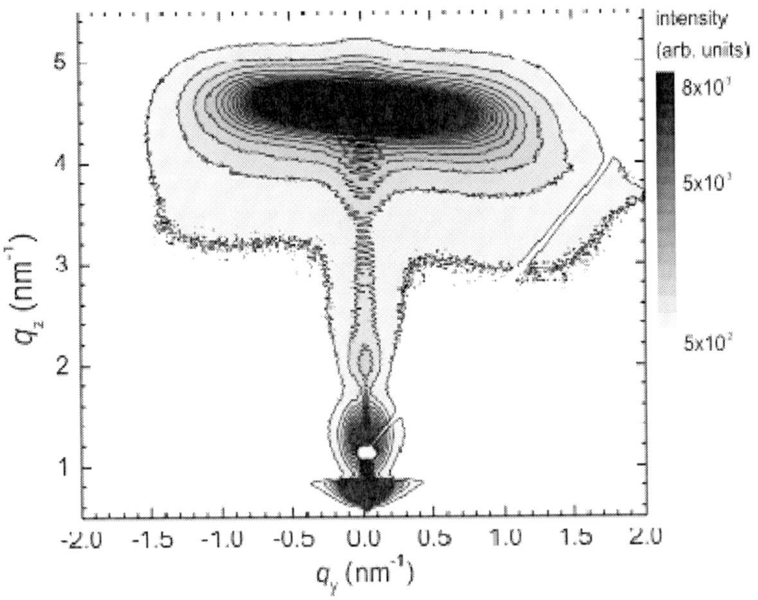

Figure 14. The GISAXS reciprocal space map of the W/B_4C mirror.

The W/B_4C multi-layer mirror was prepared by magnetron sputtering at the Incoatec GmbH. The number of bilayers was set to 80 and the nominal period was 1.38 nm. Figure 13a shows a TEM micrograph of the cross-section prepared by ion-beam milling technique. The assessment of the particular mean autocorrelation function of interfaces based on the TEM analysis is very complex due to the local character of the TEM inspection technique. However, the high reproducibility and precision of the employed magnetron sputtering technique permits a simple description of all of the interfaces with a mean universal autocorrelation function. This can be further verified by a deposition of mirror with a higher number of bilayers with no significant profile change of the Bragg sheet along the q_y direction in reciprocal space map. Figure 13b shows the measured and simulated X-ray reflectivity. The simulation confirms a comparable low interface roughness of 0.2 nm for both layers. The modulation of the X-ray reflectivity curve at the low q_z values is due to partially oxidized layers near the top of the multi-layer stack that was taken into account in the X-ray simulation. The GISAXS reciprocal space map was measured at the BW4 beamline [*38*] in HASYLAB described in more detail in the previous paragraph.

The GISAXS reciprocal space map shows the first Bragg sheet located at the identical q_z value of the first Bragg peak measured by X-ray reflectivity (Fig. 13b). The Kiessig fringes corresponding to the total multi-layer thickness of some 110 nm are clearly visible as the instrumental resolution derived from the specularly reflected beam is approximately 200 nm. The fingerprint of the oxidized top layers is also visible as a slow modulation of the scattered intensity for q_z values below 2 nm^{-1}. A slight inclination of the Bragg sheet with respect to the q_z normal direction is due to a lateral phase shift in the vertical roughness replication process and will be discussed in more detail later in this paragraph. The extracted widths of the first Bragg sheet along with the effective number of vertically correlated bilayers N_{eff} as a function of lateral roughness frequency are shown in Figure 15a and Figure 15b, respectively. Generally, the vertical correlation length $L_v\left(q_y\right)$ that is inversely proportional to the width of the Bragg sheet $\delta q_z\left(q_y\right)$ as defined by equation (23) is found to scale with the lateral frequency q_y as follows [*3, 99-101*]

$$L_v\left(q_y\right) \propto q_y^{-n} \tag{25}$$

where n depends on the relaxation process. The value n=2 implies a relaxation process described by the EW growth model and defined by the equation (14). The value n=4 implies a different relaxation process described by Mullins equation [*102*].

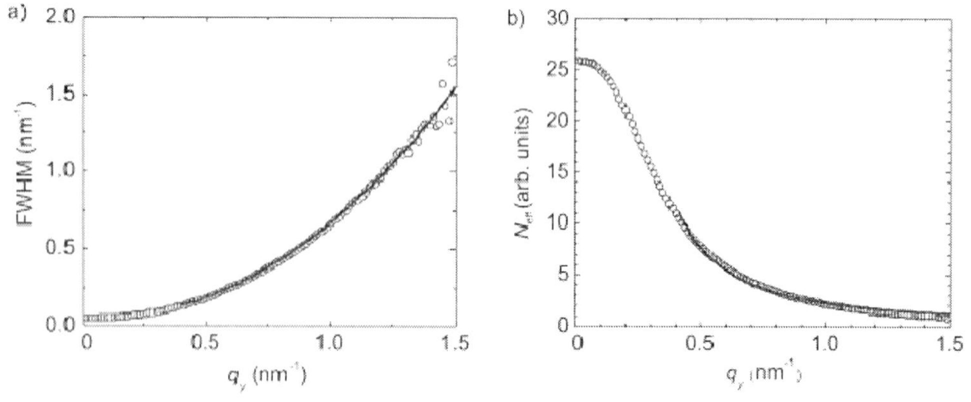

Figure 15. a) The width and b) the effective number of vertically correlated bilayers derived from the first Bragg sheet as a function of lateral roughness frequency.

The best fit of the Bragg sheet width shown in Figure 15a using the equation (25) gives the value n=2.19±0.07, which implies the EW growth model. On the basis of the EW growth model, we find the relaxation parameter ν=0.41±0.1 by fitting the effective number of vertically correlated bilayers N_{eff} (Fig. 15b) by equation (24). The saturated low-q_y region, where the finite number of vertically correlated bilayers and absorption effects play a role, was excluded from the fit. The intensity profile along the q_y value extracted from the first Bragg sheet is shown in Figure 16. The best simulation using the Ming replication model with the vertical correlation length L_v=36 nm, lateral correlation length ξ=2.3 nm and Hurst parameter H=0.1 is shown in Figure 16. Despite a very good fit to the experimental data, the simulation using the Ming model is inappropriate as we ignored the fact of the frequency dependent vertical correlation length. Another indicator of the inappropriateness of the Ming model is the value of the Hurst parameter. In particular, the EW growth equation for self-affine interfaces leads to the Hurst parameter H=0 [*1*]. A complex simulation using frequency dependent replication has to be taken into account as defined by equations (15) and (16). Figure 16 shows two simulations within the framework of the Born approximation using already obtained relaxation parameter ν and the number of bilayers N=10 and 20. It is noteworthy that the shape of the q_y cut resembles the PSD shape typical of a logarithmic autocorrelation function. Nevertheless, the high frequency tail could not be reproduced well, even for a very small Hurst parameter H=0.01. Comparison of the simulations also shows that a still larger discrepancy can be expected for a calculation with the total number of bilayers N=80 unless the H value is further reduced. Considering the 2(H+1) PSD slope for a logarithmic autocorrelation function as given by the equation (10), the calculation with H=0 is the only possibility to raise the high-frequency tail and level off the calculated intensity on the experimental curve. However, the numerical integration in equation (21) for H=0 fails because of the very steep decay of the K-correlation function at small r values for $H\rightarrow 0$ when it becomes a logarithmic function. Nevertheless, a tendency for $H\rightarrow 0$ to yield a best simulation of our GISAXS data is evident, as predicted for the EW model.

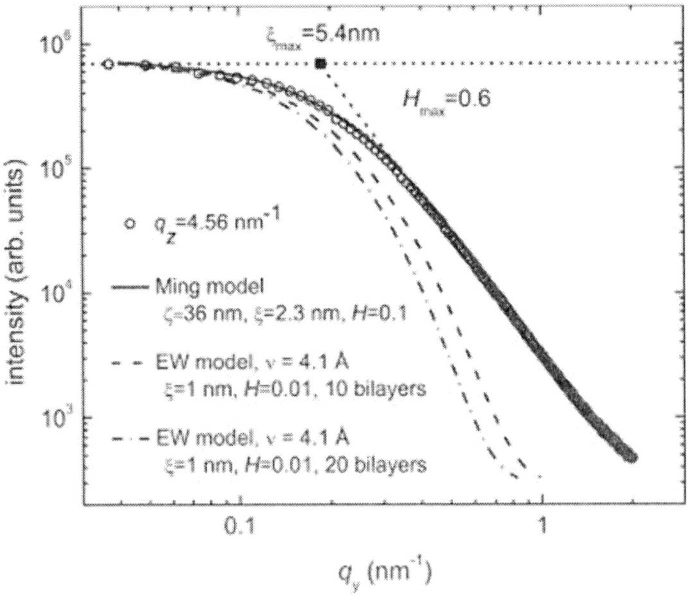

Figure 16. The extracted cut of the first Bragg sheet along the q_y direction. The graph shows also three simulations using the Ming and Edward-Willkinson model described in text.

A direct approach on how to assess the lateral correlation length ξ and the Hurst parameter H is to use the equation (10). The values ξ_{max}=5.4 nm and H_{max}=0.6 can then be derived from the crossover point indicated in Figure 6, and the decay slope identified with $2(H+1)$. However, it should be kept in mind that the values estimated in this way do not take into account the replication effects within the multi-layer stack and constitute the upper limits of the real values only. A large difference between H_{max} and the actual $H\rightarrow 0$ value confirms the strong effect of interface replication at high roughness frequencies.

The GISAXS reciprocal space map depicted in Figure 14 shows the first Bragg sheet with a certain tilt with respect to the q_y direction that is collinear with the normal to the sample surface. The interface replication deviating from the growth direction along the substrate normal was identified as the origin of the Bragg sheets tilt [*14, 103*]. An elegant inclusion of the tilted interface replication into the existing ansatz is done by introducing a complex replication factor defined by equation (16). The imaginary part of the replication factor describes in general frequency dependent phase shift between the neighboring interfaces. To verify the origin of our tilt, we rotated the sample (i.e. the Bragg sheet of the interface diffuse scattering) around the surface normal and found a systematic azimuthal dependence of the sheet tilt with a maximum value of ~4.5° (Figure 17). We can attribute the tilt to a slightly oblique deposition flux with respect to the substrate during sample preparation.

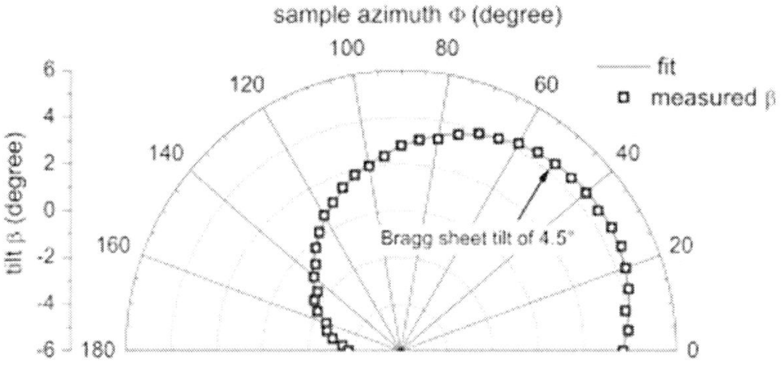

Figure 17. The azimuthal dependence of the tilt of the first Bragg sheet measured from the surface. Squares represent measured points, and the line is an elliptical fit and is a guide to the eye.

The thermal stability of a multi-layer mirror is a significant issue for practical applications. The heat load at synchrotron beamlines or intense plasma sources may reach several hundreds of watts per mm^2, and may lead to irreversible interface degradation and loss of initial optical reflectivity. Multi-layer mirrors with steep concentration gradients are thermodynamically unstable systems prone to such effects. Therefore, the behavior of the W/B_4C mirror under rapid thermal annealing is of vital importance. The deposited mirror was exposed to a rapid thermal annealing (heating rate 100 $K.s^{-1}$) in the vacuum furnance with a background pressure of 10^{-7} mbar. A series of 120 s isothermal annealings up to 1000°C simulated different thermal loads. Figure 18a shows the change in the multi-layer period extracted from the position of the first Bragg sheet. The mean multi-layer period of 1.38 nm does not exhibit any significant changes and remains stable within the $\pm$1% range up to 1000°C. For each annealing temperate, a complete X-ray reflectivity curve was recorded. The normalized X-ray reflectivity at the first Bragg peak as a function of annealing temperature is shown in Figure 18b. For the annealing temperature above 700°C, we observe a systematic decrease in the X-ray reflectivity. We attribute this drop in the X-ray reflectivity to the material interdiffusion across the interfaces. The grazing-incidence X-ray diffraction (GIXRD) was employed to track the changes in the crystalline structure of the annealed multi-layers. Initially, an amorphous multi-layer structure for the annealing temperatures above 800°C develops a broad diffraction peaks. Comparison of the diffraction pattern with X-ray standards (card Nos. 00-019-1373, 00-038-1365 and 00-043-1386; Joint Committee on Powder Diffraction Standards, 2006) suggests the presence of a poorly crystallized boron-rich W–B phase with hexagonal symmetry. Obviously, the B_4C layers start to decompose close to 1000°C and boron diffuses into tungsten. This process is presumably initiated by slight deviations from the ideal stoichiometry in very thin B_4C layers, which are known to lower the thermal stability considerably compared with the stoichiometric bulk.

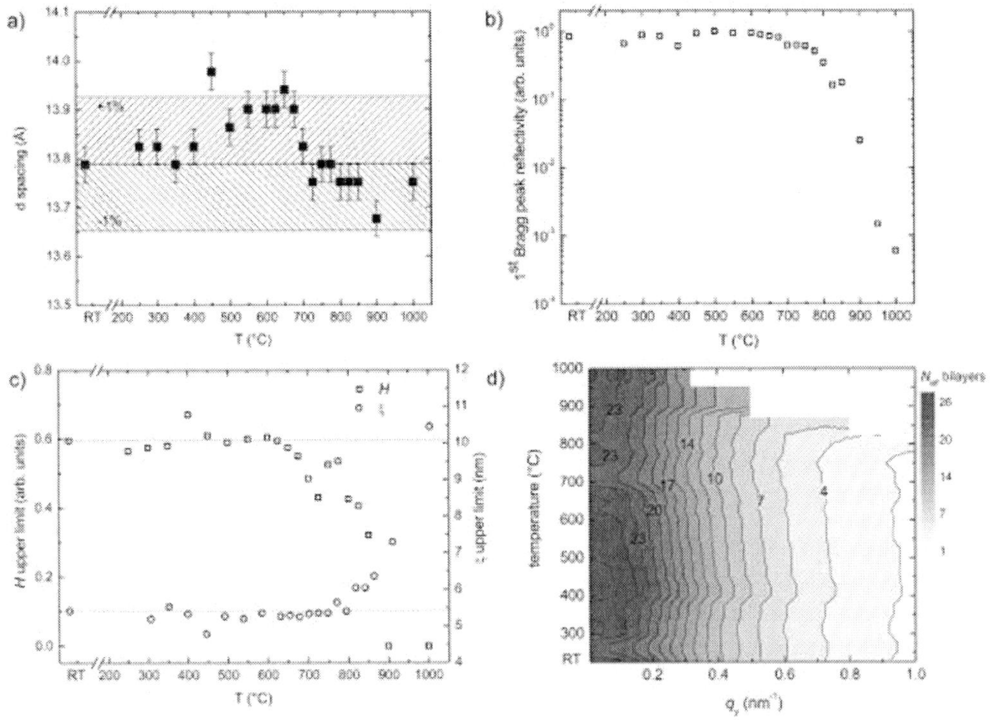

Figure 18. The graphs show the following measured parameters as a function of annealing temperature: a) the multi-layer period, b) X-ray reflectivity in the first Bragg peak, c) upper limits of lateral correlation length and Hurst parameter, d) the number of effectively correlated bilayers.

Figure 18c shows the evolution of the upper limits for the lateral correlation length and the Hurst paramaters with the growing annealing temperature. The noticeable changes in the interface parameters are visible for the temperatures above 700°C which are analogous to the observed changes in the X-ray reflectivity at the first Bragg peak. The value of the Hurst parameter is decreasing whereas the lateral correlation length is increasing with the growing annealing temperature. The number of effectively vertically correlated bilayers N_{eff} as a function of annealing temperature is shown in Figure 18d. The N_{eff} is mostly independent of annealing temperature at all evaluated lateral roughness frequencies. This suggests that material interdiffusion across the interfaces is rather uniform at the corresponding lateral length scales.

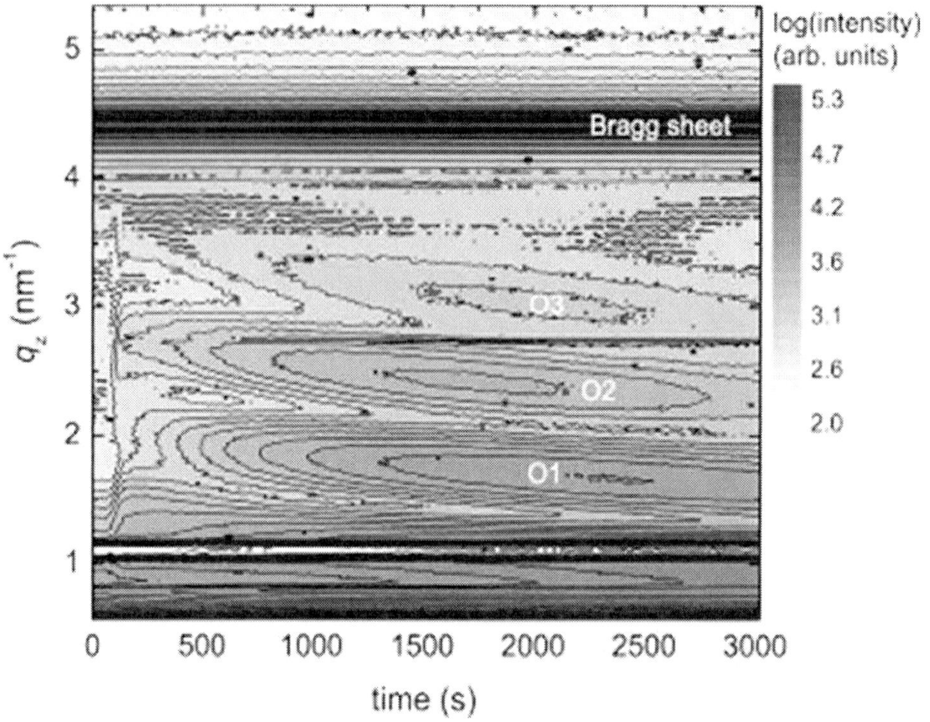

Figure 19. Temporal evolution of the GISAXS line cut at the constant q_y=0 nm^{-1} after expose in UV-ozone reactor.

Another important issue for the practical applications is the mirror stability in adverse environments, such as an oxidizing atmosphere. It is known that the intense synchrotron radiation produces ozone, which attacks optical elements and fine slits along a beamline or at laboratory equipment. In order to simulate the effect of ozone on the stability of deposited multi-layer, the GISAXS reciprocal space maps were measured in-situ in a custom-designed UV-ozone reactor equipped with an ozone-generating low-pressure mercury lamp (hν=4.9 and 6.7 eV). The total UV power at the sample surface was 2 mW.cm^{-2}. The ozone concentration was 150 mg.m^{-3} in the interaction volume. The temporal evolution of the q_z cut at q_y=0 nm^{-1} of the GISAXS reciprocal space maps after exposure in the ozone reactor for up to 3000 s is shown in Figure 19. This cut represents what is called a "detector scan" in coplanar measurements. Two new stripes between the critical angle and the first Bragg peak denoted by O1 and O2 emerge after ~250 s, followed after some delay by a third indicated as O3. The intensity of these new features increases with time, which reflects the increasing oxygen content, with a corresponding change in electron density in the near-surface oxidized region. The stripe positions shift to lower q_z values as the oxidized region spreads into the bulk. The temporal evolution of the thickness of the oxidized region was estimated independently from the O1–O2 and O2–O3 pairs of stripes, with nearly equal results (Fig. 20a).

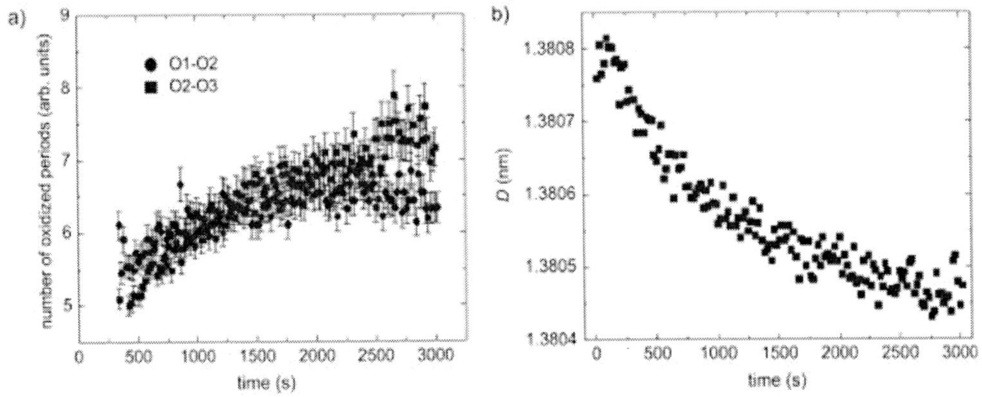

Figure 20. Temporal evolution of a) the number of oxidized periods and b) the mean multi-layer period after exposure in UV-ozone reactor.

This confirms a common origin for all of the stripes. A tendency to saturation is observed at the end of the ozone treatment, when the oxidized region reaches a depth of ~10% of the total multi-layer thickness. The effect of the ozone on the average multi-layer period can be extracted from the evolution of the first Bragg sheet position, which reveals a small but systematic decrease by 0.03% in total (Fig. 20b). When we attribute this change to the oxidized region only and consider its ~10% volume fraction, the total period reduction due to ozone treatment amounts to ~0.3% in the exposed region.

5.3. Vertical Replication Effects in La/B4C Multi-layers

The La/B_4C multi-layer mirrors were recognized as a promising candidate for the near normal incidence reflective optics of novel X-ray sources like FEL and the next generation of EUV lithography [*104-106*]. The X-ray multi-layer mirrors based on La/B_4C are suitable for the boron detection using X-ray fluorescence as well [*107-110*]. In this paragraph, we study the effects of vertical correlation between the interfaces of La/B_4C multi-layer mirrors in detail. We show the effect of ion-beam polishing on the PSD of interface roughness and its vertical correlation across the multi-layer stack. The influence of the thermal annealing on the interface stability of La/B_4C mirrors is discussed as well.

The La/B_4C multi-layer mirrors were deposited by ion-beam assisted e-beam evaporation in the ultra-high vacuum chamber [*111*]. Each La layer was polished by a Kr ion beam in order to lower the layer roughness and enhance the material density. For the study of vertical roughness correlation, multi-layer mirrors with the nominal period D=3.8 nm and N=20 bilayers were deposited. The analysis of the thermal stability was performed on a multi-layer mirror with the nominal period D=3.5 nm and N=110 bilayers. The mirror cross-sections prepared by ion-beam milling technique were inspected by TEM at various magnifications (inset of Fig. 21). The tracking of boundaries between the layers from TEM images is prone to numerical instabilities due to noise and blurry gradients. We used computation method that correlates the image intensities in the vicinity of the interfaces. The numerical simulations show that the autocorrelation function based on the intensity autocorrelation is very close to the real interface autocorrelation. The numerically processed unbiased autocorrelation functions [112] shown in Figure 21 were obtained by averaging over all interfaces within the TEM image. The autocorrelation function calculated from the image at the highest magnification (Figure 21c) is rather noisy due to insufficient scanning range available. On the other hand, the autocorrelation functions calculated from the images with the lower magnifications (Figure 21a,b) lack the information at the small *r* displacement values because of finite sampling step.

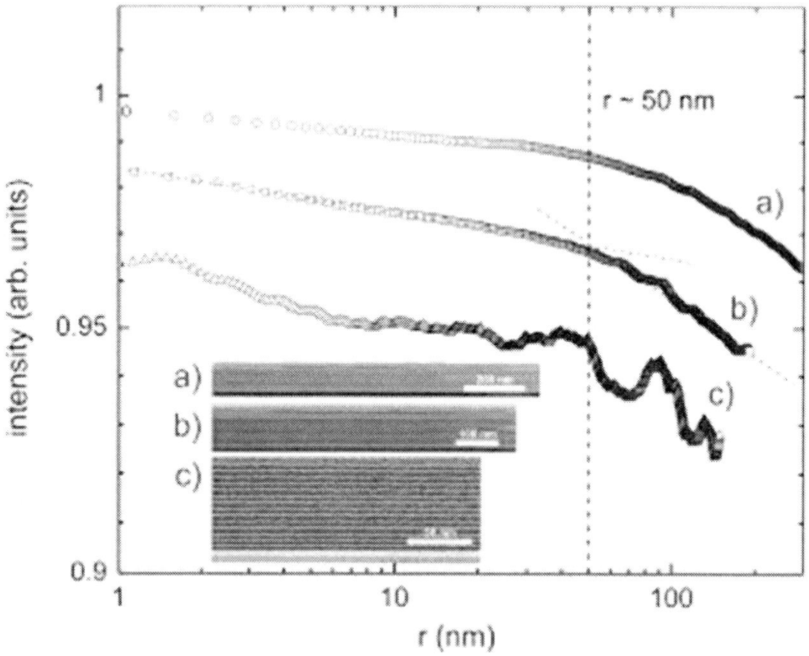

Figure 21. The unbiased autocorrelation functions of interfaces calculated from the TEM images with different magnifications.

Despite the fact that TEM is a local imaging technique, a characteristic lateral correlation length of approximately 50 nm was obtained. This rather long lateral correlation length opens doubts if the interfaces are not poorly developed mounded interfaces with a characteristic wavelength of approximately 50 nm. Consequently, it is difficult to assign whether the inspected interfaces belong to the class of self-affine or mounded interfaces. Furthermore, the vertical cross-correlations at different lateral frequencies including the lateral phase shifts between the interfaces are very difficult to study as well. The GISAXAS as a non-local reciprocal space technique is able to clearly address these issues.

For further discussions, it is convenient to use a simplified interpretation of the scattered intensity by multi-layered structure defined in equation (20). As already pointed out in paragraph 3 in the case of small interface roughness and identical interface autocorrelation functions, the integral in equation (21) will reduce to a simple Fourier transform of the autocorrelation function. If we further ignore refraction and absorption effects then the total scattered intensity can be written in a simple form as a product of the interface PSD function and of the interference function $S\left(q_{\|}, q_z\right)$ as follows:

$$I\left(q_{\|},q_z\right)\propto PSD\left(q_{\|}\right).S\left(q_{\|},q_z\right) \tag{26}$$

The equation (26) is equivalent to the principal equation of the small-angle X-ray scattering theory with the PSD function replaced by the particle form factor [*18, 20*]. A simple expression for the interference function $S\left(q_{\|},q_z\right)$ can read

$$S(q_{\|},q_z)=\sum_{j,k}^{N} e^{iq_z(z_j-z_k)}.e^{-q_{\|}^2\nu\left|z_j-z_k\right|}.e^{i\delta_{\|}(q_{\|},j,k)} \tag{27}$$

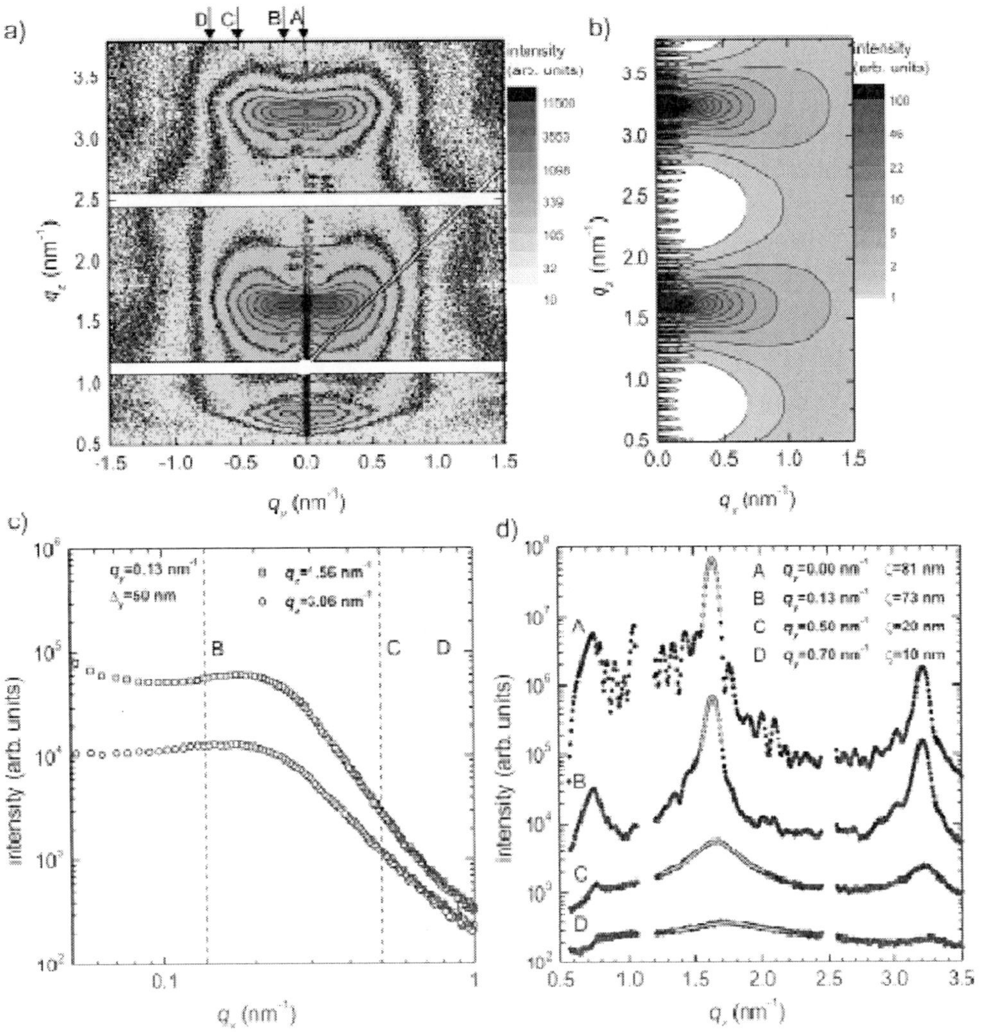

Figure 22. a) GISAXS reciprocal space map. b) Simulated interference function. c) Horizontal and d) vertical cuts through the reciprocal space map shown in graph a).

The first term describes the interference of the waves scattered by the j-th and k-th interface and is identical to the term in equation (20). The second term is responsible for the attenuation of wave interference from two interfaces separated by a distance $d = |z_j - z_k|$ due to relaxation effects in vertical correlation in the sense of the EW growth equation. This is a very simplified term as it does not include effects of roughness propagation and accumulation as defined in equation (15). Nevertheless, it will be shown as a useful approximation in the description of real multi-layered system. The last, third term, describes the lateral phase shift between the two interfaces.

The GISAXS reciprocal space maps were measured at the BW4 beamline, HASYLAB [38]. Figure 22a shows a reciprocal space map of La/B_4C multi-layer mirror with the 1st and 2nd Bragg sheets visible. The two horizontal white stripes correspond to the dead area of the two-dimensional X-ray detector composed of three modules (Pilatus 300K). The visibility of the Bragg sheets in the reciprocal space map can be interpreted in terms of interference function defined by the equation (27). The first term is independent of the q_y component and is responsible for the redistribution of the scattered intensity into Bragg sheets along q_z direction. The second term is a function of q_y component and reflects the loss of the vertical correlation with the increasing lateral frequency of replicated roughness. As a result, the Bragg sheets are broadening with the increasing q_y value. As the scattered intensity is concentrated within the Bragg sheets, no additional lateral phase shifts between the interfaces are required to reconstruct the observed reciprocal space map. For clarity, the simulated interference term is shown in Figure 22b. The following parameters were used in simulation: multi-layer period D=3.9 nm, number of bilayers N=20, relaxation parameter ν =0.5 nm and phase shift $\delta_{\|}(q_{\|}, j, k) = 0$.

Figure 22c shows the intensity profiles along the q_y direction of the 1st and 2nd Bragg sheet. Discernible broad maxima located at approximately q_y~0.2 nm^{-1} indicate vertically correlated mounded interfaces with a mean period of some 30 nm. In order to study the vertical correlation, we selected a set of representative lateral frequencies at $q_y \in \{0\,\text{nm}^{-1},\ 0.13\,\text{nm}^{-1},\ 0.5\,\text{nm}^{-1}, 0.7\,\text{nm}^{-1}\}$ denoted by the capital letters $\{\text{A, B, C, D}\}$, respectively. The vertical cuts at chosen frequencies along with the fits at the position of the 1st Bragg peak are shown in Figure 22c. The Bragg sheet width was used to calculate the vertical correlation length using equation (22). At q_y=0 nm^{-1} the roughness is fully correlated as the thickness of the stack is approximately 78 nm. A negligible decrease of the vertical correlation length at the q_y=0.13 nm^{-1} is followed by a rapid drop down to some 10 nm for spatial frequency at q_y=0.7 nm^{-1}. It is noteworthy to remark that for this multi-layers stack, the scattered intensity is confined to the Bragg sheet at the constant q_z value.

Figure 23a shows a reciprocal space map of the La/B_4C multi-layer mirror with a different vertical correlation scheme as described by the interference function in Figure 22b. For low q_y values, the scattered intensity is confined within the Bragg sheets as in the previous case. However, with the increasing q_y, the scattered intensity starts to spread more rapidly, culminating at the value q_y~0.5 nm^{-1} with nearly constant intensity distribution throughout the reciprocal space. For the values q_y>0.5 nm^{-1} , we observe an intensity increase between the Bragg sheets.

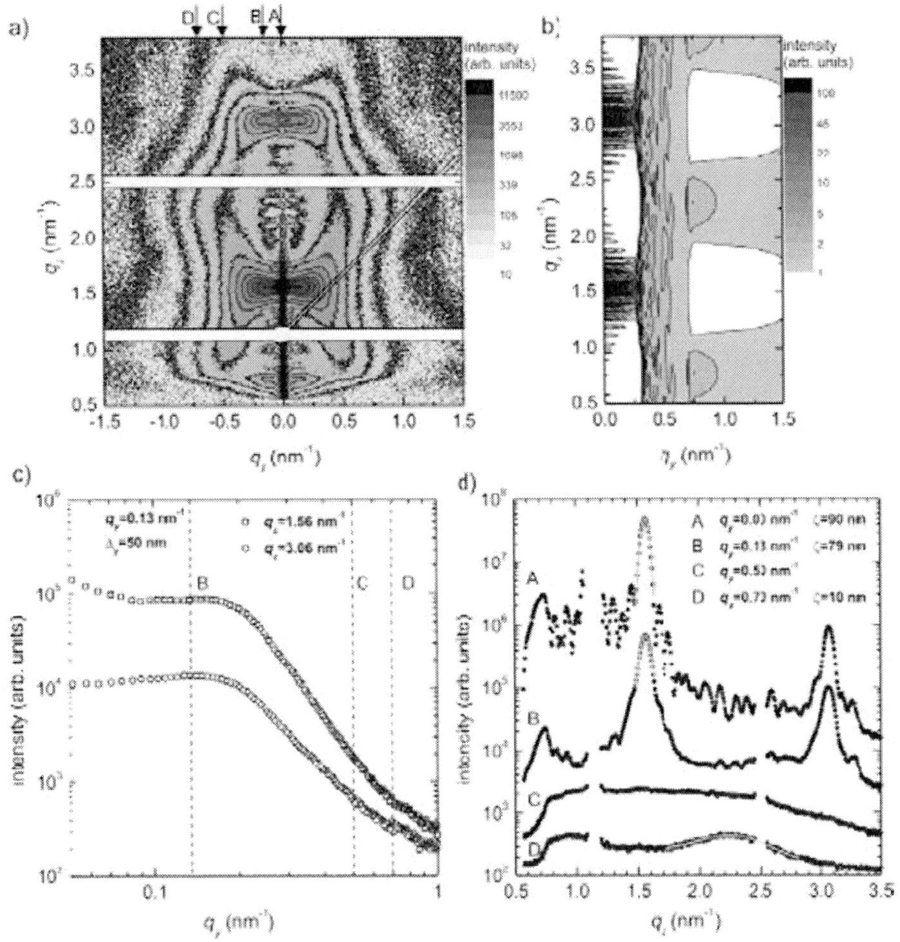

Figure 23. a) GISAXS reciprocal space map. b) Simulated interference function. c) Horizontal and d) vertical cuts through the reciprocal space map shown in graph a).

In order to understand the intensity distribution, we performed a simulation of the interference function shown in Figure 23b with the identical parameters except for the phase shift function $\delta_{\|}(q_{\|}, j, k)$. The phase shift was set to zero for the values q_y<0.4 nm^{-1}. A systematic phase shift of π between the neighboring interfaces was introduced for the values q_y>0.6 nm^{-1}. For the values $q_y \in$ <0.4 nm^{-1}, 0.6 nm^{-1}>, we introduced a random phase shifts indicating a loss in the vertical correlation between the interfaces. The product of the simulated interference function with the PSD function results in the observed GISAXS reciprocal space map. The line cuts across the Bragg sheets shown in Figure 23c indicate vertically correlated mounded interfaces with a similar characteristic wavelength as those in Figure 22c. The vertical cuts at the selected q_y values denoted by a set of capital letters $\{A, B, C, D\}$ are identical to the extracted cuts shown in Figure 23d. At lateral frequencies given by q_y=0 nm^{-1} and q_y=0.13 nm^{-1} the scattered intensity is confined within the Bragg sheets and the roughness at these lateral frequencies is fully vertically correlated across the whole multi-layer stack as indicated by the extracted vertical correlation length. In contrary to the previous multi-layer analysis, the scattered intensity line cut at the q_y=0.5 nm^{-1} is nearly constant that implies vertically uncorrelated roughness at this lateral frequency as verified by the simulation. The vertical cut at q_y=0.7 nm^{-1} shows a distinct peak with a maximum located at the position between the Bragg peaks. The vertical correlation length of approximately 10 nm at q_y=0.7 nm^{-1} is identical to the vertical correlation length of the previously inspected mirror at the same lateral frequency. The difference is in the replication process. Whereas in the previous case, the replication of the roughness was in phase between the neighboring interfaces; in this case, a phase shift of π is present between adjacent interfaces. The present analysis demonstrates that GISAXS provides a platform for study of subtle effects in vertical correlation in multi-layers.

For the fabrication of La/B_4C multi-layer mirrors, the technique of in-situ sputtering of La layers with noble gas atoms was employed. The ion-beam polishing technique used is known to reduce the layer roughness and increase the layer density. In order to understand the influence of ion-beam polishing on the interface roughness and vertical roughness correlation, we measured GISAXS reciprocal space maps of two multi-layer mirrors. The first mirror was deposited without polishing of La layers. For the second mirror, ion-beam polishing of La layer was utilized using the Kr ion-beam of E=50 eV kinetic energy. Figure 24a shows the line cuts through the 1st Bragg sheets along the q_y direction. It is evident that the effect of ion-beam polishing of La layers is influencing the high frequency tail of the scattered intensity distribution.

As the measured profile is a product of the roughness PSD function and the interference function according to equation (26), we cannot distinguish between the possible contributions. Figure 24b shows the widths of the 1st Bragg sheets as a function of q_y wave vector. The width of the Bragg sheet is inversely proportional to the vertical correlation length at given frequency. For low spatial frequencies up to q_y~0.2 nm^{-1} , the vertical correlation length for both polish and unpolished mirrors, is identical. For the spatial frequencies above q_y>0.2 nm^{-1} , the polished mirror exhibits a shorter vertical correlation length in contrast to the unpolished one. Based on this result, we can assume that the polishing of the La layers is primarily reducing the vertical roughness replication effects at the high spatial frequencies. It was already pointed out by the Spiller et al. [*67*] that the ion beam assisted deposition is reducing the replication of substrate roughness which allows non-superpolished substrates to be used with a significant cost reduction in X-ray mirror fabrication.

The thermal stability of the La/B_4C mirrors was studied ex-situ by annealing the samples up to 900°C in high vacuum furnace with a base pressure of 10^{-7} mbar. For each temperature, we heated up one mirror sample by a rate of 100 K.s^{-1} and held constant for 120 s which was followed by a spontaneous cool down to the room temperature. For each mirror sample, a GISAXS reciprocal space map was recorded. Figure 25a shows the vertical cuts across the reciprocal space at constant q_y=0 nm^{-1} value, also called the "detector scans", when measured in co-planar geometry [*9*].

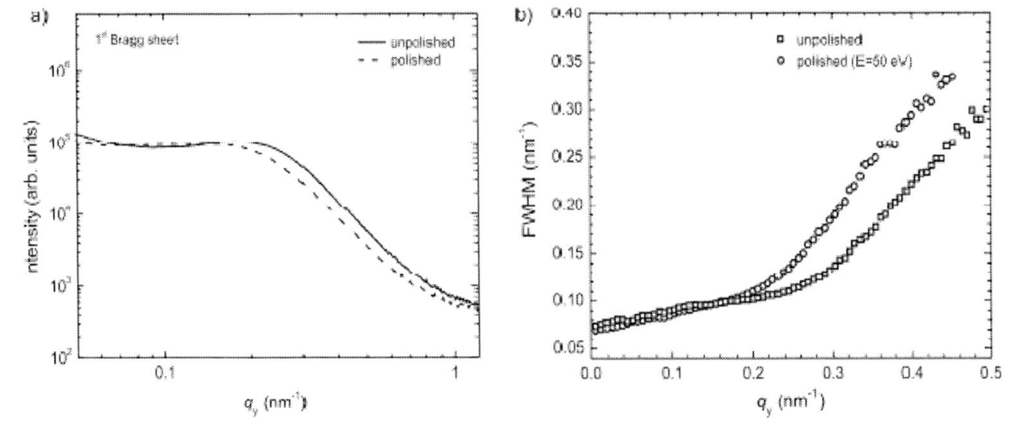

Figure 24. a) The horizontal cuts across the 1st Bragg sheets and b) the width of 1st Bragg sheet as a function of q_y for the mirrors polished and unpolished by Kr ion-beam during the deposition process.

The mult-ilayer structure is stable up to T=300°C as no significant shift in the position of the 1st and 2nd Bragg peak is visible. Above this temperature, a steady shift of the Bragg peak positions to the high q_z value is visible. This corresponds to the decreasing effective period of the multi-layer mirror due to the material inter-diffusion across the multi-layer interfaces. The considerable broadening of the 2nd Bragg peak for the annealing temperatures above T=600°C implies increasing disorder in the layering due to material intermixing at interfaces. The material intermixing is perpendicular to the multi-layer interfaces as it is driven by the gradient in chemical potential. Consequently, minor or no changes in the vertical roughness replication should be observed with the increasing annealing temperatures. This is confirmed experimentally by the inspection of horizontal cuts in reciprocal space at the 1st Bragg sheet (Fig. 25c) and between the 1st and 2nd Bragg sheet (Fig. 25d). The horizontal line cuts at the 1st Bragg sheet shown in Figure 25b display no changes in the profiles apart from the steady decay of integral intensity due to lower material density contrast of bilayers at highest annealing temperatures. The same behavior exhibits the horizontal line cuts stemming from the reciprocal space between the Bragg sheets shown in Figure 25c. The first peak located approximately at q_y=0.1 nm^{-1} is originating from the vertically correlated roughness at the characteristic mound frequency of interfaces. The second peak reflects the π-shifted vertical replication of interface roughness at high spatial frequencies described above. We observe no major shifts or changes in the scattering profiles up to annealing temperature of T=600°C. Above this temperature, diffusion induced disorder is responsible for low scattered intensity in this particular region of reciprocal space. The number of effectively correlated bilayers N_{eff} evaluated from the 1st Bragg sheet is nearly constant for the annealing temperatures up to T=600°C. Above this temperature, the number is progressively decreasing due to strong intermixing at interfaces. The same universal behavior of N_{eff} with the increasing annealing temperature was already observed for the W/B_4C multi-layer mirrors discussed in paragraph 5.2.

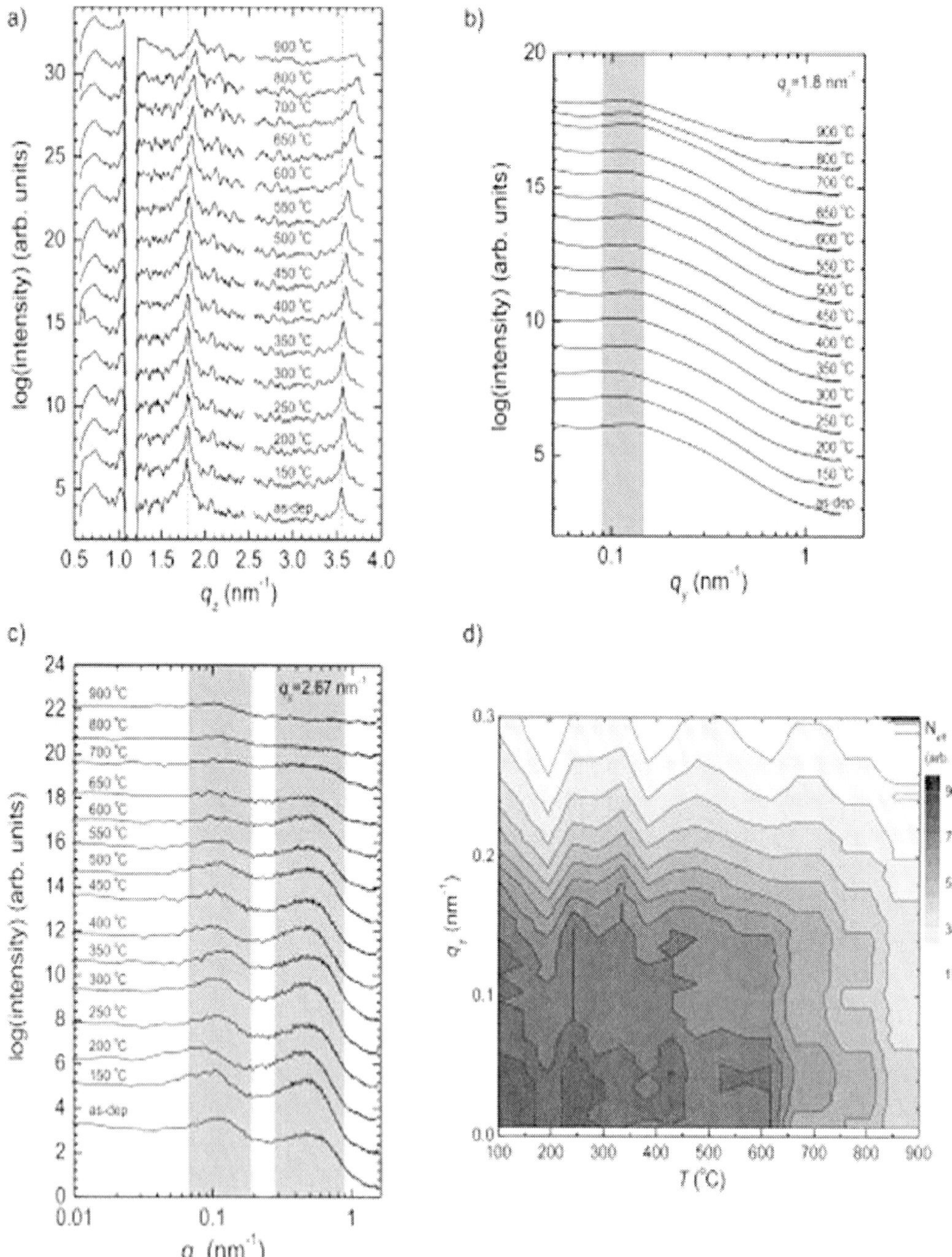

Figure 25. a) The vertical cuts of reciprocal space at constant q_y=0 nm^{-1} for different annealing temperatures. The horizontal cuts of reciprocal space at b) 1st Bragg peak and c) between the 1st and 2nd Bragg peak for different annealing temperatures. d) The number of effectively correlated bilayers as a function annealing temperaute and lateral frequency q_y

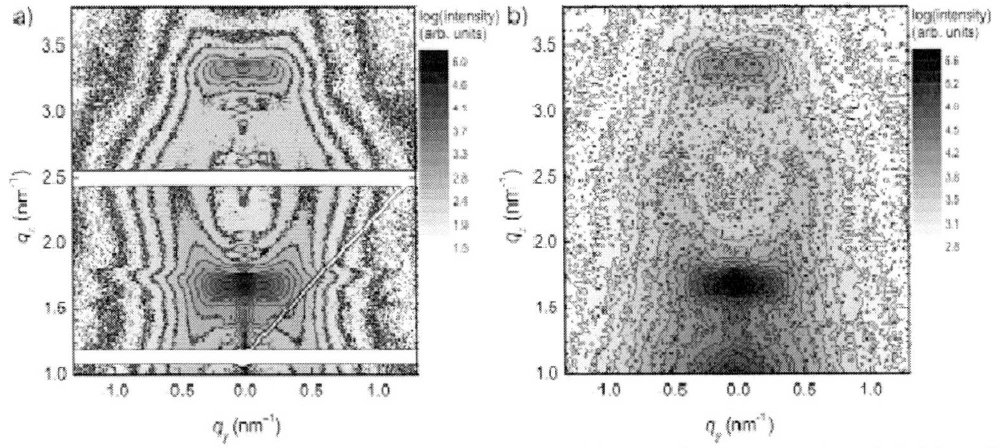

Figure 26. GISAXS reciprocal space map of La/B_4C multilayer mirror measured at a) B4 beamline (HASYLAB) and b) SAXS/GISAXS Nanostar laboratory platform.

The high brilliance synchrotron X-ray beam is the first choice for measurement and analysis of reciprocal space of multi-layer mirrors. Although the flux and resolution of the table-top laboratory GISAXS systems cannot match the synchrotron quality, the basic features of the reciprocal space can be obtained. Figure 26 shows the two measurements of La/B_4C mirror measured at BW4 beamline (HASYLAB) and using high flux mode of laboratory SAXS/GISAXS Nanostar system. Despite the Kiessig fringes at q_y=0 nm^{-1} are not resolved by the laboratory system, the important features like Bragg sheets are visible. Using the high resolution mode of Nanostar system, the reciprocal space resolution of Δq=0.03 nm^{-1} can be reached at the expense of lower flux in the primary beam.

5.4. Clusters in Ultrathin Co/C, W/B4C, Mo/B4C Multi-layers

The X-ray mirrors working at the near normal incidence condition or at very short X-ray wavelength have the individual layer thickness approaching 1 nm and less. Thin films consisting of only few atomic layers are seldom continuous due to nucleation processes controlling the layer growth at early stages [*113-115*]. The GISAXS technique provides an efficient and non-destructive way of monitoring the early stages of thin film growth [*16*]. In this paragraph, we show how the GISAXS reciprocal space maps are used to monitor the distribution of clusters in various ultrathin multi-layers composed of Co/C, W/B_4C and Mo/B_4C couples. Furthermore, we discuss the coalescence of the clusters at elevated annealing temperatures.

The advantage of the non-coplanar (GISAXS) geometry is the ability to address lateral correlations far below 50 nm that is rather difficult in coplanar geometry [*9*]. To compare the mapping capabilities of reciprocal space using the coplanar and non-coplanar geometries, we measured two Co/C multi-layer mirrors with identical periods but different absorber (Co layer) thicknesses. The Co/C multi-layers were prepared by ion-beam sputtering deposition at the Tohoku University, Japan [*82, 83*]. The multi-layer period was D=2.32 nm and the number of layers N=200. The two mirrors with Γ=0.7 and Γ=0.5 were deposited. The gamma value is defined as $\Gamma = d_{abs}/D$ where d_{abs} is the thickness of the absorber layer, in this case, the thickness of Co layer. The X-ray reciprocal space mapping in coplanar geometry was measured on an X-ray diffractometer (Bruker D8 Discover) equipped with Cu X-ray rotating anode. The GISAXS reciprocal space maps were recorded by laboratory SAXS/GISAXS system Nanostar (Bruker AXS) equipped with a collimated microfocus X-ray source (IμS, Incoatec). The coplanar reciprocal space maps of Co/C multi-layer mirror with Γ=0.7 and Γ=0.5 are shown in Figure 27a and Figure 27b, respectively. A single Bragg sheet located at q_z=2.7 nm^{-1} is visible in both reciprocal space maps. The maximum range of the q_x wave vector is ±0.1 nm^{-1} at the first Bragg sheet and ±0.3 nm^{-1} at the maximum q_z. The higher integral intensity of the first Bragg sheet for Γ=0.5 is due to lower absorption of the X-ray radiation in the thinner Co layers which was verified by measurements and simulations of the X-ray reflectivity curves. The analysis of reciprocal space map measured in coplanar geometry does not provide any insight into possible discontinuities of thin Co layers for the Γ=0.5 multi-layer mirror.

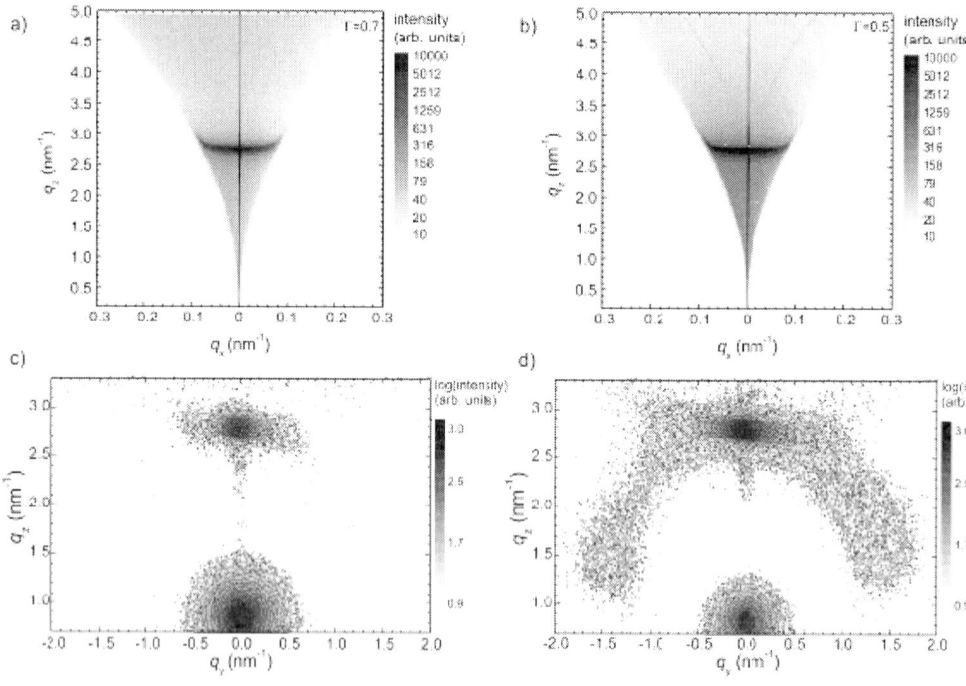

Figure 27. The reciprocal space maps measured in coplanar geometry for the Co/C multi-layer mirrors with a) Γ=0.7 and b) Γ=0.5. The corresponding reciprocal space maps measured in non-coplanar geometry for the identical mirrors with c) Γ=0.7 and d) Γ=0.5.

The reciprocal space maps of the Co/C multi-layer mirrors measured in the non-coplanar geometry are shown in Figure 27c and Figure 27d for the values Γ=0.7 and Γ=0.5, respectively. During the measurements no beamstop for the specularly reflected beam was used as evidenced by a strong circular reflection at q_z<1 nm^{-1} visible in both reciprocal space maps. The reciprocal space map of the Co/C mirror with Γ=0.7 shows one laterally extended Bragg sheet at the exact location as measured in the coplanar geometry. On the other hand the reciprocal space map of the Co/C mirror with Γ=0.5 displays a broad ring indicating a presence of correlated Co clusters. The ring is truncated due to geometrical constraints at the sample horizon. The analysis reveals the mean in-plane cluster-to-cluster distance of approximately 4.5 nm. The Co clusters do not exhibit any vertical correlation. As demonstrated by this example the extended range of the q_y wave vector component in the non-coplanar geometry is very useful when examining the short range correlations in ultra-thin multilayered films.

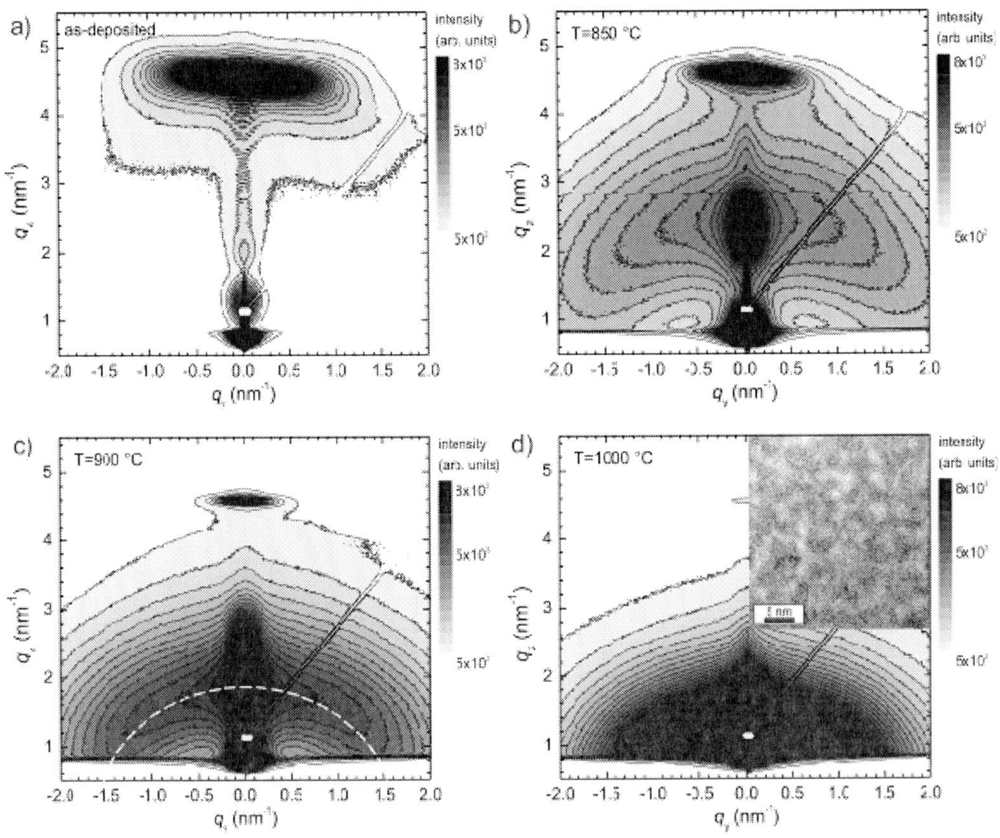

Figure 28. The GISAXS reciprocal space maps of a) as-deposited and annealed W/B4C multilayer mirrors at b) T=850°C, c) T=900°C and d) T=1000°C. The inset of graph d) shows the cross-section TEM image.

In paragraph 5.2, we took a deeper insight into interface properties and replication effects in W/B_4C multi-layer mirrors. The study of thermal stability of W/B_4C mirrors demonstrated their extraordinary structural stability up to 700°C. Above this temperature, the material intermixing through the interface boundaries took place. Figure 28 shows a series of GISAXS reciprocal space maps of the samples annealed at 850°C, 900°C and 1000°C. The reciprocal space map of the as-deposited state shown in Figure 28a displays the first Bragg sheet and small modulations below q_z<2 nm^{-1} corresponding to oxidized near-surface layers. The reciprocal space map of the sample annealed at T=850°C is shown in Figure 28b. The reciprocal space map shows the initial stage of cluster formation as indicated by a truncated broad ring. A distinct peak at q_z~2.4 nm^{-1} belongs to the interfacial oxidized layers. At

T=900°C, the reciprocal space map shows a narrower ring corresponding to a well developed distribution of clusters with a mean cluster-to-cluster distance of ~3.6 nm as indicated by a dashed white line in Fig. 28c. At T=1000°C (Fig. 28d) the mean distance between the adjacent clusters and the width of their distribution increase. The inset of Figure 28c shows the cross-section TEM micrograph of the annealed sample that clearly demonstrates the presence and distribution of the clusters.

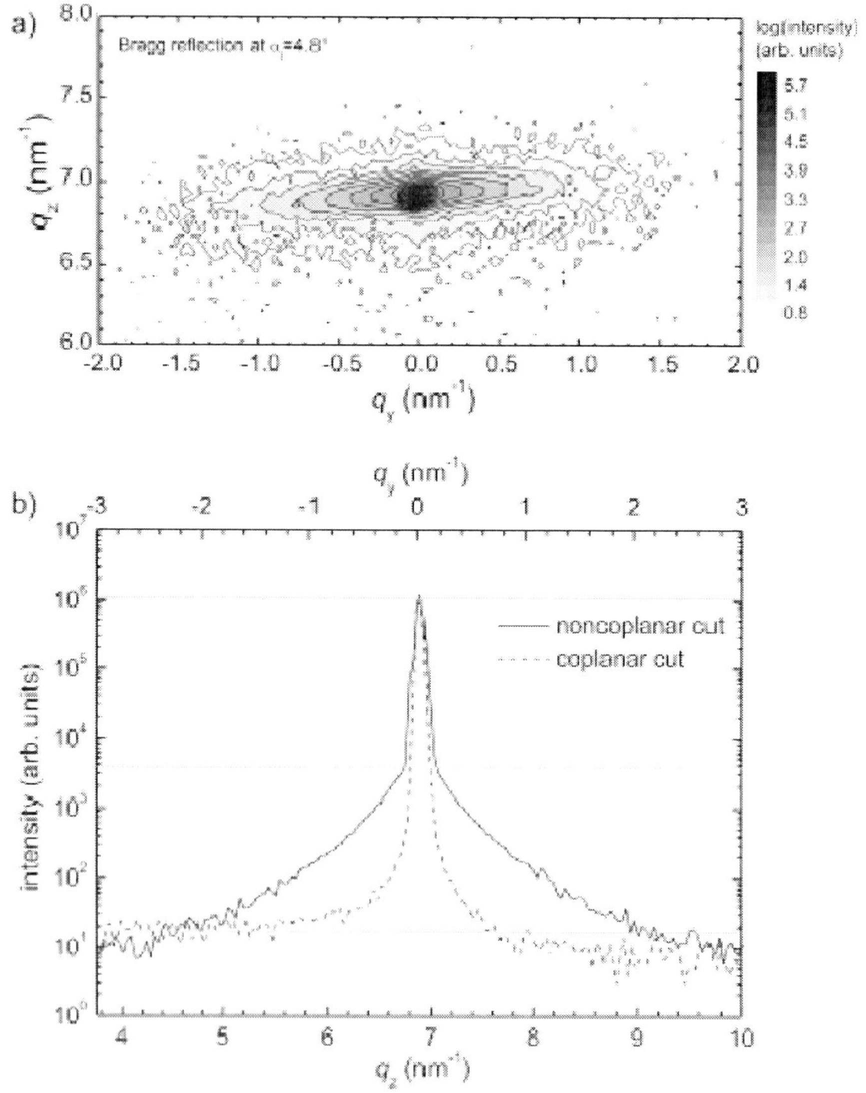

Figure 29. a) The reflected primary beam from the Mo/B_4C multi-layer mirror in resonant Bragg condition. b) The non-coplanar ($q_y \neq 0$) and coplanar ($q_y=0$) line cuts documenting the distribution of scattered intensity.

The layer discontinuities due to the cluster formation reduce the X-ray reflectivity to such an extent that its proper simulation within the model of continuous layers is impossible [*116, 117*]. We demonstrated that the GISAXS technique is very efficient in finding the critical threshold thickness above which the layer shows the properties of continuous layer. In the following, we describe another example of the multi-layered system with discontinuities in the absorber layers. We studied a Mo/B_4C multi-layer mirror with a period D=0.93 nm and number of bilayers N=400. The thickness of the Mo layer was set to 0.41 nm. The mirror was deposited by magnetron sputtering on Si substrate at Incoatec GmbH (Geesthacht, Germany). The reflection from the Mo/B_4C X-ray mirror at the Bragg condition (α_i=4.8°, λ=0.154 nm) recorded by a two-dimensional detector is shown in Figure 29a. The high dynamic range of the recorded intensities allows detailed inspection of the mirror performance. The circular spot of the reflected beam is enveloped by a diffuse scattering originating from the vertically correlated multi-layer interface roughness analogous to the Bragg sheet obtained in the non-resonant GISAXS configuration. The non-coplanar and coplanar line cuts across the reflected beam are shown in Figure 29b. The differences between the levels of the diffuse scattering around the specular ridge are due to different lateral roughness frequencies probed in the two geometries. In particular, the coplanar one collects the diffuse scattering only from small frequencies as explained in previous sections. This difference has to be taken into account when planning the high performance X-ray optics and additional pair of vertical slit has to be introduced to eliminate this parasitic scattering.

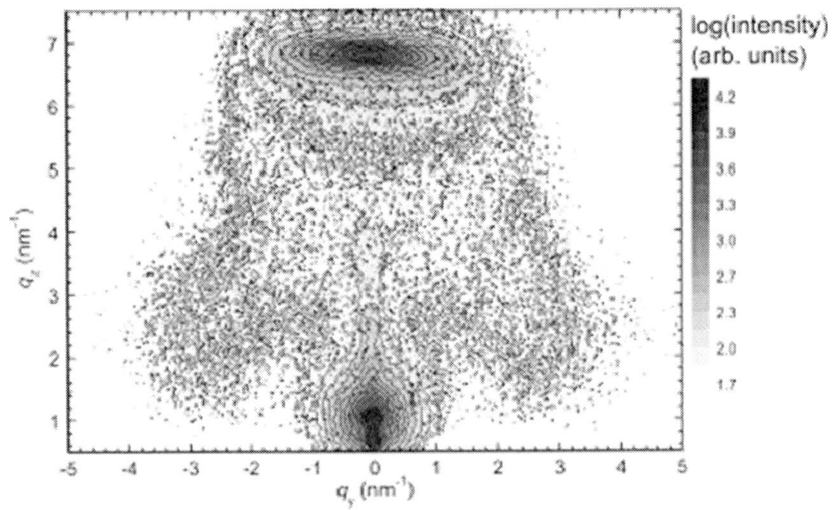

Figure 30. The GISAXS reciprocal space map of Mo/B_4C multilayer mirror indicating the presence of correlated clusters with the mean inter-cluster distance of approximately 1.8 nm.

The GISAXS reciprocal space map of the Mo/B_4C multi-layer mirror is shown in Figure 30. The first Bragg sheet is visible along the specularly reflected beam at the bottom of the map. The presence of correlated clusters in as-deposited multi-layer mirror is indicated by a truncated ring at q~3.5 nm^{-1} that corresponds to mean cluster distance of ~1.8 nm. The uppermost oxidized layers are responsible for the appearance of a minor Kiessig oscillation below the Bragg sheet. The thermal stability of the Mo/B_4C multi-layer mirrors was studied in the same way as the W/B_4C multi-layer system. The GISAXS reciprocal space maps were recorded by a laboratory SAXS/GISAXS system Nanostar (Bruker AXS). A series of reciprocal space maps obtained from the samples annealed at different temperatures is shown in Figure 31.

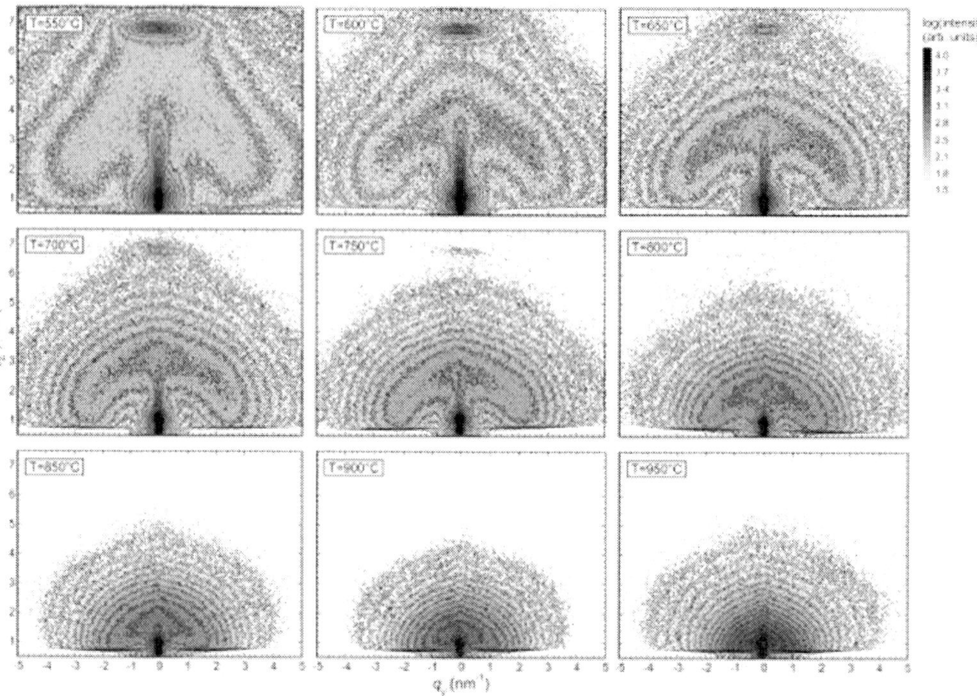

Figure 31. The evolution of the GISAXS reciprocal space maps of Mo/B_4C multilayer mirror with increasing annealing temperature.

Starting from the annealing temperature T=550°C, we observe a gradual increase of the scattered intensity within the truncated ring corresponding to the correlated Mo clusters. With the increasing annealing temperature, the coalescence of the smaller clusters leads to the steadily growing inter-cluster distance as indicated by the shrinking ring in reciprocal space. At the annealing temperature T=750°C, the last sign of the layered structure given by the barely visible Bragg sheet disappears. Above this temperature, a further growth of the mean inter-cluster distance is observed. To quantify the change in the inter-cluster distance, a pie integration (inner angle 30°) for interpretation of the scattering data was employed in analogy with SAXS data treatment [*86*]. The mean cluster distance shown in Figure 32 was determined by fitting the data with log-normal distribution. The graph shows a slow increase of the mean intercluster distance up to T=750°C followed by a more rapid growth. In order to study the cluster distribution within the volume of the multi-layer sample, we performed an angle-resolved GISAXS measurement. The Mo/B_4C multi-layer mirror annealed at T=700°C was measured at four different incident angles (0.2°, 0.3°, 0.4° and 0.7°) below and above the critical angle for X-ray reflection at λ=0.154 nm (Fig. 33a). For the GISAXS measurement, we used a laboratory setup designed and built at the Institute of Physics SAS (Bratislava, Slovakia) described in paragraph 4.

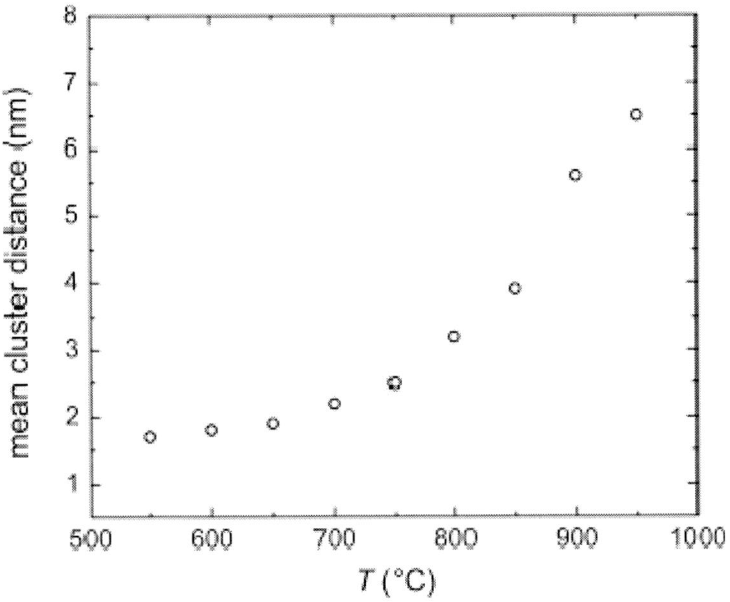

Figure 32. The growth of mean cluster distance as a function of annealing temperature.

Using the angle of incidence below and near the critical angle, we probe the surface and near surface regions due the total reflection effect [*17, 24*]. Here, only the evanescent exponentially decaying wave is refracted into the bulk [*26*].

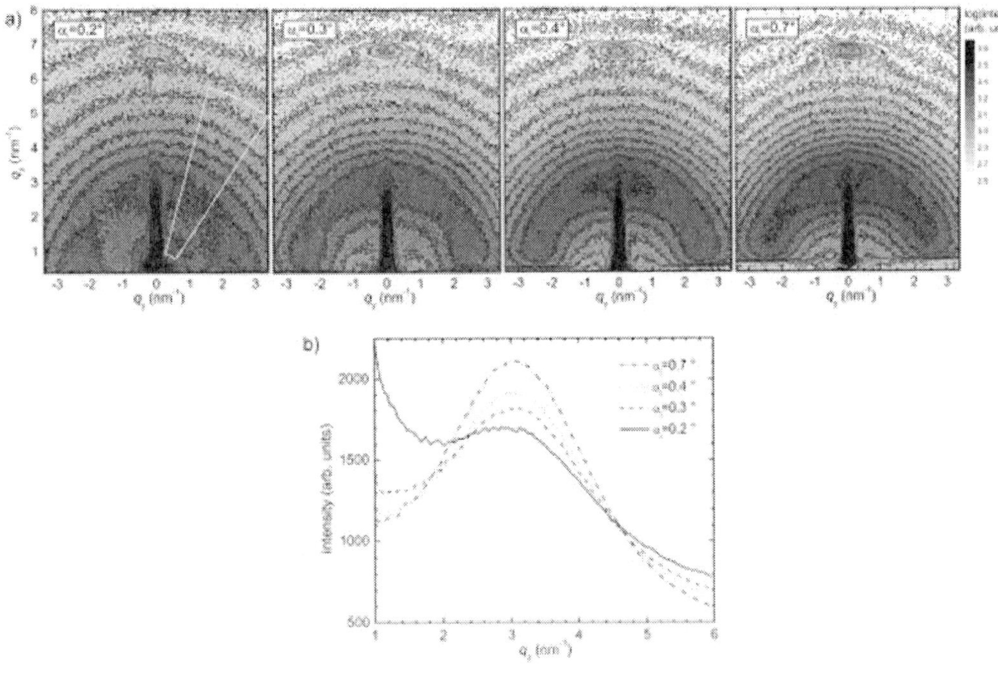

Figure 33. a) The GISAXS reciprocal space maps of Mo/B_4C multi-layer mirror obtained at different incident angles. b) The integrated profiles across the ring corresponding to the correlated Mo clusters.

With the increasing angle of incidence, the intensity of the first Bragg sheet is steadily growing due to the increasing penetration depth of the incident X-ray radiation into the multi-layer structure. Simultaneously, with the increasing intensity of the first Bragg sheet, the ring corresponding to the correlated clusters gains in intensity as well. The pie integration with an opening angle of 20° as depicted in Figure 33a was used to collect the data on cluster distribution. Figure 33b shows the extracted profiles for the four different incident angles. With the increasing penetration depth into the multi-layer bulk, the integral intensity of peak corresponding to the correlated clusters is steadily growing. The experimental results confirm the assumption that the clusters are homogenously distributed within the volume of the thermally treated multi-layer mirror.

5.5. Interface Properties and Thermal Stability Nife/Au/Co/Au Multi-layers

The morphology of interfaces of the giant magnetoresistance (GMR) [*118, 119*] devices influences directly their performance, the magnetostatic coupling, also known as Néel coupling, being a typical example [*120, 121*]. In this paragraph, we study the interface replication phenomena in the GMR thin films with perpendicular magnetic anisotropy of Co layers [*122, 123*] using the GISAXS technique. The thermal stability of the GMR multi-layer structure and the impact of thermal treatment on the vertical correlation of interface roughness are reviewed.

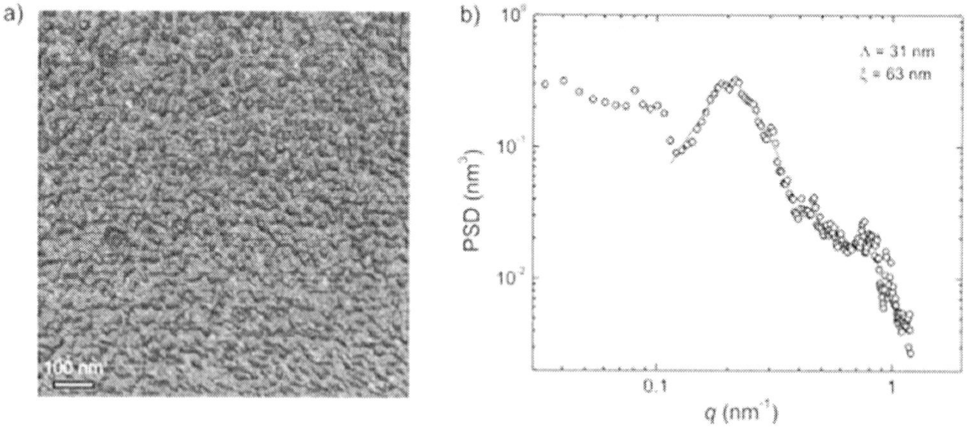

Figure 34. a) The height AFM image of the GMR multilayer taken in tapping mode. b) The calculated radially averaged PSD of surface height function measured by AFM.

The multi-layers composed of N=10 periods of [$Ni_{80}Fe_{20}$ 2nm/Au 2nm/Co 0.7nm/Au 2nm] were deposited at room temperature by magnetron sputtering onto Si(100) substrates at the Institute of Molecular Physics PAS (Poznań, Poland). The thermal stability of GMR multi-layers was studied on samples with the total number of periods N=20. The AFM image of the multi-layer surface shown in Figure 34a is a typical example of mounded surface that can be quantified by a radially averaged PSD function. The calculated PSD function depicted in Fig. 34b belongs to the class of mounded surfaces with a characteristic wavelength Λ=31 nm and a lateral correlation length ξ=63 nm. Using the angle-resolved GISAXS technique, we can track the replication of the surface morphology from the sample surface to the buried interfaces beneath. A series of reciprocal space maps recorded with the table-top GISAXS instrument designed at the Institute of Physics SAS (Bratislava, Slovakia) described in paragraph 4 is shown in Figure 35.

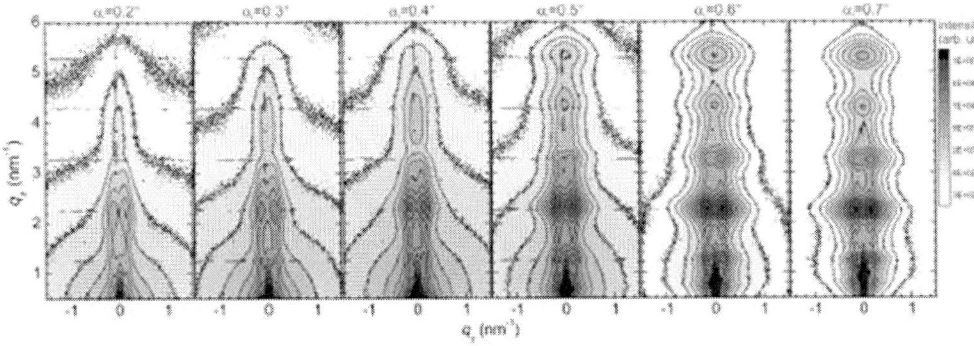

Figure 35. The evolution of scattered intensity in the reciprocal space with the increasing incident angle of primary beam.

The critical angle of the GMR multi-layer stack is α_c~0.5° at the X-ray wavelength λ=0.154 nm used. At the lowest incident angle α_i=0.2°, we probe approximately 10 nm beneath the surface. The two visible side maxima (q_y~±0.13 nm^{-1}) in the 2nd Bragg sheet located at q_z~2.2 nm^{-1} clearly indicate vertical replication (conformality) of the mounded interfaces across the multi-layer stack. If we assume minor effects of the vertical replication on the damping of the scattered intensity profile of the 2nd Bragg sheet for q_y<0.2 nm^{-1} , we can estimate the mounded wavelength Λ=48 nm and the lateral correlation length ξ=35 nm from the position and width of the side maxima, respectively. As the probing penetration depth increases with the growing incident angle, more interfaces can coherently scatter the incoming radiation which results in narrowing the observed Bragg sheets. At the incident angle α_i=0.7°, well above the critical angle α_c, a series of five Bragg sheets is visible. The first three Bragg sheets show the modulations due to vertical replication of the mounded interfaces while in the fourth and fifth Bragg sheet this modulation is absent. The broadening of the Bragg sheet width with the increasing diffraction order is associated with the vertical disorder in roughness replication process [*19*].

This second-type disorder of the vertical roughness replication smears out the Bragg sheet modulation at higher diffraction orders. The width of the Bragg sheets as a function of the q_z scattering wave vector was fitted with a linear function at two different lateral wave vectors q_y=0 nm^{-1} and q_y=0.13 nm^{-1} [*84, 86*]. The vertical correlation length was derived from the intercept of the linear fit at q_z=0. For the both inspected roughness lateral frequencies, the vertical correlation length equals to the total thickness of the GMR multi-layer stack. The angle-resolved GISAXS measurements confirmed the presence of fully vertically correlated mounded interfaces. The origin of vertically correlated mounded interfaces was found to be related to the columnar crystalline growth of the metallic GMR multi-layers [*124-127*].

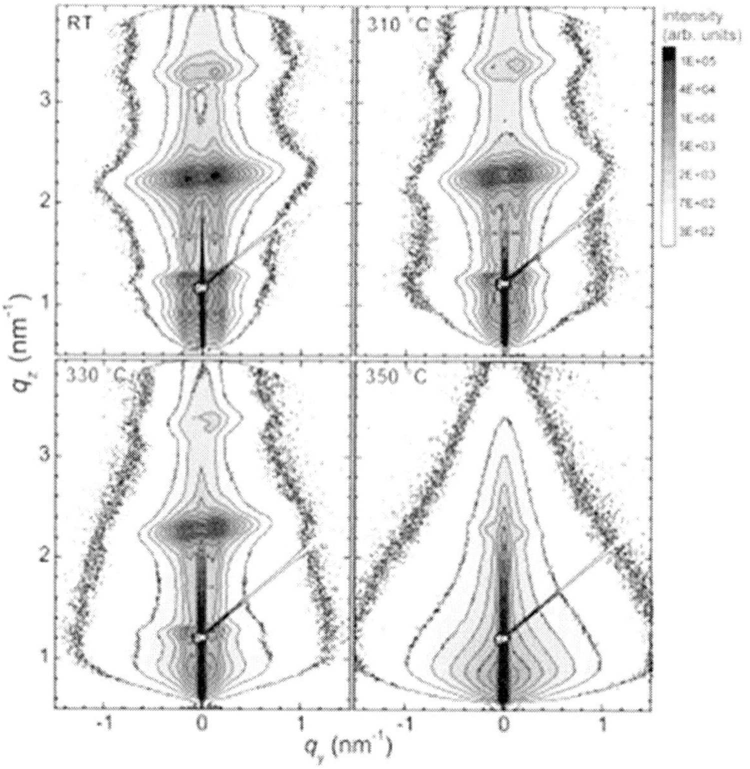

Figure 36. The GISAXS reciprocal space maps of GMR multi-layers annealed at different temperatures.

To study the thermal stability of GMR multi-layers, the samples were vacuum annealed for 120 s in the vacuum chamber with the base pressure of 10^{-7} mbar. The GISAXS reciprocal space maps were measured at synchrotron beamline BW4 [*38*] in HASYLAB (Hamburg, Germany). The GISAXS reciprocal space maps of the as-deposited and annealed samples at T=310°C, T=330°C and T=350°C are shown in Figure 36. The GMR multi-layers are stable up to approximately T=330°C. Below this temperature, no significant changes in the Bragg sheets are observed. Above this temperature, rapid intermixing across the multi-layer interfaces is responsible for the quenching of integral intensity of the 2nd Bragg sheet as shown in Figure 36 for the sample annealed at T=350°C. More detailed information on the material interdiffusion across the multi-layer interfaces can be learned from the vertical line cuts of the GISAXS reciprocal space maps. The vertical line cuts at the q_y=0 nm^{-1} and q_y=0.13 nm^{-1} are shown in Figure 37a and Figure 37b, respectively.

The line cuts at q_y=0 nm^{-1} correspond to the vertical roughness correlations at the long lateral frequencies in order of hundreds of nanometers. The first noticeable change in the scattered intensity profile is observed in the Yoneda region [*128*] at T=310°C. The initially sharp Yoneda peak corresponding to the top Au layer at T=310°C broadens which indicates the changes in the X-ray critical angle due to material intermixing in the near-surface region. At T=330°C, a new Yoneda peak appeared at the lower exit angles suggesting a different material composition of the uppermost layer. Simultaneously, the Kiessig fringes appeared originating from the interference of partially scattered waves from the top and bottom of the multi-layer stack. This implies the material intermixing at the interfaces proceeded by a continuous formation of a single layer with the thickness of the whole multi-layer and an averaged density. The final annealed state at T=350°C show absence of any layering as the Bragg peak is no longer visible. The small shifts in the position of Bragg peaks for the samples annealed at different temperatures are due to inhomogeneous deposition resulting in the slightly different multi-layer periods of different samples. The vertical roughness replication at the mean characteristic mound wavelength (q_y=0.13 nm^{-1}) is very similar except for the T>330°C. The sample annealed at this temperature still exhibits vertical correlation of mounds indicated by a presence of minor Bragg peak. The observed stability of the vertical correlation near the characteristic mound frequency is due to the crystalline columnar growth of the metallic multi-layers that controls the interface morphology.

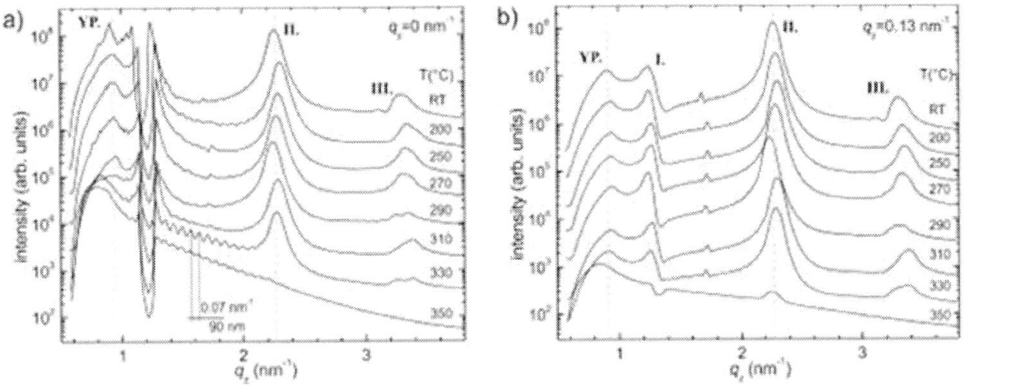

Figure 37. The vertical line cuts of the GISAXS reciprocal space maps at a) q_y=0 nm^{-1} and b) q_y=0.13 nm^{-1} for different annealing temperatures.

CONCLUSIONS

The goal of this chapter was to demonstrate the capabilities of the GISAXS reciprocal space mapping technique for inspection of morphology of buried interfaces in terms of correlation and scaling properties. The GISAXS provides a unique non-destructive tool to study vertical replication of roughness within the volume of multi-layered materials. Another benefit of GISAXS analysis consists in the non-local statistical character of the obtained data in contrast to the local imaging scanning techniques. The GISAXS has been traditionally confined to high brilliance synchrotron beamlines. However, due to recent progress in the microfocus X-ray tubes and collimating/focusing X-ray optics GISAXS is also moving to laboratories. The first commercial SAXS/GISAXS devices like Nanostar (Bruker AXS) are available for the ex-situ and in-situ studies of multi-layered thin films. In the near future, the in-situ GISAXS monitoring/control of the multi-layer growth in laboratory conditions may be expected. We would like to thank to the following people who considerably contributed to this work: K. Vegso, L. Chitu, Y. Halahovets, S. Luby, I. Matko, A. Timmann, J. Perlich, I. A. Makhotkin, S. V. Roth, G. A. Maier, J. Keckes, T. Tsuru, T. Harada, M. Yamamoto, S. Hendel, M. Lass, M. D. Sacher, W. Hachmann, U. Heinzmann, A. Hembd, F. Hertlein, J. Wiesmann, F. Stobiecki, and B. Szymanski.

This publication is the result of the project implementation: *Center of Applied Nanoparticle research,* ITMS code 26240220011, supported by the Research & Development Operational Program funded by the ERDF.

REFERENCES

[1] M. Pelliccione, T. M. Lu, *Evolution of thin film morphology : modeling and simulations*. Springer series in materials science (Springer, New York ; London, 2008), pp. xi, 204 p.
[2] J. M. Bennett, L. Mattsson, *Introduction to surface roughness and scattering*. (Optical Society of America, Washington, D.C., 1989), pp. viii, 110 p.
[3] A.-L. Barabási, H. E. Stanley, *Fractal concepts in surface growth*. (Press Syndicate of the University of Cambridge, New York, NY, USA, 1995), pp. xx, 366 p.
[4] S. K. Sinha, E. B. Sirota, S. Garoff, H. B. Stanley, X-Ray and Neutron-Scattering from Rough Surfaces. *Phys Rev B* 38, 2297 (Aug 1, 1988).
[5] G. Palasantzas, Roughness Spectrum and Surface Width of Self-Affine Fractal Surfaces Via the K-Correlation Model. *Phys Rev B* 48, 14472 (Nov 15, 1993).
[6] G. Palasantzas, J. Krim, Effect of the Form of the Height-Height Correlation-Function on Diffuse-X-Ray Scattering from a Self-Affine Surface. *Phys Rev B* 48, 2873 (Aug 1, 1993).
[7] T. Salditt, T. H. Metzger, C. Brandt, U. Klemradt, J. Peisl, Determination of the Static Scaling Exponent of Self-Affine Interfaces by Non-specular X-Ray-Scattering. *Phys Rev B* 51, 5617 (Mar 1, 1995).
[8] T. Salditt *et al.*, Determination of the Height-Height Correlation-Function of Rough Surfaces from Diffuse-X-Ray Scattering. *Europhys Lett* 32, 331 (Nov 1, 1995).
[9] V. Holý, U. Pietsch, T. Baumbach, *High-resolution X-ray scattering from thin films and multi-layers*. Springer tracts in modern physics, (Springer, Berlin ; New York, 1999), pp. xi, 256 p.
[10] Z. H. Ming *et al.*, Microscopic Structure of Interfaces in Si1-Xge/Si Heterostructures and Superlattices Studied by X-Ray-Scattering and Fluorescence Yield. *Phys Rev B* 47, 16373 (Jun 15, 1993).
[11] S. F. Edwards, D. R. Wilkinson, The Surface Statistics of a Granular Aggregate. *Proceedings of the Royal Society A* 381, (1982, 1982).
[12] D. G. Stearns, X-Ray-Scattering from Interfacial Roughness in Multi-layer Structures. *J Appl Phys* 71, 4286 (May 1, 1992).
[13] E. Spiller, D. Stearns, M. Krumrey, Multi-layer X-Ray Mirrors - Interfacial Roughness, Scattering, and Image Quality. *J Appl Phys* 74, 107 (Jul 1, 1993).
[14] D. G. Stearns, E. M. Gullikson, Non-specular scattering from extreme ultraviolet multi-layer coatings. *Physica B* 283, 84 (Jun, 2000).
[15] J. Daillant, A. Gibaud, *X-ray and neutron reflectivity : principles and applications*. Lecture notes in physics Monographs, (Springer, Berlin Heidelberg, 2009), pp. xxiii, 331 p.
[16] G. Renaud, R. Lazzari, F. Leroy, Probing surface and interface morphology with Grazing Incidence Small Angle X-Ray Scattering. *Surf Sci Rep* 64, 255 (Aug 31, 2009).

[17] M. Born, E. Wolf, *Principles of optics : electromagnetic theory of propagation, interference and diffraction of light.* (Cambridge University Press, Cambridge ; New York, ed. 7th expanded, 1999), pp. xxxiii, 952 p.

[18] O. Glatter, O. Kratky, *Small angle x-ray scattering.* (Academic Press, London ; New York, 1982), pp. 515 p.

[19] A. Guinier, *X-ray diffraction in crystals, imperfect crystals, and amorphous bodies.* A Series of books in physics (W.H. Freeman, San Francisco,, 1963), pp. x, 378 p.

[20] A. Guinier, G. Fournet, *Small-angle scattering of X-rays.* Structure of matter series (Wiley, New York,, 1955), pp. 268 p.

[21] L. A. Feigin, D. I. Svergun, G. W. Taylor, *Structure analysis by small-angle X-ray and neutron scattering.* (Plenum Press, New York [etc.], 1987), pp. XIII, 335 S.

[22] S. K. Sinha, Reflectivity Using Neutrons or X-Rays - a Critical Comparison. *Physica B* 173, 25 (Aug, 1991).

[23] S. K. Sinha, X-Ray Diffuse-Scattering as a Probe for Thin-Film and Interface Structure. *J Phys Iii* 4, 1543 (Sep, 1994).

[24] J. Als-Nielsen, D. McMorrow, *Elements of modern X-ray physics.* (Wiley, New York, 2001), pp. xi, 318 p.

[25] D. T. Attwood, *Soft x-rays and extreme ultraviolet radiation : principles and applications.* (Cambridge University Press, Cambridge ; New York, 2000), pp. xvi, 470 p.

[26] M. Birkholz, P. F. Fewster, C. Genzel, *Thin film analysis by X-ray scattering.* (Wiley-VCH, Weinheim, 2006), pp. xxii, 356 p.

[27] E. Spiller, *Soft X-ray optics.* (SPIE Optical Engineering Press, Bellingham, Wash., USA, 1994), pp. x, 278 p.

[28] V. Holy, T. Baumbach, Nonspecular X-Ray Reflection from Rough Multi-layers. *Phys Rev B* 49, 10668 (Apr 15, 1994).

[29] V. Holy, J. Kubena, I. Ohlidal, K. Lischka, W. Plotz, X-Ray Reflection from Rough Layered Systems. *Phys Rev B* 47, 15896 (Jun 15, 1993).

[30] S. Dietrich, A. Haase, Scattering of X-Rays and Neutrons at Interfaces. *Phys Rep* 260, 1 (Sep, 1995).

[31] I. A. Vartanyants, D. Grigoniev, A. V. Zozulya, Coherent X-ray imaging of individual islands in GISAXS geometry. *Thin Solid Films* 515, 5546 (May 23, 2007).

[32] A. V. Zozulya *et al.*, Imaging of nanoislands in coherent grazing-incidence small-angle X-ray scattering experiments. *Phys Rev B* 78, (Sep, 2008).

[33] H. Kiessig, Interferenz von Röntgenstrahlen an dünnen Schichten. *Ann. Phys. (Leipzig)* 10, 769 (1931).

[34] V. M. Kaganer, S. A. Stepanov, ouml, R. hler, Bragg diffraction peaks in x-ray diffuse scattering from multi-layers with rough interfaces. *Phys Rev B* 52, 16369 (1995).

[35] T. Salditt, T. H. Metzger, J. Peisl, Kinetic Roughness of Amorphous Multi-layers Studied by Diffuse-X-Ray Scattering. *Phys Rev Lett* 73, 2228 (Oct 17, 1994).

[36] G. Renaud, M. Ducruet, O. Ulrich, R. Lazzari, Apparatus for real time in situ quantitative studies of growing nanoparticles by grazing incidence small angle X-ray scattering and surface differential reflectance spectroscopy. *Nucl Instrum Meth B* 222, 667 (Aug, 2004).

[37] D. M. Smilgies, N. Boudet, B. Struth, O. Konovalov, Troika II: a versatile beamline for the study of liquid and solid interfaces. *J Synchrotron Radiat* 12, 329 (May, 2005).

[38] S. V. Roth *et al.*, Small-angle options of the upgraded ultrasmall-angle x-ray scattering beamline BW4 at HASYLAB. *Rev Sci Instrum* 77, 085106 (2006).
[39] J. R. Levine, L. B. Cohen, Y. W. Chung, P. Georgopoulos, Grazing-incidence small-angle X-ray scattering - new tool for studying thin-film growth. *J Appl Crystallogr* 22, 528 (DEC 1 1989, 1989).
[40] P. Siffalovic *et al.*, Characterization of Mo/Si soft X-ray multi-layer mirrors by grazing-incidence small-angle X-ray scattering. *Vacuum* 84, 19 (Aug 25, 2009).
[41] P. Laggner, M. Kriechbaum. (ICDD, 2007), vol. 22, pp. 190.
[42] H. Onuki, P. Elleaume, *Undulators, wigglers, and their applications.* (Taylor & Francis, London ; New York, NY, 2003), pp. x, 438 p.
[43] V. K. Pecharsky, P. Y. Zavalij, *Fundamentals of powder diffraction and structural characterization of materials.* (Springer, Berlin, ed. 2nd, 2009).
[44] J. Pedersen, A flux- and background-optimized version of the NanoSTAR small-angle X-ray scattering camera for solution scattering. *J Appl Crystallogr* 37, 369 (2004).
[45] R. Caciuffo, S. Melone, F. Rustichelli, A. Boeuf, Monochromators for x-ray synchrotron radiation. *Physics Reports* 152, 1 (1987).
[46] M. Montel, *The X-ray microscope with catamegonic roof-shape objective.* V. E. Cosslett, A. Engstrom, H. H. Pattee, Eds., X-Ray Microscopy and Microradiography (Elsevier, Amsterdam, 1957).
[47] P. Kirkpatrick, A. V. Baez, Formation of Optical Images by X-Rays. *J. Opt. Soc. Am.* 38, 766 (1948).
[48] U. Shymanovich *et al.*, Characterization and comparison of X-ray focusing optics for ultrafast X-ray diffraction experiments. *Applied Physics B* 92, 493 (2008).
[49] C. G. Schroer *et al.*, Hard x-ray nanoprobe based on refractive x-ray lenses. *Appl Phys Lett* 87, (Sep 19, 2005).
[50] A. Timmann *et al.*, Small angle x-ray scattering with a beryllium compound refractive lens as focusing optic. *Rev Sci Instrum* 80, 046103 (2009).
[51] P. Siffalovic *et al.*, Measurement of nanopatterned surfaces by real and reciprocal space techniques. *Meas Sci Rev* xx, xx (2010).
[52] B. B. He, *Two-dimensional x-ray diffraction.* (Wiley, Hoboken, N.J., 2009), pp. xiv, 426 p.
[53] S. V. Roth *et al.*, In situ observation of nanoparticle ordering at the air-water-substrate boundary in colloidal solutions using x-ray nanobeams. *Appl Phys Lett* 91, 091915 (2007).
[54] P. Siffalovic *et al.*, Real-Time Tracking of Superparamagnetic Nanoparticle Self-Assembly. *Small* 4, 2222 (Dec, 2008).
[55] H. Mimura *et al.*, Breaking the 10 nm barrier in hard-X-ray focusing. *Nature Physics* 6, 57 (Feb, 2010).
[56] B. L. Morgan, *Photoelectronic image devices 1991 the McGee Symposium proceedings of the 10th Symposium on Photoelectronic Image Devices held at Imperial College of Science, Technology and Medicine, London, 2-6 September 1991.* Institute of Physics conference series (Institute of Physics, Bristol [etc.], 1992), pp. XI, 440 S.
[57] M. W. Tate *et al.*, A Large-Format High-Resolution Area X-Ray-Detector Based on a Fiberoptically Bonded Charge-Coupled-Device (Ccd). *J Appl Crystallogr* 28, 196 (Apr 1, 1995).

[58] D. M. Khazins *et al.*, A parallel-plate resistive-anode gaseous detector for X-ray imaging. *Ieee T Nucl Sci* 51, 943 (Jun, 2004).
[59] D. Bourgeois, J. P. Moy, S. O. Svensson, A. Kvick, The point-spread function of X-ray image-intensifiers/CCD-camera and imaging-plate systems in crystallography: assessment and consequences for the dynamic range. *J Appl Crystallogr* 27, 868 (1994).
[60] P. Kraft *et al.*, Performance of single-photon-counting PILATUS detector modules. *J Synchrotron Radiat* 16, 368 (2009).
[61] C. M. Schlepütz *et al.*, Improved data acquisition in grazing-incidence X-ray scattering experiments using a pixel detector. *Acta Crystallographica Section A Foundations of Crystallography* 61, 418 (2005).
[62] T. Weber *et al.*, Reciprocal-space imaging of a real quasicrystal. A feasibility study with PILATUS 6M. *J Appl Crystallogr* 41, 669 (Aug, 2008).
[63] P. Kraft, Dissertation , Eidgenössische Technische Hochschule ETH Zürich, Nr 18466, 2010, ETH (2010).
[64] F. Hertlein, A. Oehr, C. Hoffmann, C. Michaelsen, J. Wiesmann, State-of-the-art of multi-layer optics for laboratory X-ray devices. *Part Part Syst Char* 22, 378 (May, 2006).
[65] C. Michaelsen *et al.*, Optimized performance of graded multi-layer optics for X-ray single crystal diffraction. *Advances in Mirror Technology for X-Ray, Euv Lithography, Laser, and Other Applications* 5193, 211 (2003).
[66] A. E. Yakshin, I. V. Kozhevnikov, E. Zoethout, E. Louis, F. Bijkerk, Properties of broadband depth-graded multi-layer mirrors for EUV optical systems. *Opt Express* 18, 6957 (Mar 29, 2010).
[67] E. Spiller *et al.*, High-performance Mo-Si multi-layer coatings for extreme-ultraviolet lithography by ion-beam deposition. *Appl Optics* 42, 4049 (Jul 1, 2003).
[68] E. Louis *et al.*, Progress in Mo/Si multi-layer coating technology for EUVL optics. *Emerging Lithographic Technologies Iv* 3997, 406 (2000).
[69] E. Louis *et al.*, Multi-layer coatings for the EUVL process development tool. *Emerging Lithographic Technologies IX, Pts 1 and 2* 5751, 1170 (2005).
[70] H. N. Chapman *et al.*, Femtosecond time-delay X-ray holography. *Nature* 448, 676 (Aug 9, 2007).
[71] E. Goulielmakis *et al.*, Single-cycle non-linear optics. *Science* 320, 1614 (Jun 20, 2008).
[72] A. Wonisch *et al.*, Design, fabrication, and analysis of chirped multi-layer mirrors for reflection of extreme-ultraviolet attosecond pulses. *Appl Optics* 45, 4147 (Jun 10, 2006).
[73] A. Wonisch *et al.*, Aperiodic nanometer multi-layer systems as optical key components for attosecond electron spectroscopy. *Thin Solid Films* 464-65, 473 (Oct, 2004).
[74] B. Schmiedeskamp *et al.*, Electron-Beam-Deposited Mo/Si and Mo(X)Si(Y)/Si Multi-layer X-Ray Mirrors and Gratings. *Opt Eng* 33, 1314 (Apr, 1994).
[75] U. Kleineberg *et al.*, Effect of substrate roughness on Mo/Si multi-layer optics for EUVL produced by UHV-e-beam evaporation and ion polishing. *Thin Solid Films* 433, 230 (Jun 2, 2003).
[76] A. Ulyanenkov *et al.*, X-ray scattering study of interfacial roughness correlation in Mo/Si multi-layers fabricated by ion beam sputtering. *J Appl Phys* 87, 7255 (May 15, 2000).

[77] S. Bajt *et al.*, Improved reflectance and stability of Mo-Si multi-layers. *Opt Eng* 41, 1797 (Aug, 2002).
[78] S. Braun *et al.*, Mo/Si-multi-layers for EUV applications prepared by Pulsed Laser Deposition (PLD). *Microelectron Eng* 57-8, 9 (Sep, 2001).
[79] A. Ulyanenkov, S. Sobolewski, Extended genetic algorithm: application to x-ray analysis. *J Phys D Appl Phys* 38, A235 (May 21, 2005).
[80] U. Kleineberg, Effect of substrate roughness on Mo/Si multi-layer optics for EUVL produced by UHV-e-beam evaporation and ion polishing. *Thin Solid Films* 433, 230 (2003).
[81] A. Anopchenko *et al.*, Effect of substrate heating and ion beam polishing on the interface quality in Mo/Si multi-layers - X-ray comparative study. *Physica B* 305, 14 (Oct, 2001).
[82] T. Tsuru, M. Yamamoto, In-situ ellipsometric monitor with layer-by-layer analysis for precise thickness control of EUV multilayer optics. *Thin Solid Films* 515, 947 (Nov 25, 2006).
[83] M. Yamamoto, Y. Hotta, M. Sato, A tracking ellipsometer of picometer sensitivity enabling 0.1 % sputtering-rate monitoring of EUV nanometer multi-layer fabrication. *Thin Solid Films* 433, 224 (Jun 2, 2003).
[84] A. V. Petukhov *et al.*, Microradian X-ray diffraction in colloidal photonic crystals. *J Appl Crystallogr* 39, 137 (Apr, 2006).
[85] R. P. A. Dullens, A. V. Petukhov, Second-type disorder in colloidal crystals. *Epl-Europhys Lett* 77, 58003 (2007).
[86] N. Stribeck, *X-ray scattering of soft matter*. Springer laboratory manuals in polymer science (Springer, Berlin, 2007), pp. 238 S.
[87] J. M. Slaughter *et al.*, Structure and performance of Si/Mo multi-layer mirrors for the extreme ultraviolet. *J Appl Phys* 76, 2144 (1994).
[88] M. Ishino *et al.*, Boundary structure of Mo/Si multi-layers for soft X-ray mirrors. *Jpn J Appl Phys 1* 41, 3052 (May, 2002).
[89] S. Braun, H. Mai, M. Moss, R. Scholz, A. Leson, Mo/Si multi-layers with different barrier layers for applications as extreme ultraviolet mirrors. *Jpn J Appl Phys 1* 41, 4074 (Jun, 2002).
[90] A. K. Srivastava, P. Tripathi, M. Nayak, G. S. Lodha, R. V. Nandedkar, Formation of Mo5Si3 phase in Mo/Si multi-layers. *J Appl Phys* 92, 5119 (Nov 1, 2002).
[91] S. Bajt, D. G. Stearns, P. A. Kearney, Investigation of the amorphous-to-crystalline transition in Mo/Si multi-layers. *J Appl Phys* 90, 1017 (Jul 15, 2001).
[92] X. M. Jiang, T. H. Metzger, J. Peisl, Non-specular X-Ray-Scattering from the Amorphous State in W/C Multi-layers. *Appl Phys Lett* 61, 904 (Aug 24, 1992).
[93] A. F. Jankowski, L. R. Schrawyer, M. A. Wall, Structural Stability of Heat-Treated W/C and W/B4c Multi-layers. *J Appl Phys* 68, 5162 (Nov 15, 1990).
[94] A. E. Yakshin, I. I. Khodos, I. M. Zhelezniak, A. I. Erko, Fabrication, Structure and Reflectivity of W/C and W/B4c Multi-layers for Hard X-Ray. *Opt Commun* 118, 133 (Jul 1, 1995).
[95] J. Wiesmann, C. Michaelsen, F. Hertlein, M. Stormer, A. Seifert, State-of-the-art thin film X-ray optics for conventional synchrotrons and FEL sources. *Synchrotron Radiation Instrumentation, Pts 1 and 2* 879, 774 (2007).

[96] F. Hertlein, S. Kroth, C. Michaelsen, A. Oehr, J. Wiesmann, Nanoscaled multi-layer coatings for X-ray optics. *Adv Eng Mater* 10, 686 (Jul, 2008).

[97] M. Schuster, H. Gobel, Parallel-Beam Coupling into Channel-Cut Monochromators Using Curved Graded Multilayers. *J Phys D Appl Phys* 28, A270 (Apr 14, 1995).

[98] P. Siffalovic *et al.*, Interface study of a high-performance W/B4C X-ray mirror. *J Appl Crystallogr* 43, 1431 (2010).

[99] T. Salditt *et al.*, Characterization of interface roughness in W/Si multi-layers by high resolution diffuse X-ray scattering. *Physica B* 221, 13 (Apr, 1996).

[100] T. Salditt *et al.*, Observation of the Huygens-principle growth mechanism in sputtered W/Si multi-layers. *Europhys Lett* 36, 565 (Dec 10, 1996).

[101] T. Salditt *et al.*, Interfacial roughness and related growth mechanisms in sputtered W/Si multi-layers. *Phys Rev B* 54, 5860 (Aug 15, 1996).

[102] W. W. Mullins, Theory of Thermal Grooving. *J Appl Phys* 28, 333 (1957).

[103] M. Schmidbauer, R. Opitz, T. Wiebach, R. Kohler, Inclined inheritance of interface roughness in semiconductor superlattices as characterized by x-ray reciprocal space mapping. *Phys Rev B* 64, art. no. (Nov 15, 2001).

[104] T. Tsarfati, R. W. E. van de Kruijs, E. Zoethout, E. Louis, F. Bijkerk, Nitridation and contrast of B4C/La interfaces and X-ray multi-layer optics. *Thin Solid Films* 518, 7249 (Oct 1, 2010).

[105] T. Tsarfati, R. W. E. V. de Kruijs, E. Zoethout, E. Louis, F. Bijkerk, Reflective multi-layer optics for 6.7 nm wavelength radiation sources and next generation lithography. *Thin Solid Films* 518, 1365 (Dec 31, 2009).

[106] D. Ksenzov *et al.*, Reflection of femtosecond pulses from soft X-ray free-electron laser by periodical multi-layers. *physica status solidi (a)* 206, 1875 (2009).

[107] C. Michaelsen *et al.*, Multi-layer mirror for x rays below 190 eV. *Opt Lett* 26, 792 (Jun 1, 2001).

[108] C. Michaelsen *et al.*, Recent developments of multi-layer mirror optics for laboratory x-ray instrumentation. *X-Ray Mirrors, Crystals, and Multi-layers Ii* 4782, 143 (2002).

[109] J. M. Andre *et al.*, La/B4C small period multi-layer interferential mirror for the analysis of boron. *X-Ray Spectrometry* 34, 203 (May-Jun, 2005).

[110] P. Ricardo, J. Wiesmann, C. Nowak, C. Michaelsen, R. Bormann, Improved Analyzer Multi-layers for Aluminium and Boron Detection with X-Ray Fluorescence. *Appl. Opt.* 40, 2747 (2001).

[111] S. Hendel, Bielefeld University (2009).

[112] J. O. Smith, *Mathematics of the discrete fourier transform (DFT) with audio applications*. (BookSurge Publishing, [S.l.], ed. 2nd, 2007), pp. 306 p.

[113] H. Lüth, *Solid surfaces, interfaces and thin films*. Advanced texts in physics (Springer, Berlin, ed. 4th, rev. and extended, 2001), pp. 559 S.

[114] J. A. Venables, *Introduction to surface and thin film processes*. (Cambridge University Press, Cambridge, 2000), pp. 372 S.

[115] M. Ohring, *Materials science of thin films deposition and structure*. (Academic Press, San Diego, ed. 2nd, 2002), pp. 794 S.

[116] C. C. Walton, G. Thomas, J. B. Kortright, X-ray optical multi-layers: Microstructure limits on reflectivity at ultra-short periods. *Acta Mater* 46, 3767 (Jul 1, 1998).

[117] A. F. Jankowski, D. M. Makowiecki, M. A. Wall, M. A. McKernan, Subnanometer multi-layers for x-ray mirrors: Amorphous crystals. *J Appl Phys* 65, 4450 (1989).

[118] M. N. Baibich *et al.*, Giant Magnetoresistance of (001)Fe/(001)Cr Magnetic Superlattices. *Phys Rev Lett* 61, 2472 (1988).
[119] M. Ziese, *Spin electronics*. Lecture notes in physics (Springer, Berlin, 2001), pp. 493 S.
[120] J. C. S. Kools, W. Kula, D. Mauri, T. Lin, Effect of finite magnetic film thickness on N[e-acute]el coupling in spin valves. *J Appl Phys* 85, 4466 (1999).
[121] S. Tegen, I. Mönch, J. Schumann, H. Vinzelberg, C. M. Schneider, Effect of Néel coupling on magnetic tunnel junctions. *J Appl Phys* 89, 8169 (2001).
[122] F. J. A. Denbroeder, D. Kuiper, A. P. Vandemosselaer, W. Hoving, Perpendicular Magnetic-Anisotropy of Co-Au Multi-layers Induced by Interface Sharpening. *Phys Rev Lett* 60, 2769 (Jun 27, 1988).
[123] B. Szymanski, F. Stobiecki, M. Urbaniak, Changes in magnetic and magnetoresistive characteristics of Ni-Fe/Au/Co/Au multi-layers induced by annealing. *Phys Status Solidi B* 243, 235 (Jan, 2006).
[124] S. Stavroyiannis *et al.*, Low-field giant magnetoresistance in (111)-textured Co/Au multi-layers prepared with magnetron sputtering. *J Appl Phys* 84, 6221 (Dec 1, 1998).
[125] H. P. Sun, Z. Zhang, W. D. Wang, H. W. Jiang, W. Y. Lai, Microstructure of columnar crystallites in Ni80Fe20/Cu magnetic multi-layers. *J Appl Phys* 87, 2835 (Mar 15, 2000).
[126] T. Kehagias *et al.*, Growth of fcc Co in sputter-deposited Co/Au multi-layers with (111) texture. *J Cryst Growth* 208, 401 (Jan, 2000).
[127] M. Hecker *et al.*, Thermally induced modification of GMR in Co/Cu multi-layers: correlation among structural, transport, and magnetic properties. *J Phys D Appl Phys* 36, 564 (Mar 7, 2003).
[128] Y. Yoneda, Anomalous Surface Reflection of X Rays. *Physical Review* 131, 2010 (1963).

In: X-Ray Scattering
Editor: Christopher M. Bauwens
ISBN: 978-1-61324-326-8

Chapter 2

In Situ, Real-Time Synchrotron X-Ray Scattering

Bridget Ingham
Industrial Research Ltd., Lower Hutt, New Zealand

Abstract

Advances in synchrotron X-ray scattering techniques and X-ray detector technology in recent years are allowing more and more possibilities for *in situ*, real-time experiments to be conducted. Many of these experiments require the construction of a specialised cell that can survive and/or contain the *in situ* environment (e.g. solution composition, temperature, gas pressure, etc.). In this chapter, cell design considerations will be discussed and examples given of two *in situ* cells that have been constructed and used in real-time experiments: nanoparticle synthesis, and electrochemical deposition of thin films.

1. Introduction

The ongoing development of synchrotron sources offers opportunities to explore materials systems on smaller length scales and faster time scales than ever before. Synchrotron sources are becoming increasingly accessible to researchers worldwide, as new facilities are being constructed and older facilities upgraded and expanded. Synchrotrons offer many advantages over laboratory instruments. Synchrotron light is broad spectrum electromagnetic radiation, extending from the infrared to hard X-rays. It is several orders of magnitude brighter than laboratory sources, and the X-ray energy is tuneable, through the use of monochromators to select the desired wavelength. Undulator insertion devices in third generation synchrotron sources produce highly coherent light that is orders of magnitude brighter again.

This intense brightness increases the signal-to-noise ratio of a measurement dramatically. Synchrotron sources, therefore, have great value for studying nanoscale, dilute, and weakly scattering systems (including thin films, nanoparticles, nanocrystals in a matrix, polymers, liquids, organic films, etc.). They also enable the measurement time required to be greatly reduced. The ability to choose the wavelength of the X-ray source is also advantageous. High energy X-rays are able to penetrate reasonably thick cell windows and solution layers without being greatly absorbed by them. In addition, the tuneability of the X-rays allows anomalous diffraction and anomalous scattering experiments to be performed, where measurements are recorded at X-ray energies spanning an absorption edge of one or more of the elements present in a sample. This yields element-specific information about the system. Such techniques are beyond the scope of this chapter.

In some situations, the measurement time can be made sufficiently short that chemical and physical processes are able to be followed in real-time. For example, if measurements can be recorded every minute, and the process takes an hour, this gives sixty points in a time series. There are many benefits to being able to follow these processes in real time, without stopping a reaction: (a) taking the sample out of its *in situ* environment (e.g. out of solution) can often change its structure; (b) some events proceed too quickly to quench the system for *ex situ* measurements, and/or quenching may not even be possible; (c) intermediate or metastable phases may be formed; (d) observing the same sample throughout the measurement provides continuity; etc.

The variety of *in situ* apparatuses that can be built or purchased is immense. Experiments have been reported using high pressure diamond anvil cells, vacuum cells; high temperature, cryogenic temperature, controlled gas mixtures, magnetic fields, Langmuir troughs, etc. These all enable new regions of parameter space to be explored for a wide variety of systems.

1.1. Detector Considerations

While point detectors such as photodiodes, scintillators, and photon counters are common X-ray detectors, they must be operated in scanning mode in angular space. Thus it takes several minutes to collect a pattern with a point detector, even with a synchrotron X-ray source, and one must always compromise between the counting time at each point and the angular range and resolution required.

The use of 1D strip and 2D area detectors is one major factor in reducing the time required for a measurement, since they enable an entire diffraction pattern or scattering profile to be collected in a single exposure. There are several different types of area detectors available, each with advantages and disadvantages. Choosing the best detector for an experiment is a critical part of experiment planning. Factors to bear in mind are:

- What is the minimum angular resolution required? For nanoparticles or disordered systems, diffraction peaks are broad and the angular resolution will not need to be as high as for systems containing larger crystallites.
- What is the minimum measurement time required? For some detectors, the read-out time can greatly limit the achievable time resolution of the experiment.
- What is the maximum count rate expected? Detectors can be overloaded or behave in a non-linear fashion if the count rate is too high.

– What level of energy discrimination is required? A sample irradiated with X-rays of an energy just above its absorption edge will fluoresce; some detectors are better than others for disregarding lower energy photons.

There are several classes of detector available. These are summarised in a general sense in Table I.

Table I. Typical detector properties

	CCD	Image Plate	Pilatus	Mythen
Pixel size (μm)	20-100	100-150	170	0.004°
# pixels	Up to 4000 x 3000	Up to 2300 x 2300	2463 x 2527 (6M)	80° = 20,000 (1D strip)
Maximum counts	60,000 (16 bit)	130,000 (17 bit)	1,000,000 (20 bit)	250,000 (18 bit)
Readout time	~ 1 s	30-80 s	5 ms	0.3 ms
Detection method	Scintillator crystal	Photo-stimulated luminescence	Single photon counting	Single photon counting
Noise level	10	1	~0	1

CCDs were first developed in the 1970s and used in many devices today (both in research and personal use, e.g. digital cameras). X-ray detectors are scintillator devices, and the amplification and reading of the electrical signal introduces noise. The maximum count rate per pixel is usually around 60,000 and the readout time is usually a few seconds, depending on the size of the detector. The advantage of some CCD detectors is that they currently offer the smallest pixel size of the area detector options; this can be as small as 20 μm.

Image plates, such as the Mar345, generally have lower noise than a CCD, a higher dynamic range (up to 130,000 counts per pixel), and larger pixel size. They operate, as the name implies, with a plate made of an X-ray storage phosphor material. X-rays create excited electrons in the storage phosphor which become trapped in metastable states. The plate is then read by illuminating each pixel with light which enables the electrons to return to its ground state, releasing light of another wavelength. This process is called photostimulated luminescence, and the intensity of the luminescence is proportional to the initial X-ray dose. The major disadvantage of image plates for use in *in situ* experiments is the long readout time, which can take several minutes depending on the size of the plate.

In the last 5-10 years, single photon counting detectors have been developed by the Paul Scherrer Institute in Switzerland. These include the Pilatus (2D) and Mythen (1D) detectors, which are available in a range of sizes. These have virtually no noise and very fast readout times. At present the pixel size of the Pilatus area detectors is limited to around 170 μm. They also have a large dynamic range - up to 1,000,000 counts per pixel [1-6]. The Mythen strip detector spans several tens of degrees in arc with a step size of 0.004° [7-9].

1.2. Geometrical Considerations

It is worthwhile at this point to introduce the scattering vector, **Q**. This is the difference between the incident and exit X-ray vectors, **k** and **k'**, as shown graphically in Fig. 1. The magnitude of **Q** and its direction in relation to the sample or substrate is controlled by setting the substrate and detector angles relative to the incidence beam, which is obviously fixed for synchrotron instruments. The crystal planes that are probed are those for which the vector normal to those planes is in the direction of **Q**. The magnitude of **Q** is given by $Q = \frac{4\pi}{\lambda}\sin\theta$, where λ is the X-ray wavelength and 2θ is the scattering angle.

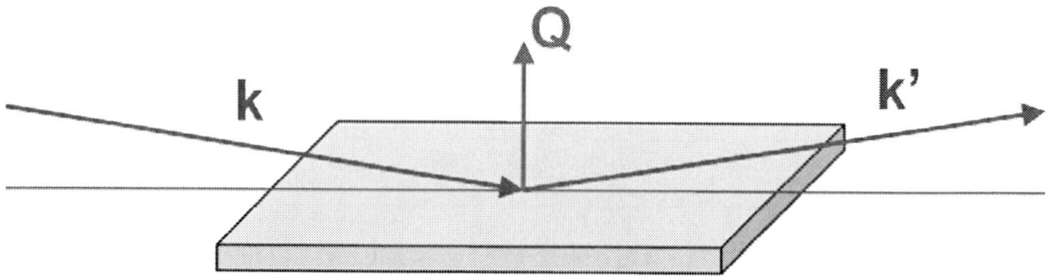

Figure 1. Relation of the scattering vector, **Q**, to the incident and exit X-ray vectors **k** and **k'**.

Experimental geometries used for X-ray scattering broadly fall into two kinds: transmission, where the X-ray beam passes through the sample; and reflection, where the X-ray beam is reflected off the sample. These are shown graphically in Fig. 2.

Conventional laboratory X-ray diffraction (XRD) experiments typically operate in a symmetric Bragg (θ-2θ) geometry; i.e. the incident and exit angles are both equal. In this case, the scattering vector **Q** is always normal to the sample surface (in reflection geometry). However, when an area detector is used, X-rays are detected at many exit angles for a single incidence angle. Therefore the direction of **Q** relative to the sample is not the same at all points. This is not an issue when the sample is isotropic, but it can pose a problem in the interpretation of X-ray diffraction patterns collected from textured samples (samples showing preferential orientation). This will be discussed in more detail in Section 3.2.

Transmission geometries (Fig. 2a) are frequently used for studying colloidal samples in a solution. In these experiments it is desirable to keep the solution scattering and therefore the X-ray path length through the solution to a minimum, both in the design of the cell and the angle the cell is placed at during the measurement. When area detectors are in use, the small X-ray path length through solution will also limit the peak broadening introduced by the sample thickness (as diffraction will arise from the front and back of the sample).

Reflection geometries (Fig. 2b) are used to study bulk samples and thin film samples on substrates. In the case of thin films, often the use of grazing incidence reflection is desirable in order to minimise the substrate contribution to the diffraction pattern. However, one needs to take care when attempting to record patterns using an area detector when the sample is in a grazing incidence geometry, as the size of the beam footprint on the sample can be large, the diffraction peaks will be broadened because of this, since there is no post-sample collimation.

Small-angle X-ray scattering (SAXS) experiments are notoriously difficult to perform using a reflection geometry (also called grazing incidence SAXS (GISAXS)). The sample must be flat over a large area (tens of millimetres), and background subtraction can be difficult. While successful *in situ* GISAXS experiments have been reported, the vast majority of *in situ* SAXS work is performed in a transmission geometry.

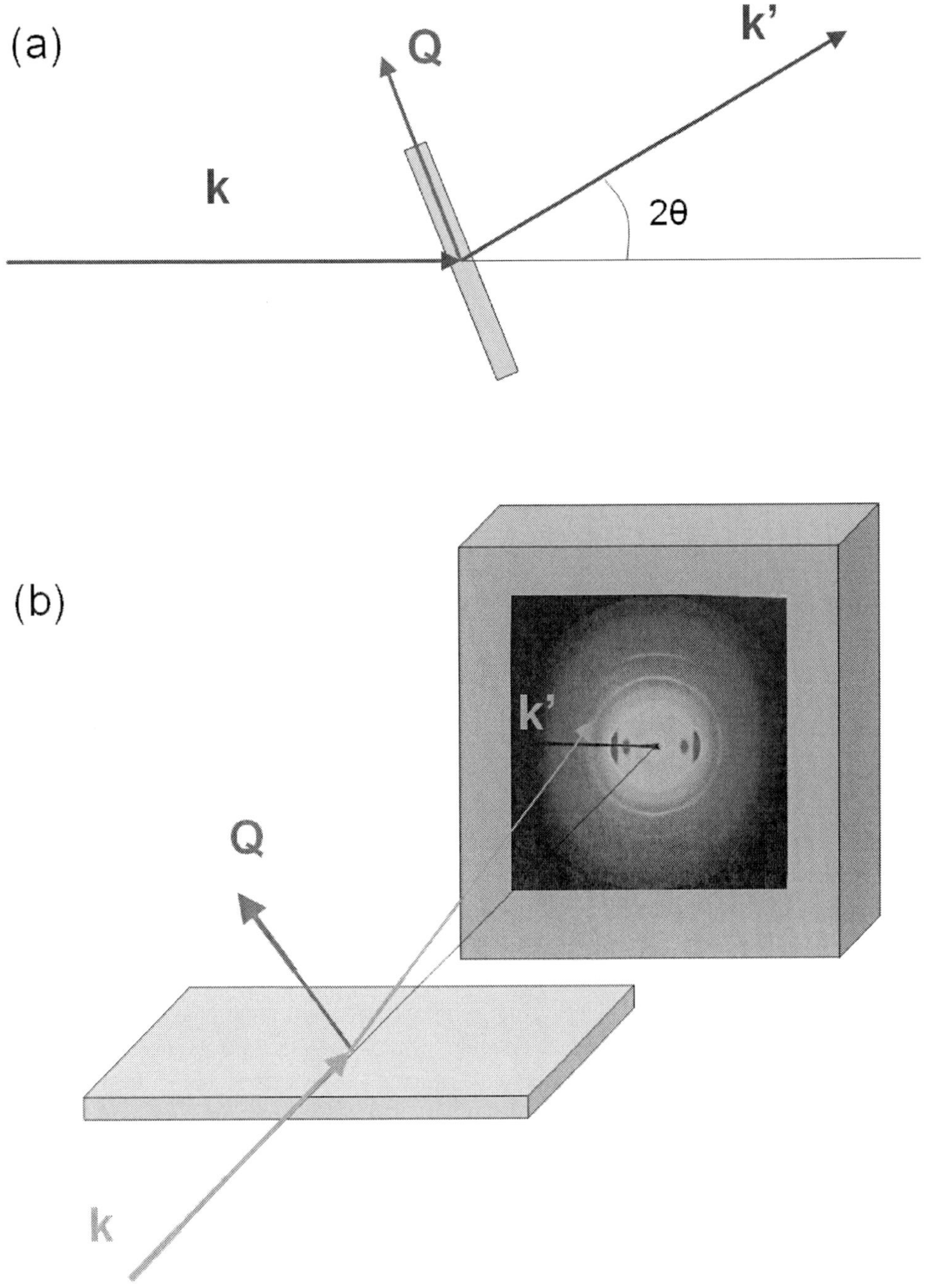

Figure 2. (a) Transmission and (b) reflection geometries used in X-ray scattering experiments.

The remainder of this chapter will describe two experimental setups used for various *in situ* X-ray scattering experiments. The first is a chemical reaction vessel used to study nanoparticle formation from a precursor solution in a transmission geometry. The second is an electrochemical cell with heating capability, used to study the formation of iron carbonate corrosion product films on mild steel, and the electrochemical deposition of ZnO nanostructured films.

2. Nanoparticle Formation Cell

Solution synthesis of nanoparticles has advantages over other methods such as chemical vapour deposition, sputtering, inert gas aggregation, and so forth. Samples with a high degree of size monodispersity can be formed by careful selection of an appropriate surfactant, with the size of the particle controlled by the relative concentrations of the surfactant and metal precursor salt. Non-spherically-shaped particles are often reported. These may be highly faceted [10-15], have high aspect ratios (e.g. nanorods and nanowires) [16-18], or have dendritic structures with high surface areas [19-26]. Natural questions arise as to how these structures nucleate and grow, since structural observations are often limited to post-synthesis analysis techniques, such as *ex situ* transmission electron microscopy (TEM), XRD, etc. It can be difficult to quench the reaction at pre-selected times in order to make *ex situ* measurements on the nucleating structures. Therefore *in situ* investigations of the nanoparticles' size and shape and how they nucleate and develop, are desirable.

In the laboratory, synthesis of noble metal nanostructures (for use in catalysis) can be conducted using a Fischer-Porter bottle, pressurised with hydrogen gas. The metal precursor and surfactant solution is placed in the bottle, which is then evacuated and backfilled with hydrogen, sealed, and placed in an oven for several hours. Pt [23, 27], Pd [19, 28], Ru [29], Ni [30] and Sn [31] nanoparticles have all been produced in this way.

The *in situ* cell needs to mimic the attributes of the Fischer-Porter bottle: to withstand temperatures up to 150°C, to be evacuated and backfilled with hydrogen, to withstand a slight overpressure, and to not react with the solution. In addition, it needs to be compatible with the X-ray scattering setup: low X-ray background windows, minimal solution thickness (to minimise the solution scattering and absorption of the X-rays), wide enough exit angles on the windows to enable the necessary diffraction peaks to be detected (this latter requirement is obviously less critical for a small-angle X-ray scattering experiment). It is clear that a transmission geometry will be the most efficient for these experiments.

The cell we have designed and used with great success is made of stainless steel with a valve adaptor attachment, as shown in Fig. 3.

The cell is assembled by clamping the windows in place, injecting the solution using a syringe, then attaching the valve piece. The windows are circular pieces of Kapton (for XRD) or mica (for SAXS) and are held in place with silicone O-rings while the window supports are screwed into position. When the cell is assembled, the reaction volume is around 0.3 cm^3. The incident aperture is 5 mm in size and straight, while the exit aperture is tapered to allow an acceptable angular range for diffraction patterns to be recorded over a 2θ range of -24 to 24°. Using an X-ray energy of 16 keV this gives a Q_{max} of 3.35 $Å^{-1}$, which is sufficient to observe the (111) and (200) peaks of Pt. The valve has standard Swagelok fittings to attach it to the gas system.

Once the solution has been inserted and the valve piece attached, the cell is flushed with hydrogen, sealed, and disconnected from the gas supply. It can then be transported to the diffractometer (in this case, the synchrotron XRD or SAXS beam line).

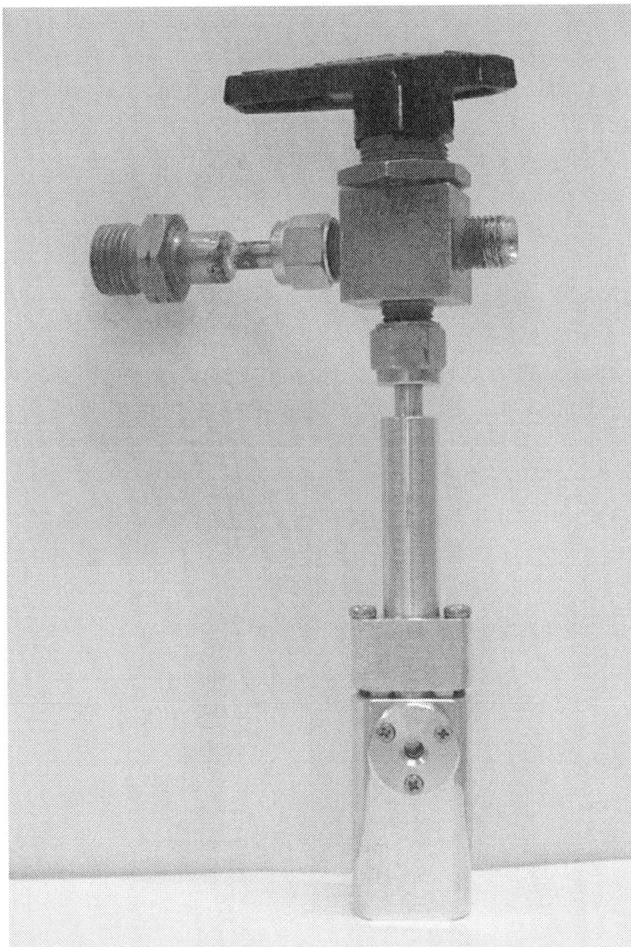

Figure 3. Photograph of the *in situ* cell developed for X-ray scattering experiments.

The reaction cell is inserted into a heating unit mounted on the diffractometer. This heating unit consists of an aluminium chamber with entrance and exit apertures, of which the exit aperture is tapered. The aluminium chamber is surrounded by a layer of thermal insulation material. Heating is supplied by a number of (8-10) cartridge heaters similar to those found in soldering irons, inserted into holes drilled into the walls of the chamber. These easily permit heating up to 300°C. A thermocouple is mounted in the aluminium block and the heating unit is controlled by a standard temperature controller. The heating unit is versatile and can be installed on both SAXS and XRD diffractometers. Being able to use the same equipment to perform two different experiments allows the results to be compared directly; differences in the time scale of the reaction due to the cell setup, are eliminated.

Experiment planning (in terms of the detector, data collection time, step size etc.) strongly depends on the time scale of the reaction and the scattering strength of the material being formed. Pt nanoparticles form over a time scale of roughly 8 hours [32], whereas Pd nanoparticles form over a time scale of 1 hour or less [19]. A strip or area detector could be used in either case, and for the Pt formation experiments, measurement times of 5-10 minutes using a point detector in continuous scanning mode would be acceptable. There will always need to be compromises made between the desired angular range, the angular resolution (step size or pixel spacing), and measurement time. Since these are nanoparticles, the peaks are expected to be broad and so the pixel size of an area detector would generally not be a limiting factor. In addition, the particles are expected to be isotropic since they are suspended in solution, meaning that line scans (such as those obtained using a point or strip detector) will capture all the necessary information.

2.1. Pt Nanoparticles

As mentioned above, Pt nanoparticles can be formed in the *in situ* reaction cell over a period of about 8 hours at 70°C [32]. Experiments were conducted at the Stanford Synchrotron Radiation Lightsource (SSRL) on beam lines 7-2 (XRD) and 4-2 (SAXS). For the XRD experiments, an X-ray energy of 16 keV was used (λ = 0.775 Å). A point detector was used with 1 mrad Soller slits for post-sample collimation. XRD patterns were recorded in a continuous scanning fashion over a 2θ range of 18-24°, with a step size of 0.05° and a counting time of 2-3 seconds per point. This enables a scan encompassing the Pt (111) and (200) peaks to be completed every 5-10 minutes.

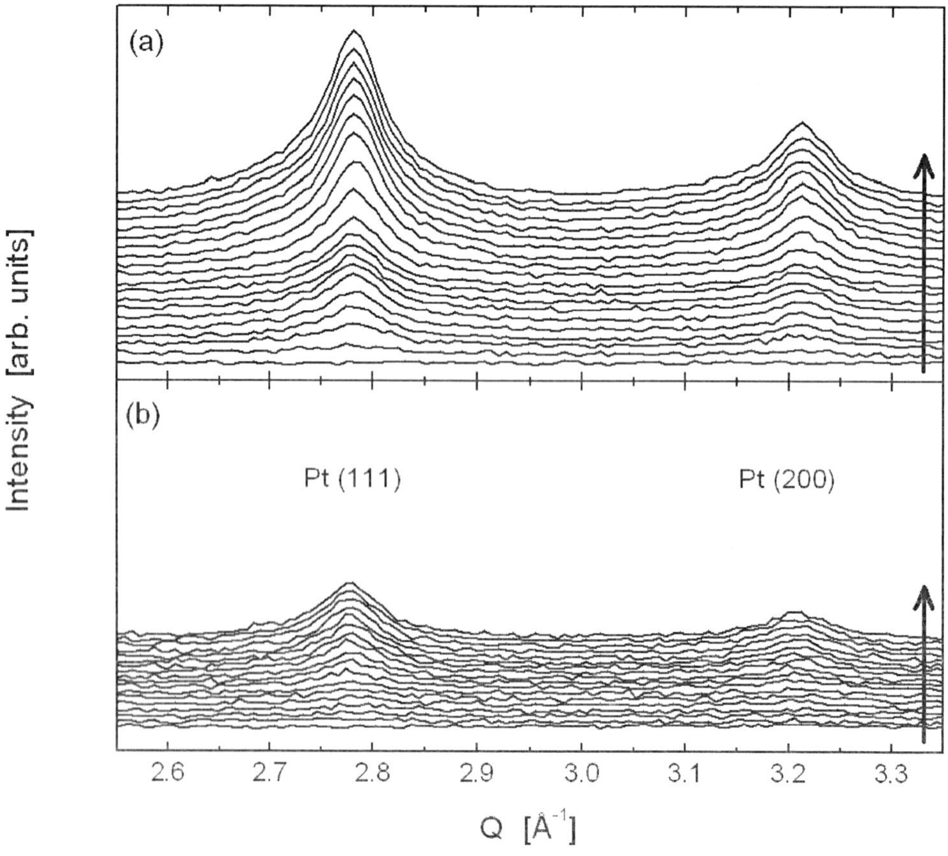

Figure 4. X-ray diffraction raw data of Pt nanoparticles formed from (a) high and (b) low concentration solutions. Data were collected for 500 minutes, time progresses as shown by the arrows.

The solution consists of Pt(II) acetylacetonate dissolved in toluene, with oleylamine added as the surfactant. Two experiment regimes were used: one with a 0.05 M Pt solution ('high') and one with a 0.005 M Pt solution ('low'). From previous work it had been shown that the 'high' Pt concentration resulted in high surface area, porous nanostructures, while the 'low' Pt concentration resulted in highly faceted, compact particles. XRD raw data for both experiments are shown in Figure 4. After about an hour, diffraction peaks are able to be observed.

The XRD patterns can be fitted using two peaks, and the peak area, full width at half-maximum (FWHM) and position obtained. These give information respectively about the relative amount of crystalline material, the average crystallite size, and the uniform lattice strain.

The peak positions do not change over time and are the same as those expected for bulk Pt, within uncertainty, indicating negligible uniform strain.

The peak FWHMs decrease over time, indicating that the crystallite size increases. These are both shown in Figure 5. For the low concentration experiment, this increase is monotonic and reaches a plateau of a crystallite size of around 8 nm at longer times. For the high concentration, the crystallite size increases to a plateau of around 8 nm, between 100 and 200 minutes, and then decreases slightly to a second plateau of around 7.5 nm at longer times.

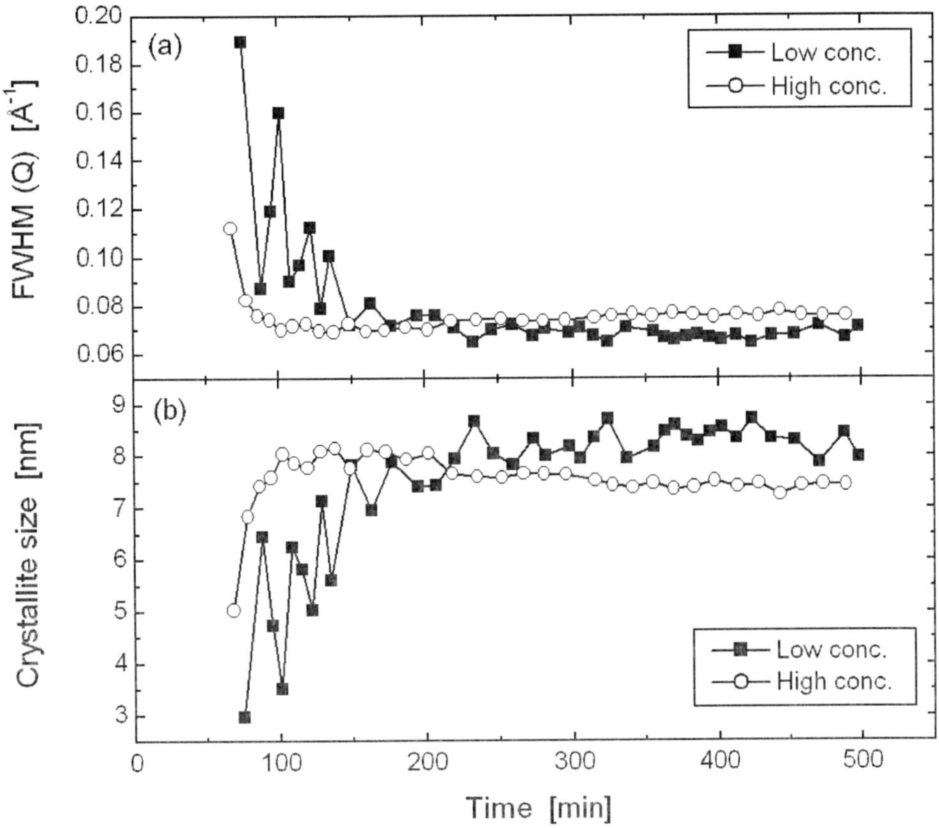

Figure 5. Pt (111) full-width at half-maximum (a) and corresponding crystallite size (b) from the data shown in Figure 4.

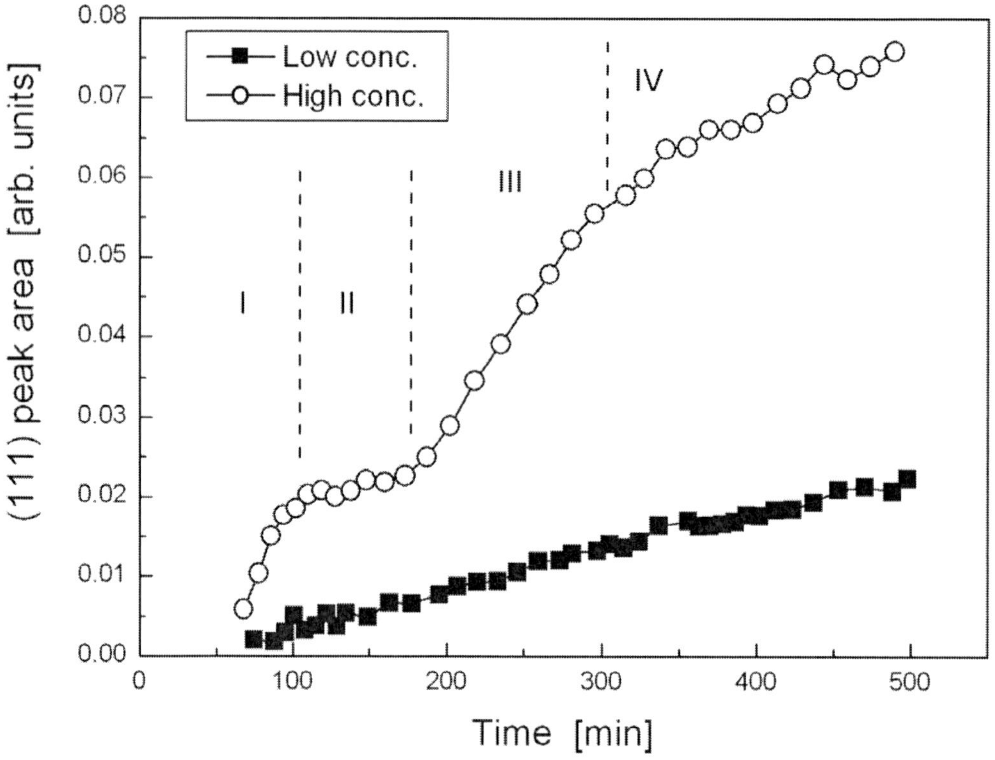

Figure 6. Pt (111) peak area as a function of time for nanoparticles prepared from high and low concentration solutions.

The peak areas in both the high and low Pt concentration experiments increase over time (Figure 6) and give insight into the kinetics of nucleation and growth. The low concentration results show that the area steadily increases in a nearly linear fashion, whereas in the high concentration experiment, several growth regimes are evident: (I) an initially high growth rate up to 100 minutes, followed by (II) a period from 100 to 180 minutes where the peak area is almost constant, then (III) a second period of rapid growth from 180 to 300 minutes, and finally (IV) a period of somewhat slower growth.

These results were complemented with *ex situ* TEM images recorded from samples where the reaction was ceased at intermediate times corresponding to these features of interest. For the low concentration experiment, samples were studied after 80 and 400 minutes (Figure 7). These show the formation of highly faceted cubic and triangular particles. This is consistent with the XRD results: the crystallite size increases monotonically and plateaus to a value of 8 nm (which is comparable with the average particle size), and the peak area increases steadily, indicating the steady growth of the crystallites. The resultant shapes and the growth kinetics indicate that these particles are formed in a thermodynamically controlled growth regime [32-35].

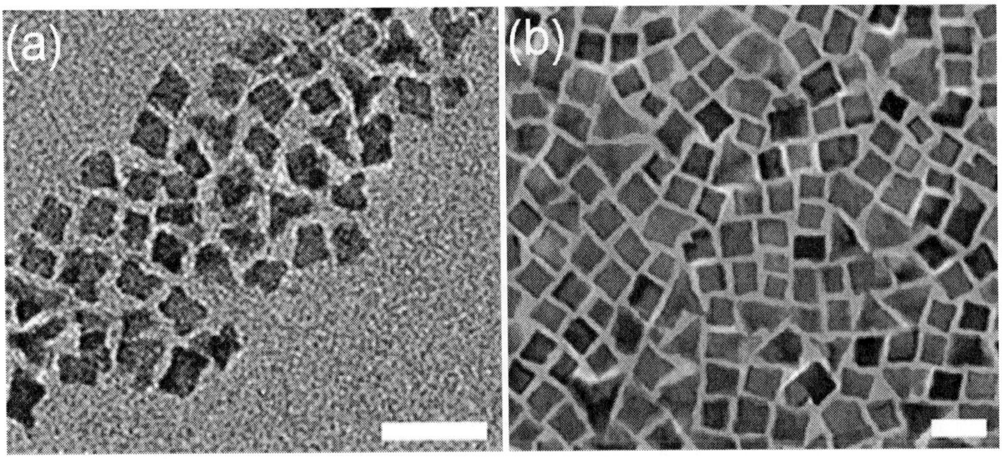

Figure 7. *Ex situ* TEM images of Pt nanoparticles formed from low concentration solutions, harvested after (a) 80 and (b) 400 minutes.

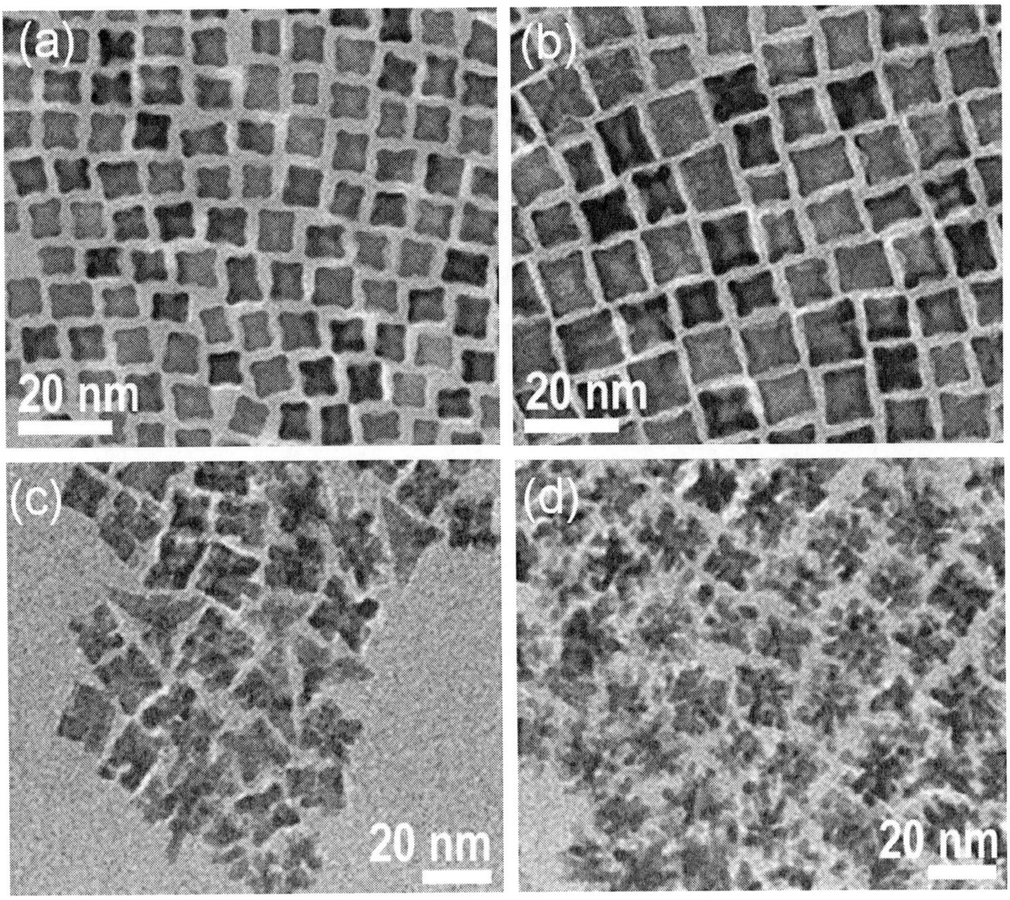

Figure 8. *Ex situ* TEM images of Pt nanoparticles formed from high concentration solutions, harvested after (a) 75, (b) 120, (c) 240, and (d) 500 minutes.

The high concentration experiment was ceased after 75, 120, 240, and 500 minutes and the harvested nanoparticles examined using TEM. The images are shown in Figure 8. They show a progression in the formation of porous, dendritic nanostructures from initial cubic, faceted nuclei. At intermediate times (120-240 minutes) the (100) facets of the initial nanocubes are preferentially dissolved away and branched structures with high surface area result. This explains the different growth stages observed in the XRD. The initial nucleation of the nanocubes is rapid during stage I. In stage II, the corners of the nanocubes continue to grow, while the faces are etched away. These two processes are nearly balanced out in terms of the amount of material being simultaneously dissolved and deposited, and so the peak area is nearly constant during this stage. In stage III, growth continues and more branches are formed. The formation of such branched structures is indicative of kinetically controlled growth [34]. In stage IV, this growth slows as the Pt^{2+} precursor is depleted.

In situ SAXS experiments have also been performed on the low concentration Pt system under the same conditions, using beam line 4-2 at SSRL with a sample-detector distance of 2.5 m and an X-ray energy of 9 keV. A MarCCD-165 detector was used. The raw data are shown in Figure 9. At early times, there is a single feature at around 0.03 $Å^{-1}$ that increases in intensity (corresponding to scattering objects of size ~ 10 nm). At later times, a second feature at 0.1 $Å^{-1}$ is observed to increase (corresponding to scattering objects of size ~ 3 nm). The data can be fitted using a hard sphere model with local monodisperse approximation [36-37] to obtain the average particle size and follow the relative volume of particles. These are shown in Figures 10 and 11 respectively. The shape of the 'scale' curve (time series progression of the total volume of particles) resembles that of the 'area' curves from the XRD results: a monotonically increasing function. The two size distributions are of interest: the first starts at about 1.5 nm and increases smoothly up to 11 nm, while the second arises after about 20 minutes and is nearly constant at 2.5 nm. This means that there is a bimodal size distribution of particles present, perhaps due to Ostwald ripening.

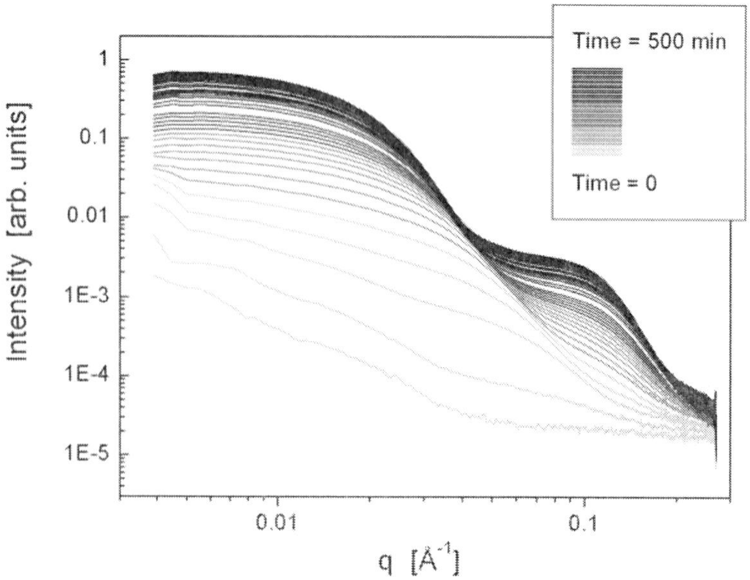

Figure 9. Raw *in situ* small-angle X-ray scattering data for Pt nanoparticles formed from a low concentration solution.

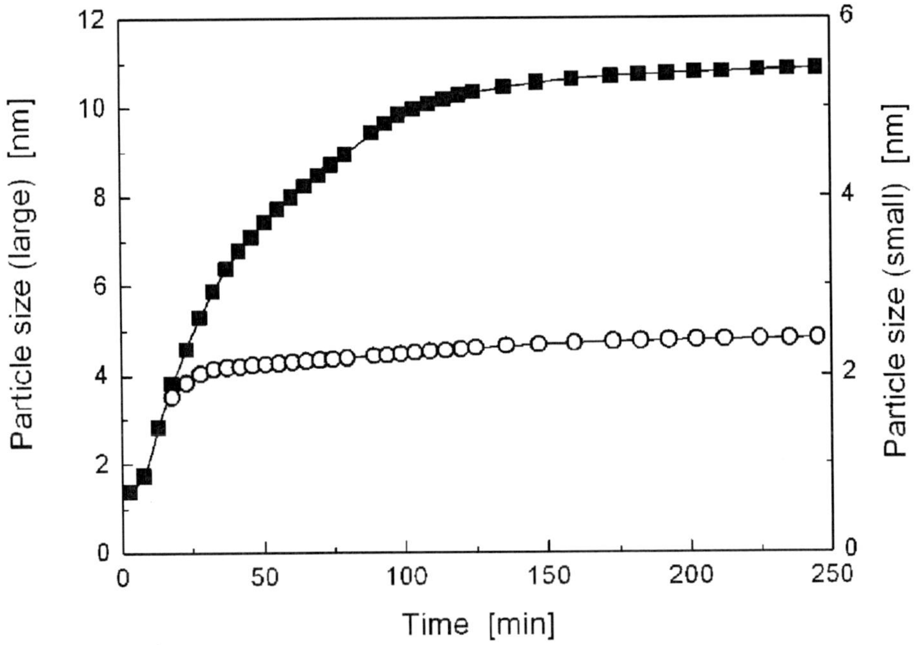

Figure 10. Respective particle sizes of the two distributions fitted to the data in Figure 9.

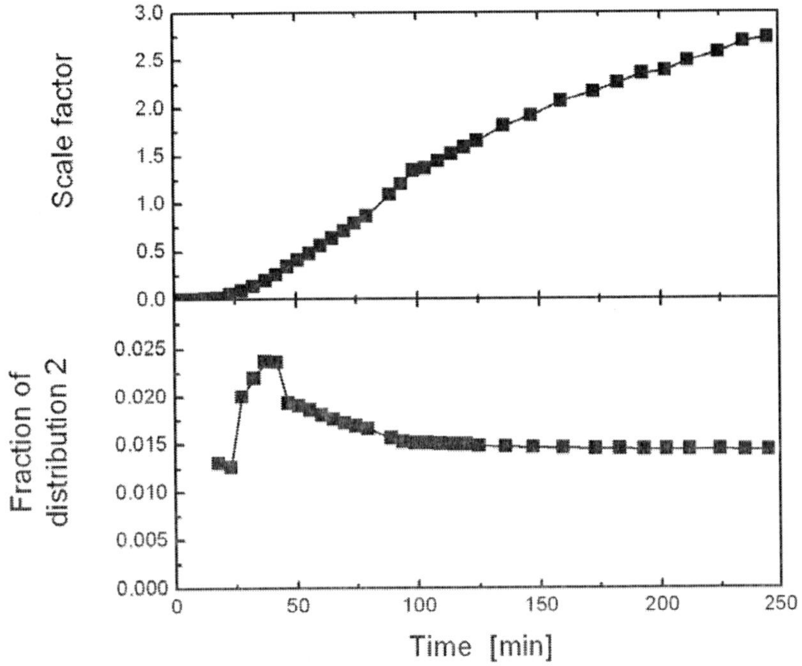

Figure 11. Scale factor (top - related to volume of scattering particles) and volume fraction of the smaller particle distribution, obtained from fitting the data in Figure 9 to a log-normal bimodal size distribution.

2.2. Pd Nanoparticles

In a similar synthesis method to the platinum nanoparticles in Section 2.1, palladium nanoparticles with a variety of morphologies can be produced by varying the chemical nature of the surfactant species, namely the ratio of oleylamine:oleic acid [19]. For instance, when 100% oleylamine is used, faceted spherical and polyhedral particles of around 10 nm in size are formed; whereas when a 1:1 mixture of oleylamine:oleic acid is used, highly branched nanostructures up to 100 nm in size can be produced [19]. The reaction time is typically about an hour. *In situ* XRD experiments were performed using beam line 7-2 at SSRL to follow the nucleation and growth. The scan range (covering the (111) peak only), step size, and measurement time per point were optimised so that each scan took 5 minutes. As for the Pt experiments, the peak in each scan was fitted to obtain the position, FWHM, and area. The peak position was found to be the same as that expected from bulk Pd, within experimental uncertainty, indicating no significant lattice strain. The peak FWHM was constant with time. The raw data and the areas obtained from the fits are shown in Figure 12.

In the experiment with 100% oleylamine, the growth rate is steady and slow compared to the 1:1 oleylamine:oleic acid experiment. Faceted spherical and polyhedral particles are produced and the growth is thermodynamically controlled.

For the 1:1 experiment, the growth rate is initially similar to that of the 100% oleylamine experiment. If the reaction is halted at this point, the same faceted particles are observed [19]. However after about 15 minutes the growth rate increases dramatically. *Ex situ* TEM shows the formation of branched nanostructures. The branching initially occurs in the [112] directions, creating tripods, with further dendritic growth in other directions at later times. Over time the structures grow to around 100 nm in size. Particles this large tend to settle in solution, out of the X-ray beam; this accounts for the drop in peak area towards the end of the experiment. However, as noted earlier, the peak FWHM (which is inversely proportional to the crystallite size) is relatively constant throughout the experiment. This indicates that the average crystallite size in the [111] direction is around 10 nm, even though the branched structures themselves are larger than this.

Palladium has the ability to absorb hydrogen into its crystal structure, which results in a noticeable lattice expansion [38-40]. This makes it appealing for applications in hydrogen gas sensing and storage. Nanostructured Pd shows size-dependent differences in the hydriding behaviour - the pressure-hydrogen concentration relation for the two primary hydride phases formed, and the size of the hydrogen concentration gap between them (known as the miscibility gap) [38, 40-42]. Experiments to explore the hydridation behaviour of these solution-synthesised Pd nanostructures [43] were performed at the Australian Synchrotron's powder diffraction beamline, using the same cell and a Mythen detector. This allows the entire diffraction pattern to be collected at once, which improves both the counting statistics and the time resolution.

Pd nanostructures were initially grown off-line by injecting the precursor solution (0.01 mM Pd^{2+} with 1 mM oleic acid in toluene) into the cell, filling with 3 bar H_2, and placing the sealed cell into an oven at 90°C for 2 hours. After this, the cell was transported to the beam line and measurements recorded every 1 minute. The X-ray wavelength was 1 Å (12.4 keV) and the accessible angular range was 2-35° (the cell was tilted in order to observe peaks at higher angles). This enabled the (111) and (200) peaks of the Pd and PdH_x phases to be observed (see Fig. 13). The peaks were fitted and the parameters for the peak position, area, and FWHM extracted for analysis.

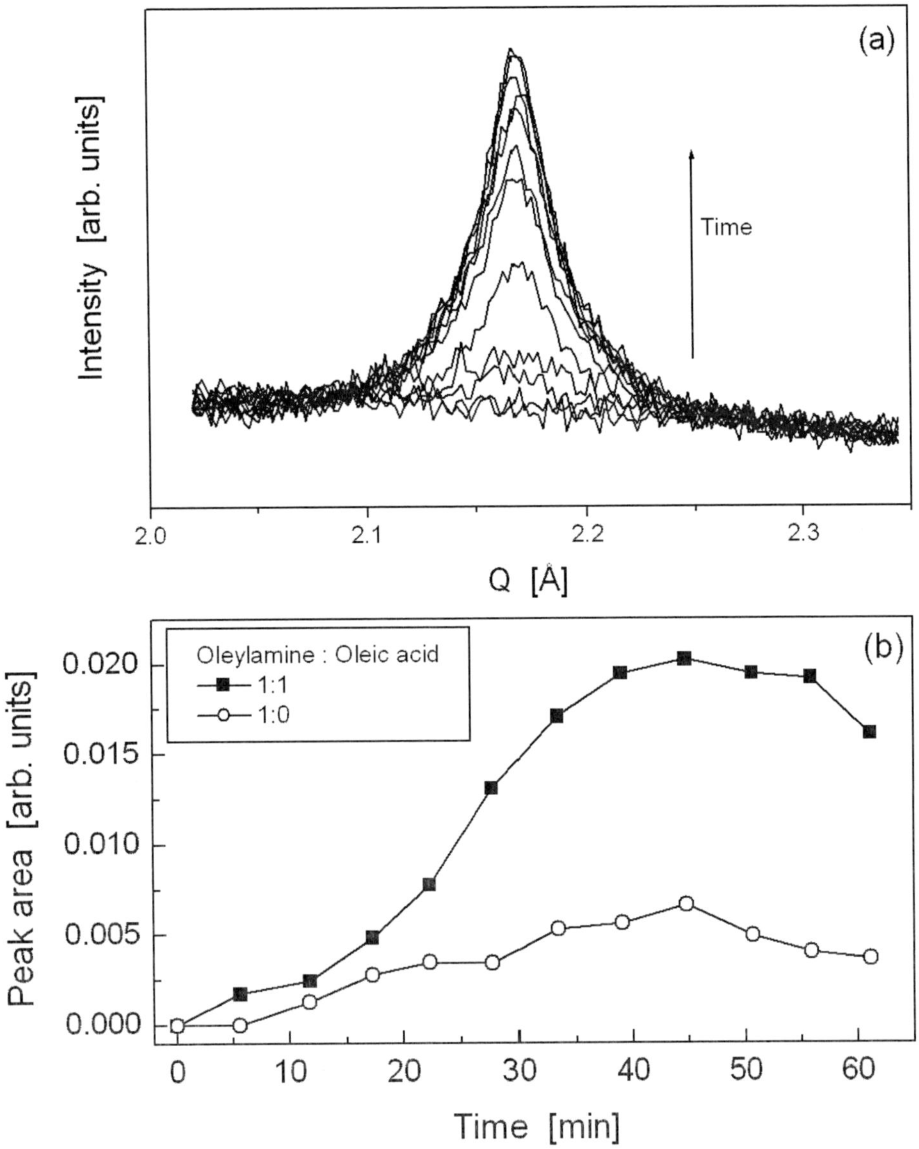

Figure 12. (a) Raw *in situ* X-ray diffraction data obtained from formation of Pd nanoparticles with 1:1 oleylamine:oleic acid surfactant; (b) Peak area versus time for formation of Pd nanoparticles from solutions with different surfactant species ratios.

From the peak positions, the lattice parameter can be extracted. Both Pd and PdH_x are cubic structures, and the lattice parameter is proportional to the hydrogen content, x [44]. The peak area ratios are proportional to the volume fraction of each phase. The peak FWHM is proportional to the crystallite size. The crystallite size was not observed to change significantly during the experiment. Figure 14 shows the correlation between the volume fractions and lattice parameter as a function of time.

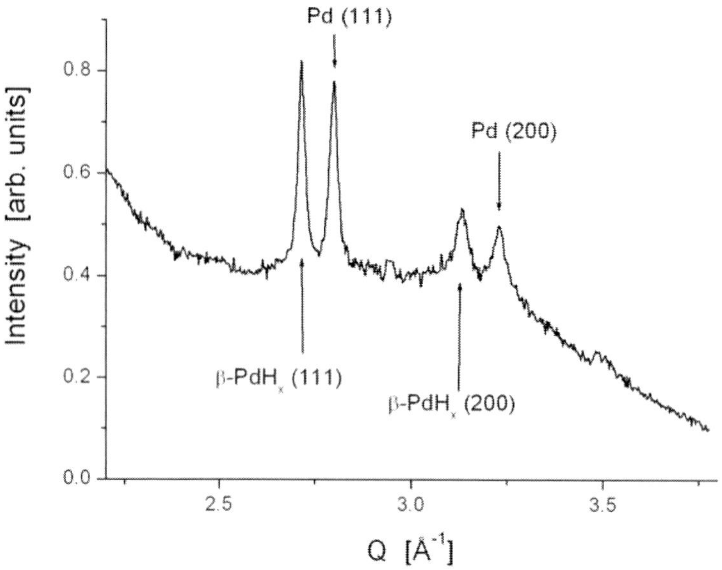

Figure 13. Example of an X-ray diffraction pattern obtained during a measurement showing hydridation of Pd nanoparticles. Peaks attributed to the Pd and PdH_x phases are labelled.

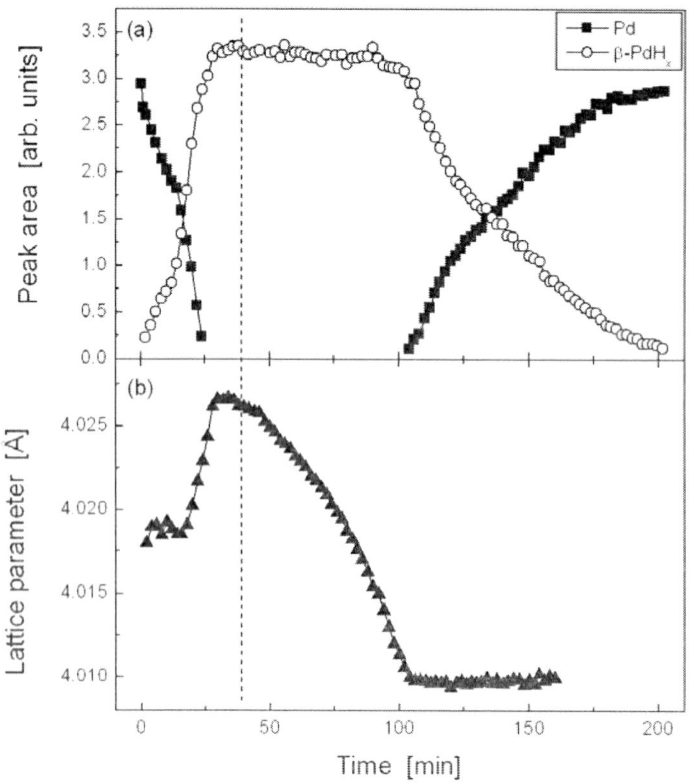

Figure 14. (a) Peak area of the Pd and PdH_x phases, proportional to the total volume of each phase, and (b) lattice parameter of the PdH_x phase, as a function of time.

Initially, the sample consists mostly of Pd. Within the first 25 minutes as the cell cools, the hydride phase forms. While the hydride phase is forming, the lattice parameter is constant (indicating a threshold concentration of hydrogen). Once all the Pd is converted to PdH_x, the lattice parameter starts to increase, indicating the hydrogen concentration is increasing. This indicates that the formation of the hydride at the threshold hydrogen concentration, is more energetically favourable than increasing the hydrogen content. Once the system appeared to stabilise (after about 45 minutes), the cell valve was opened to release the hydrogen. PdH_x is not stable under ambient conditions [38, 45-46]. Almost immediately, we start to see the lattice parameter of PdH_x decrease, as the hydrogen content decreases. PdH_x starts to convert to Pd once a lower threshold of hydrogen concentration is reached (at around 100 minutes); the lattice parameter of the hydride phase remains constant while the relative peak areas change. The threshold concentration for hydrogen desorption is lower than that for hydrogen absorption. This is a feature of the hysteretic behaviour of the miscibility gap in nanoscale Pd systems [38].

2.3. Final Comments

The solution reaction vessel described in this section has successfully been used for *in situ* synchrotron XRD and SAXS experiments to study the formation of metal nanoparticles, and subsequent processing (hydridation of Pd nanoparticles). Since the particles are suspended in solution, their scattering signal is isotropic and can be recorded using a 1D detector (strip detector, or scanning a point detector). The cell has many benefits: it mimics the laboratory setup (Fischer-Porter bottle) well, the small windows are able to withstand several bars of pressure *when* the cell is correctly assembled, and the X-ray scattering signal from the particles compared to the solution is acceptable. The reaction vessel can be placed inside a heating unit for temperature studies (we have tested it up to 150°C). The cell design could easily be applied to or adapted for use in other chemical systems.

3. Electrochemical Cell, with Heating Capacity

Electrochemistry is used in laboratories for many purposes, to study corrosion processes in a controlled manner, to dissolve and deposit materials (e.g. thin film deposition), to form or reduce oxide films on metals, etc. It enables chemical reactions to be controlled according to their known redox potentials, thus inducing or preventing certain reactions from occurring, and/or changing the reaction rates.

The simplest electrochemical cell geometry consists of three electrodes in an electrolyte solution: the counter, working, and reference electrodes. Potentials are measured or applied between the reference and working electrodes, while currents are measured or applied between the counter and working electrodes. The potential may be constant (for a potentiostatic experiment) or it may be ramped or cycled (cyclic voltammetry), while the current is monitored. Conversely, a constant current may be applied (a galvanostatic experiment) and the potential monitored.

When it comes to designing an experimental setup to be used for recording *in situ* X-ray scattering patterns during an electrochemical reaction, there are a number of factors to take into account. Once again, the cell design needs to be compatible with both the electrochemistry requirements and the X-ray scattering setup. There needs to be a reservoir of solution, three electrodes in some geometry, and a short X-ray path length through the solution. The sample is the working electrode, which is used in a reflection X-ray scattering geometry in nearly all experiments (often grazing-incidence).

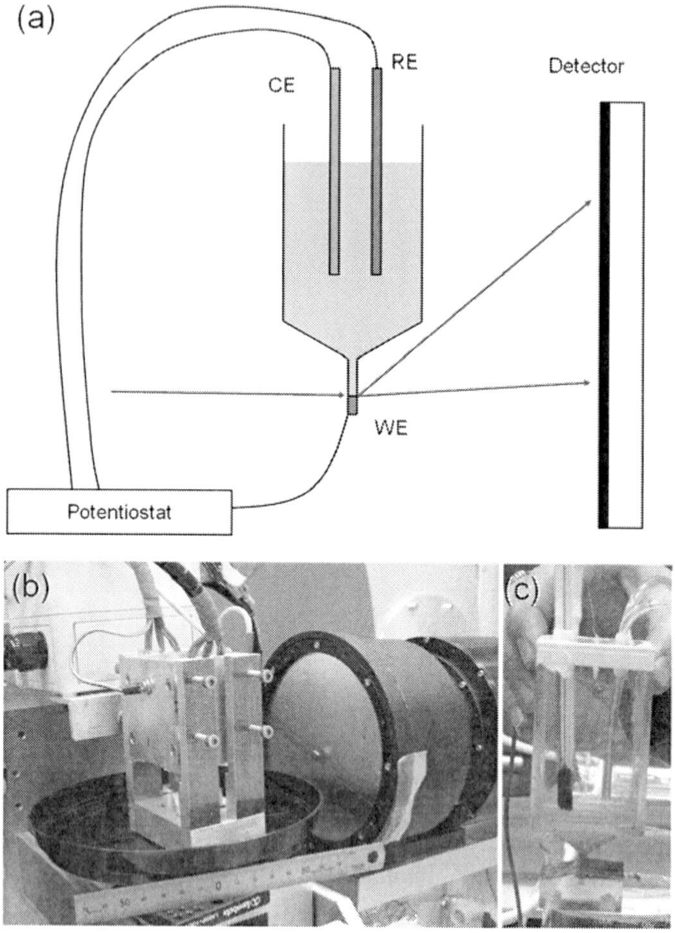

Figure 15. (a) Diagram of the electrochemical cell used for *in situ* X-ray scattering experiments. (b) Photograph of the cell on the Australian Synchrotron Powder Diffraction beam line. (c) Photograph of the Perspex cell insert, showing placement of the electrodes in the reservoir, and cross-section of the sample area at the base.

A diagram and photograph of our cell design is given in Fig. 15. This is comprised of a reservoir, into which the reference and counter electrodes are inserted, which is surrounded by a heating block (heating is provided by six cartridge heaters). The reservoir contains about 60 mL of solution and the bottom of it is tapered down to the working electrode, such that the solution thickness at this point is 2 mm. This distance was chosen to give an acceptable signal ratio of diffraction peaks to solution scattering, without being so short that ionic diffusion in the solution is greatly limited. The tapered design is necessary to allow a sufficient exit angle for the diffracted X-rays to be detected, while also attempting to minimise the distance from the bottom of the heater to the working electrode, so that the temperature drop is not too great. The cell is made of an inert material (Teflon or Perspex), and the windows of the cell in the tapered region are Kapton film, affixed to the cell using an appropriate adhesive. The reservoir is capped with a lid to minimise evaporation of solution and also enable a gas blanket to be used. The dimensions of the reservoir are sufficient to contain the two electrodes, a thermometer, and gas inlets and outlets.

The samples consist of rods which are inserted into a slot in the tapered base either from the reservoir side or from underneath. The top surface of the rod (2 mm x 2 mm) is the sample area being probed. X-rays strike the sample at a grazing incidence and diffraction patterns are recorded using a 2D area detector. The illuminated area is limited by the sample size (2 mm), so the peaks are not broadened greatly by the use of the grazing incidence geometry without post-sample collimation. Coating of the samples may be necessary, for example to minimise parasitic reactions (e.g. crevice corrosion). In electrodeposition this is generally not considered to be a problem as the small amount of solution down the sides of the rods is quickly depleted of the ions of interest, and further ionic diffusion is low. Electrical connection is made to the sample rod by a screw through the base (if the sample is inserted from the reservoir), or by direct connection to an uncoated part of the rod (if inserted from underneath).

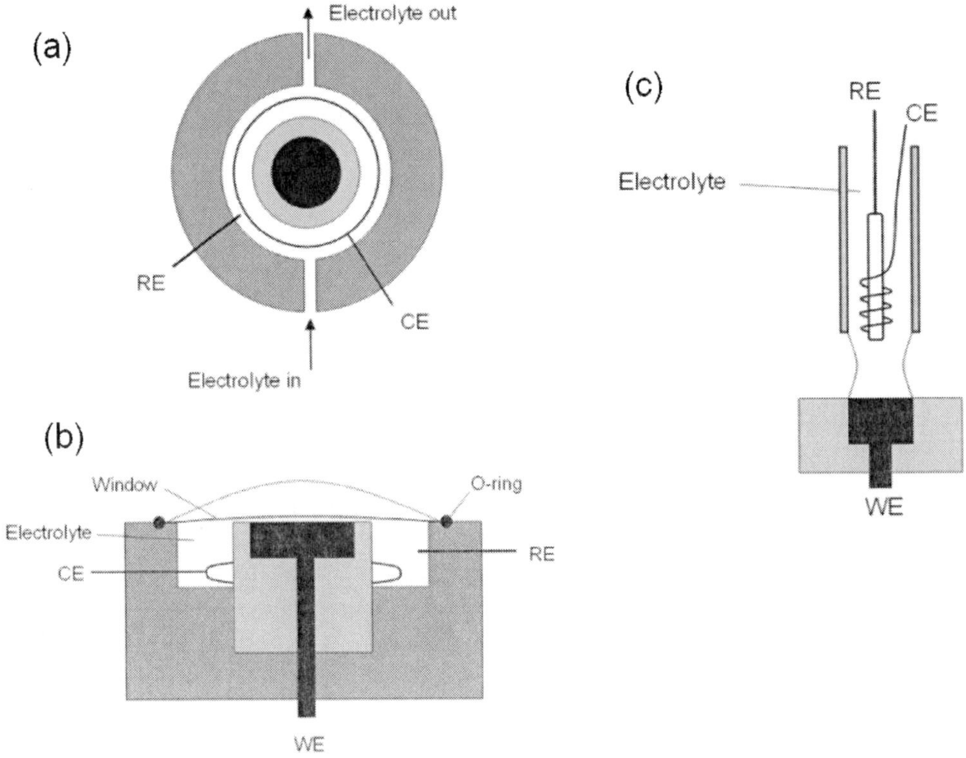

Figure 16. Diagrams of (a) top and (b) side view of a 'thin-layer' cell; (c) side view of a 'drop' cell.

There are several other designs of electrochemical cells used for *in situ* X-ray scattering experiments that have been reported in the literature. These can broadly be divided into two classes: thin-layer cells, and drop cells.

Thin-layer cells [47-55] have designs similar to Figure 16a-b. The working electrode is in the centre of the cell and the electrolyte reservoir is below it. The reference and counter electrodes are situated in the reservoir, and there are solution inlet and outlet ports. The cell window is made of a thin polymer (e.g. polypropylene, Mylar, Kapton) which is inflated by pumping solution into the cell. Electrochemistry is performed in the inflated configuration, then the cell is deflated to allow X-ray scattering measurements to be recorded. This cell design has advantages in that it allows for a large angular range to be measured, however it is not appropriate for time-resolved experiments since the solution layer above the electrolyte must be very thin while X-ray measurements are being recorded. The cells are also prone to leakage as the thin polymer window is continually inflated and deflated.

Drop cells [56-58] have designs similar to Figure 16c. These use a very small amount of solution (~ 1 mL) in a drop which is held in place by capillary forces between the working electrode at the base and the counter and reference electrode arrangement above it. They offer a fast response time and the possibility of performing *in situ*, real-time measurements. The disadvantages are in the drop size limitation (i.e. the amount of material available; there is no reservoir to replenish the reactants for electrodeposition or to allow diffusion of corrosion products away from the electrode), solution evaporation (which can be reduced by mounting the drop cell inside a humidity chamber), and the difficulty in miniaturising the electrode setup. In addition, it would be difficult to perform experiments at elevated solution temperatures using this kind of cell.

Our cell design offers advantages over both of these in that it allows for *in situ*, real-time X-ray scattering experiments to be performed. The scattering signal as a function of time can be related directly to the electrochemical response. The cell design is reasonably robust and has heating capability. It has successfully been used at X-ray scattering beam lines at both SSRL and the Australian Synchrotron [59-60]. The biggest issues with the use of the cell are three-fold. (1) In early designs, leaks formed particularly when the Teflon cell was operated at higher temperatures (80°C). This is because of the thermal stability of the adhesives used to attach the Kapton window to the cell. Use of the Perspex cell with the Kapton windows affixed using standard cyanoacrylate 'superglue' has virtually eliminated this problem. (2) Evaporation of solution when the cell is operated at high temperatures (>80°C) for long periods. If the solution level drops too far inside the reservoir, the cell can warp; and if it drops so that the electrodes are no longer immersed, the electrical circuit is broken. The lid design has been improved to attempt to minimise solution evaporation. Experiments up to 2 hours in duration have been successfully conducted using aqueous solutions at 90°C. For longer experiments, more solution could be introduced to the reservoir. (3) It has been speculated that the cell design could give rise to a large solution resistance [55]. This is because the solution width is only 2 mm in the vicinity of the working electrode. However, the tapered design means that the effective solution area quickly increases; we calculate that the solution resistance is a few tens of ohms.

As mentioned earlier, the cell is used in a grazing-incidence reflection geometry. The heating unit and support is mounted directly onto the diffractometer and the Teflon or Perspex cells can easily be inserted and withdrawn. At the beginning of each experiment, the electrode must be aligned with the X-ray beam. The sample rod (working electrode) is inserted into the cell, the cell placed into the heating block, and the working electrode connected. The solution is added and the reference and counter electrodes, thermometer, and gas lines (if required) are installed. If alignment is performed on the dry electrode (although rough alignment can be performed at this stage), introduction of the solution, placement of the electrodes, thermometer etc., and sealing the cell, causes a small amount of movement and the cell would need to be realigned. X-ray scattering patterns are recorded using the area detector, to optimise the substrate signal relative to the solution scattering. This should occur when the top of the X-ray beam is just illuminating the entire sample (Figure 17). If the beam is higher than this, the solution scattering will increase while the substrate signal remains constant. Since it is notoriously difficult to ensure that the top surface of the rod is flat and perpendicular to the sides, it is generally not possible to obtain an accurate measure of the incidence angle. One can experiment with different incidence angles by rotating the cell on the diffractometer about the θ-axis to optimise the signal. Once the sample is aligned, the experiment can begin. The sample is stationary throughout the experiment. Area detectors are preferred for these studies, since electrochemical reactions are usually quite far from chemical equilibrium and can give rise to preferentially oriented structures. This crystallographic texture results in the X-ray diffraction peaks being detected as arcs rather than rings, as is the case for isotropic systems. A 1D scan (a strip detector or a point detector used in scanning mode) may not capture all the necessary information.

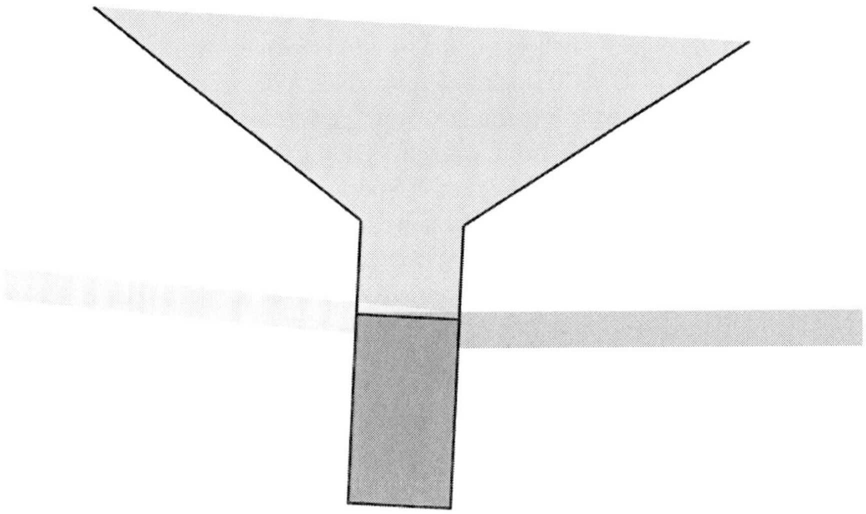

Figure 17. Diagram of alignment of the sample rod with respect to the X-ray beam, to minimise solution scattering.

The data collection parameters depend on many factors. The measurement time for each image will depend on the time scale of the reaction and the time resolution required. The minimum time resolution is limited by the detector read-out time. The X-ray wavelength needs to be sufficiently high so that the diffraction peaks of interest can be observed in the angular range offered by the design of the cell. The detector needs to be chosen according to the time resolution required (limited by its readout time), the pixel size and number of pixels (which limit the angular resolution and angular range), and in some cases the maximum count rate. These factors will now be demonstrated with two examples of *in situ*, real-time X-ray diffraction studies: of iron carbonate corrosion product films [59], and of zinc oxide nanostructures [60].

3.1. $FeCO_3$ Corrosion Product Films

Carbon dioxide corrosion of mild steel is one of the major causes of failure in production tubing, pipelines, and flowlines [61]. The normal approach used by industry today to prevent corrosion is to add corrosion inhibitor molecules to the flowstream. However, under certain conditions it is possible that the iron carbonate corrosion product itself may form a passive film that is superior in its corrosion-resistant properties [62]. Various iron carbonate, oxy-carbonate, and hydroxy-carbonate phases are observed under different conditions [53-54, 59]. *In situ*, real-time X-ray diffraction studies have been used to obtain kinetic information about the growth of $FeCO_3$ scales as a function of temperature from 40-90°C and how the addition of corrosion inhibitors affects the growth of the scale.

Sample rods of 1020 carbon steel were coated with 24-hr curing epoxy and the top surface polished to a 1 μm finish using diamond paste. The electrolyte solution consisted of 0.5 M NaCl and was saturated with CO_2 by bubbling the gas at 1 bar through the solution in the laboratory. In the experiments with corrosion inhibitors, up to 100 ppm of amino trimethylene phosphonic acid (ATMPA) was also added. The pH of the solution was then adjusted to 6.3 at room temperature using NaOH, before being heated and transferred to the cell. CO_2 continued to be gently bubbled in the solution reservoir while the experiment was in progress. An Ag/AgCl (sat. KCl) microreference electrode was used, while the counter electrode was a piece of Pt foil. Experiments were performed using the powder diffraction beam line at the Australian Synchrotron, using a VHR CCD detector (pixel size 26 μm). The X-ray wavelength was 0.8265 Å (15 keV) and the beam size was 0.2 x 0.7 mm (v x h).

Experiments were run in both galvanostatic and potentiostatic modes, with excellent reproducibility. Each experiment began by applying a potential of -1 V (vs. Ag/AgCl) to the sample for 5 minutes to remove any air-formed surface oxide, then the test potential (or current) applied for durations up to 2 hours. XRD measurements were recorded continuously, with the time taken for each measurement being 1 minute. The area of the peaks arising from the scale is proportional to the film thickness. The use of constant current results in a continuous supply of Fe^{2+} ions being provided to the solution, thus accelerating the corrosion process.

An example of the diffraction images obtained at the beginning and end of an experiment, and the processed diffractogram, are shown in Fig. 18. The presence of rings indicates that the film is not textured. The development of 'spots' in the Fe (110) ring has been attributed to the selective dissolution of small Fe grains from the steel during the experiment. The raw data were radially averaged using Fit2D [63] (after masking out the beamstop and regions of the detector occluded by the heating unit) and the sample-detector distance calibrated using LaB_6 600a standard powder. A dark scan was recorded for the same length of time as the measurements (1 min), integrated, and subtracted from each data set. No further background subtraction was applied. The most intense $FeCO_3$ diffraction ring (the (104), at Q = 2.25 $Å^{-1}$) was fitted using a Lorentzian function to obtain the peak position, FWHM, and area.

$FeCO_3$ is the primary phase observed at all temperatures from 40 to 90°C, with unidentified and different minority phases observed at 40 and 90°C. The potential responses, and the peak areas, are shown versus time in Fig. 19. The potential drops from -0.4 V (vs. Ag/AgCl), and returns to this value after some time - the time scale of which is shorter for higher temperatures, indicating the formation of a protective scale. For example, at 90°C, this occurs after about 40 minutes, which is also when the peak area stops increasing. At 80°C the film is starting to become protective, but the scale observed is thicker. At 60 and 40°C the film shows steady growth with no sign of passivation over the time scale of the experiment.

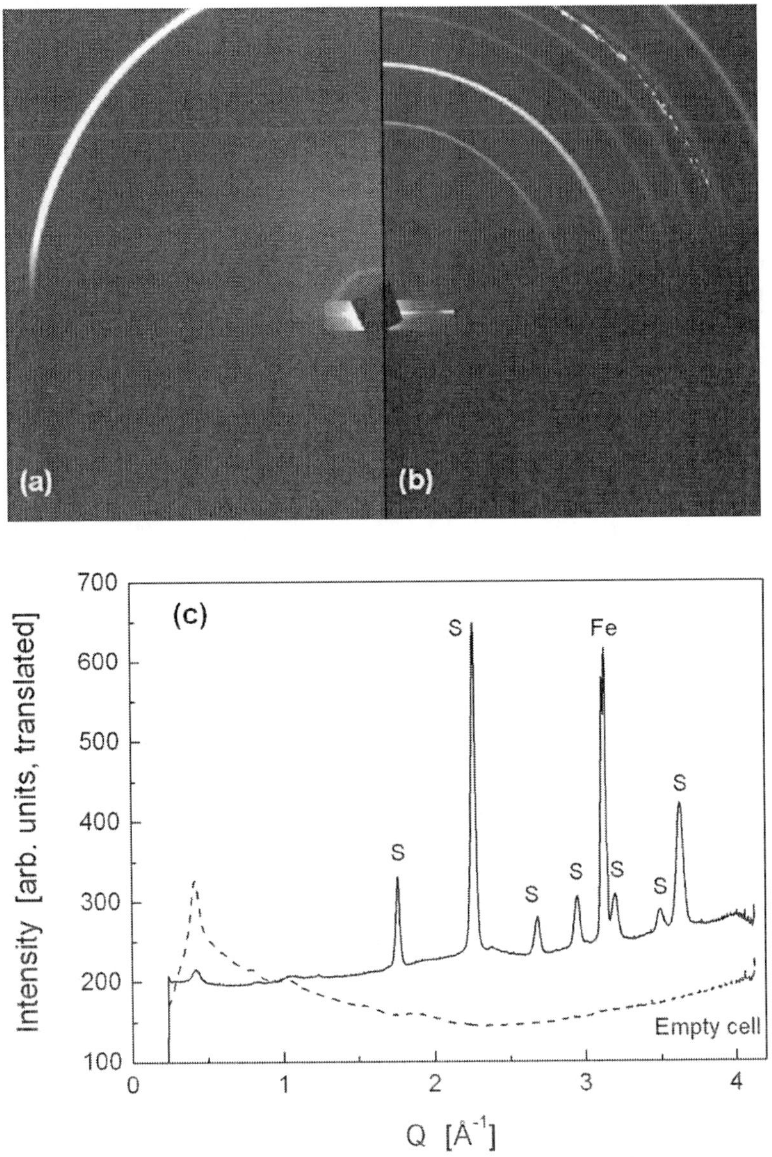

Figure 18. Raw 2D X-ray diffraction images recorded at (a) the beginning and (b) the end of a deposition, showing Fe and $FeCO_3$ diffraction rings. The processed diffractogram from image (b) is shown in (c).

When ATMPA is added to the solution, the film morphology is changed dramatically [59]. 1 ppm ATMPA has little effect, but adding 10 or 100 ppm slows the scale growth considerably. The effect is even more pronounced when the system is potentiostatically controlled [64].

These results have provided new insights into the growth mechanisms of $FeCO_3$ scale in CO_2-saturated NaCl brines and how these are changed by the addition of scale inhibitors.

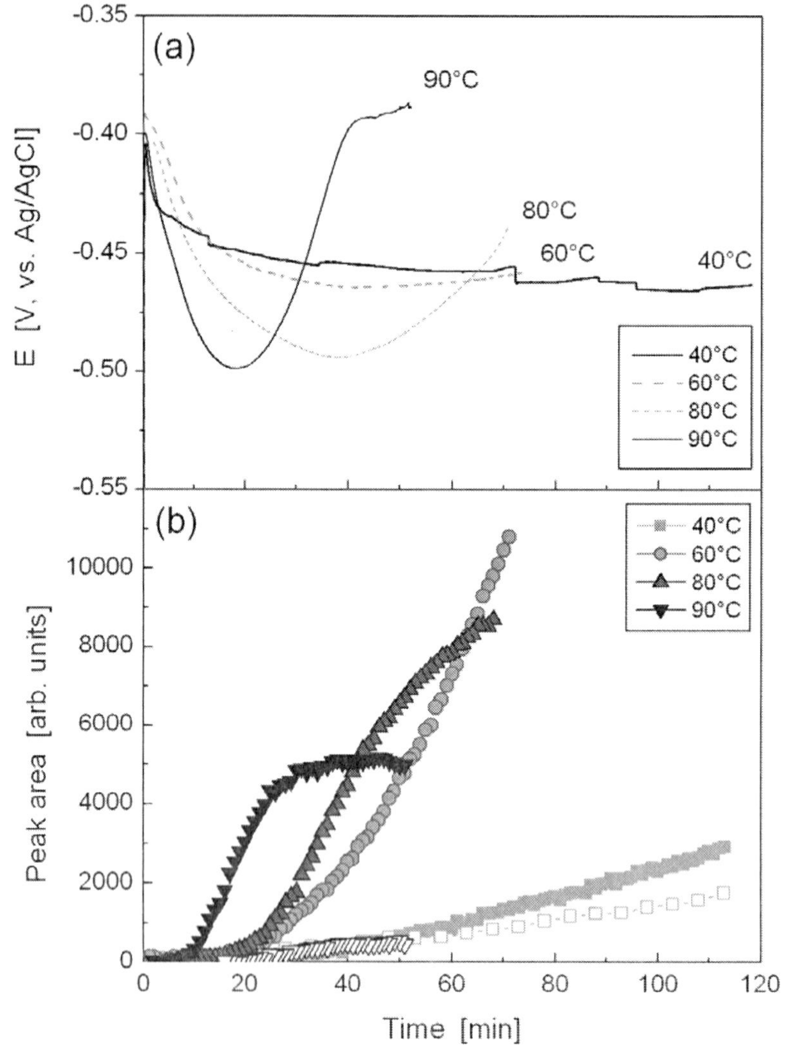

Figure 19. (a) Electrochemical potential and (b) corresponding X-ray diffraction peak area for $FeCO_3$ (filled symbols) and unidentified minority phases (open symbols) at different temperatures.

3.2. ZnO Nanostructured Films

The *in situ* electrochemical cell design described here has also been used for electrochemical deposition of ZnO from aqueous solution [60]. ZnO is a wide band gap semiconductor, which has many potential uses in photonic applications and solar cells [65-

71]. It has the ability to form many different shapes and sizes of nanostructures, and this morphological control can be used to tailor the electronic properties. Electrochemical deposition has many advantages over some other techniques, including controllable deposition rates, ease of scaling for production, and low-temperature processing using an industrially-established technology (viz. electroplating). The morphology of ZnO nanostructures can be controlled in electrochemical deposition through the control of experimental parameters such as the electrochemical potential [72-80] and concentration of Zn^{2+} in the solution [73, 81-87]. High aspect ratio nanorods can easily be formed because of the anisotropy along the [002] crystallographic direction [87-88]. While the deposition of ZnO can be monitored directly using *in situ* XAS [82, 89], this is only sensitive to the amount of Zn being deposited, and does not yield any crystallographic information. *In situ*, real time XRD experiments provide important additional information about the crystallinity and orientation of the ZnO nuclei and how these might change as the nanorods grow.

The electrolyte solution consists of 0.1 M KCl and 5 mM $Zn(NO_3)_2$. An Ag/AgCl (sat. KCl) microreference electrode was used, while the counter electrode was a piece of Pt foil. Oxygen gas was bubbled through the solution, which was held at a constant temperature (65-70°C) throughout the deposition. The depositions were controlled potentiostatically. XRD experiments were performed at beam line 11-3 at SSRL, with an X-ray wavelength of 0.9736 Å and beam size of 0.05 x 0.15 mm (v x h). A Mar345 image plate detector was used. The sample rods were made of quartz, coated with 150 nm of sputtered Au.

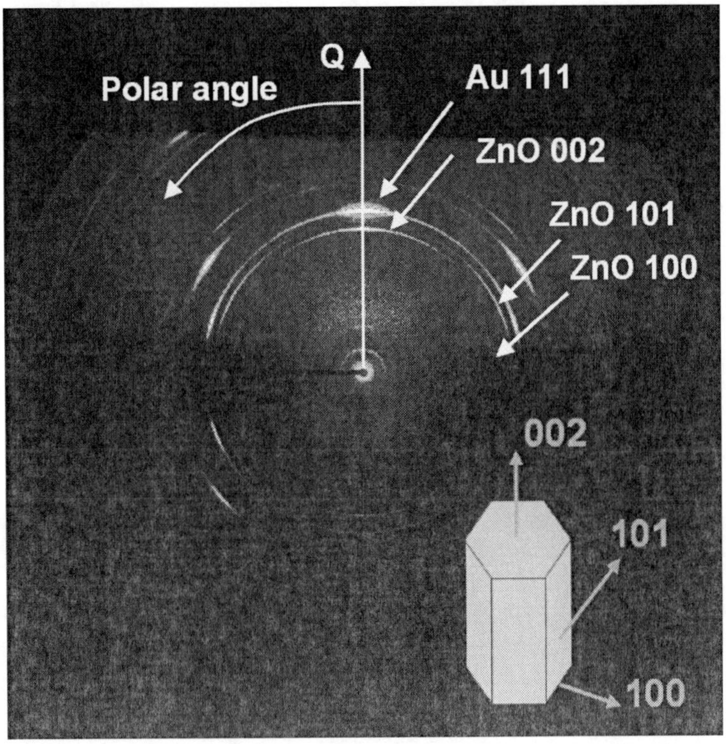

Figure 20. Example of a diffraction pattern from the end of a deposition. Diffraction arcs from ZnO and Au are as labelled, along with a diagram showing the crystallographic texture.

An example of the diffraction pattern observed at the end of an experiment is shown in Figure 20. Diffraction peaks are observed from both Au and ZnO as arcs, indicating strong preferential orientation. For ZnO the (002) reflection is normal to the surface, which is consistent with the hexagonal motif of the nanorods (being imputed from the wurtzite crystal structure). This sample exhibits *fibre texture*, that is, there is preferential alignment normal to the substrate, while the in-plane orientation is isotropic. Analysis of the time series data is more complex than simply taking a radial average of each scan. Fit2D [63] is used once again to convert the data from Cartesian to polar co-ordinates. From this data intensity plots can be extracted at constant angle or constant radius.

Since the aim of the experiment is to follow the development of crystallographic texture in the sample, intensity at constant radius as a function of polar angle is extracted and followed as a function of time. The question then becomes, which diffraction peak is the best to use for the analysis? Although the (002) (centred at a polar angle of 0°) is the most intense, it is a specular peak for which the diffraction condition (using grazing incidence) is not optimised. In contrast, the (100) peak is in-plane, so although the diffraction condition is close to optimal, the intensity drops away at a polar angle of 90°C. The peak chosen for the analysis is therefore the (101): of the three most intense peaks, it is the closest to satisfying the diffraction condition. See Figure 21.

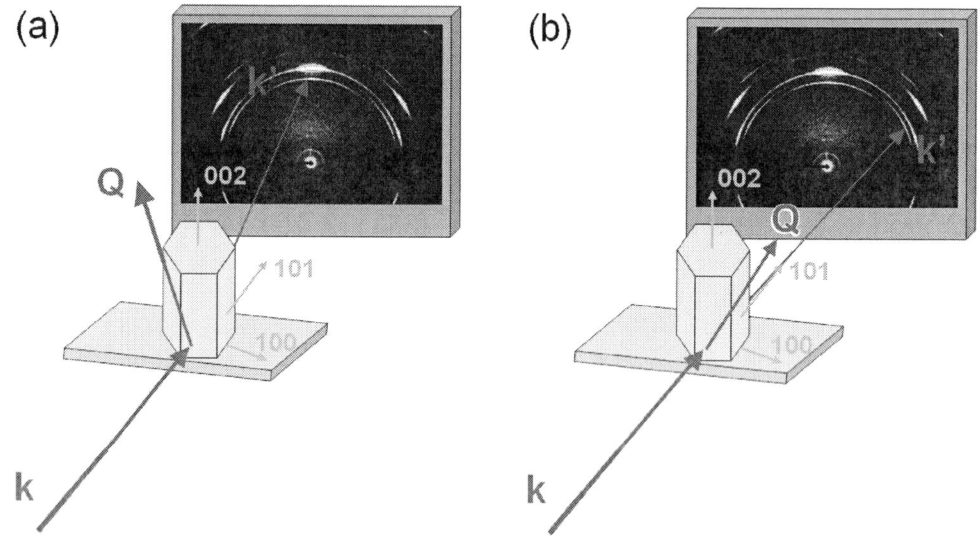

Figure 21. Diagrams showing how (a) the (002) reflection is not optimised in a grazing-incidence geometry, but (b) the (101) reflection is.

Plots of intensity versus polar angle have thus been extracted for the (101) peak, and the peak area and FWHM extracted as a function of time. These yield information about the amount of crystalline material, and the degree of texture, respectively. These are shown in Figure 22. The peak intensity increases linearly from zero to 6 minutes, at which point there is an abrupt change in slope and the intensity continues to increase linearly at a slower rate at longer times. This behaviour has previously been observed in *in situ* X-ray absorption spectroscopy (XAS) experiments [89] and coincides with features in the electrochemical current density. The current density shows an initial peak corresponding to the formation of

ZnO nuclei. These grow in three dimensions until adjacent nanorods meet and the electrode surface is covered, upon which the current density drops and the growth rate slows as the nanostructures continue to grow in one dimension. The electrochemical parameters (potential, Zn^{2+} concentration) affect the growth rate and behaviour [82, 89].

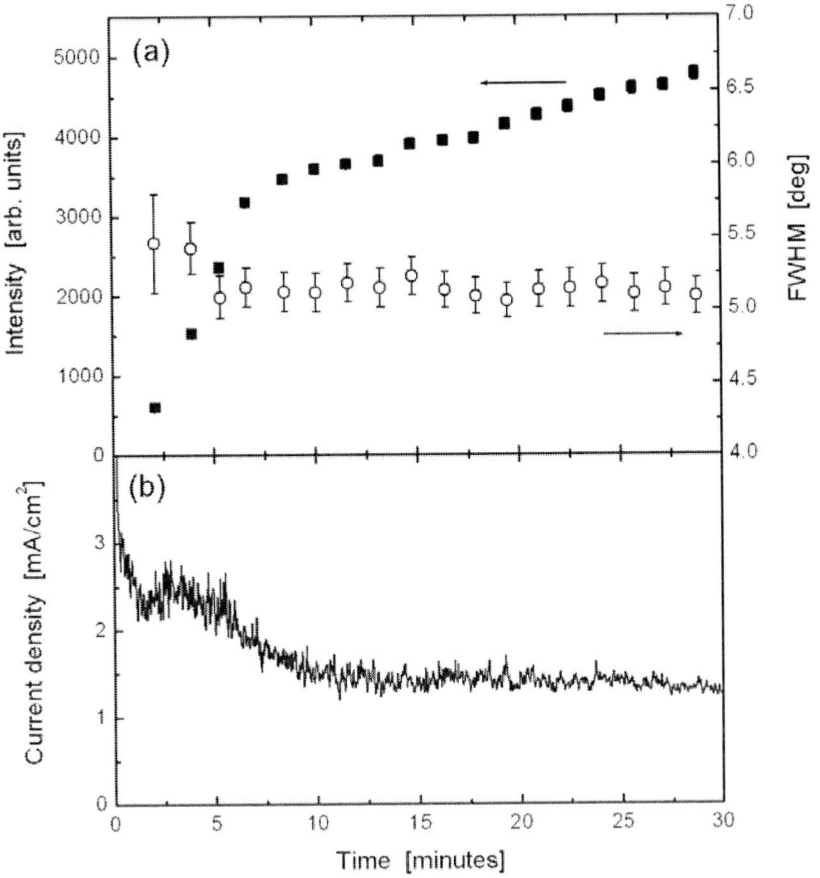

Figure 22. (a) X-ray diffraction peak intensity and polar angle width as a function of time; (b) electrochemical current density as a function of time.

If the sample was isotropic (as the $FeCO_3$ films), a plot of intensity versus polar angle would be constant. Thus the FWHM of the peak gives an indication of the width of the orientation distribution. From the earliest times, a peak is present in the intensity versus polar angle plot, with a FWHM of no more than 6°. This indicates that the nuclei themselves are preferentially oriented, even though their morphology is spherical and does not suggest any preferential direction related to the crystal structure, as the later hexagonal nanorods do.

3.3. Final Comments

The question naturally arises as to how the tapered geometry of the cell in the region of the working electrode affects the diffusion and hence the rate of reaction. For the ZnO experiments, the time scale of the peak area versus time can be compared with that of the XAS signal versus time under the same experimental conditions. In the XAS cell, the working electrode is large and the solution path is unrestricted (the X-ray signal is measured in a fluorescence geometry from outside the cell) [82, 89]. A ZnO nanostructured film deposited from 5 mM $Zn(NO_3)_2$ solution with 0.1 M $CaCl_2$ electrolyte at 65°C and an applied potential of -0.67 V (Ag/AgCl) exhibits a change in slope after 6 minutes in the *in situ* XRD cell, but only 2.5 minutes in the *in situ* XAS cell. In both cases the drop in the electrochemical current density corresponds with the change in slope.

Similarly, the electrochemical response from the *in situ* $FeCO_3$ experiments can be compared with laboratory work. For example, potentiostatic experiments (+200 mV vs. o.c.p.) performed at 70-80°C in the laboratory exhibit a peak in the current density after about 10 minutes [59]. For *in situ* synchrotron XRD experiments under the same conditions, the peak is observed at about 75 minutes [90]. Therefore the geometry does have an effect, but the time scales can be rationalised by directly comparing the XRD signal and the electrochemical response.

The cell design could easily be applied to a wide variety of electrochemical systems, in both aqueous and non-aqueous solutions (depending on the cell materials used). It is easy to use: individual sample rods can be easily changed and the sample position is reproducible. The cells themselves are interchangeable and are easily inserted into and removed from the heating block so that the sample changeover time is minimised. The major benefit of the cell is that it allows real time, *in situ* measurements to be taken without changing the cell geometry (as for other XRD cell designs). This allows direct correlations to be made between the XRD signal and the electrochemical response, instead of 'pausing' the electrochemistry in order to remove most of the solution and record an XRD pattern.

Conclusion

Design considerations for constructing an *in situ* cell for X-ray scattering experiments include:

- What environment does the cell need to withstand: temperature range, pressure, solution composition, gas? This impacts the materials that can be used and the dimensions of the cell (with maximum dimensions usually limited by the diffractometer dimensions)
- What is the minimum solution thickness needed? (e.g. to minimise diffusion-limited effects)
- What is the maximum solution thickness possible? (e.g. for 12-15 keV X-rays, 2 mm of water results in 60-70% transmission)
- Is a transmission or reflection geometry needed?
- Is gas flow needed?
- What is the angular range needed for the measurement? The use of higher X-ray energies also has the effect of compressing the angles.
- Does the cell need heating capability?

– What detector will be used? Consider the angular range, angular resolution, measurement of texture (off-axis angular information), and time resolution required - compromises will most likely be necessary.

In summary, two specific cells for X-ray scattering measurements have been described, as well as the rationale behind their design. Use of these cells have enabled the nucleation and growth of nanoparticles in solution, and the nucleation and growth of thin films on substrates by electrochemical deposition, to be followed in real time. These studies have yielded new insights into the mechanisms of formation, which can be very difficult or even impossible to obtain from *ex situ* XRD and electron microscopy observations.

ACKNOWLEDGMENTS

Portions of this research were carried out at the Stanford Synchrotron Radiation Lightsource (SSRL), a national user facility operated by Stanford University on behalf of the U. S. Department of Energy, Office of Basic Energy Sciences. Portions of this research were undertaken on the Powder Diffraction beam line at the Australian Synchrotron, Victoria, Australia. The views expressed herein are those of the authors and are not necessarily those of the owner or operator of the Australian Synchrotron. Portions of this work were funded by the New Zealand Foundation for Research, Science and Technology under contract CO8X0419. A grant from the New Zealand Synchrotron Group Ltd. to construct the electrochemical cell is gratefully acknowledged. Thanks are due also to Dr. Mike Toney (SSRL), Dr. Mary Ryan and Dr. Benoit Illy (Imperial College, U.K.), Dr. Richard Tilley, Dr. John Watt, and Dr. Soshan Cheong (Victoria University, N.Z.), Prof. David Williams (University of Auckland, N.Z.), Dr. Nick Laycock, Dr. Gareth Kear, and Monika Ko (Quest Integrity Ltd., N.Z.), Dr. Peter Kappen (Latrobe University, Australia), Dr. Kia Wallwork, Dr. Justin Kimpton and Dr. Qinfen Gu (Australian Synchrotron, Melbourne) for their contributions to the work presented.

REFERENCES

[1] Hülsen, G. *PhD thesis*, U. Erlangen-Nürnberg, Germany, 2005.

[2] Hülsen, G.; Brönnimann, C.; Eikenberry, E. F.; Wagner, A. *J. Appl. Cryst.* 2006, 39, 550-557.

[3] Brönnimann, C.; Eikenberry, E. F.; Henrich, B.; Horisberger, R.; Hülsen, G.; Pohl, E.; Schmitt, B.; Schulze-Briese, C.; Suzuki, M.; Tomizaki, T.; Toyokawa H.; Wagner, A. *J. Synch. Rad.* 2006, 13, 120-130.

[4] Henrich, B.; Bergamaschi, A.; Brönnimann, C.; Dinapoli, R.; Eikenberry, E. F.; Johnson, I.; Kobas, M.; Kraft, P.; Mozzanica, A.; Schmitt, B. *Nucl. Instr. Meth. Phys. Res. A* 2009, 607, 247-249.

[5] *http://pilatus.web.psi.ch/pilatus.htm*

[6] *http://www.dectris.com*

[7] Schmitt, B.; Brönnimann, C.; Eikenberry, E. F.; Gozzo, F.; Hörmann, C.; Horisberger R.; Patterson, B. *Nucl. Instr. Meth. Phys. Res. A* 2003, 501, 267-272.

[8] Bergamaschi, A.; Cervellino, A.; Dinapoli, R.; Gozzo, F.; Henrich, B.; Johnson, I.; Kraft, P.; Mozzanica, A.; Schmitt B.; Shi, X. *J. Synch. Rad.* 2010, 17, 653-668.
[9] http://pilatus.web.psi.ch/mythen.htm
[10] Jun, Y. W.; Choi, J. S.; Cheon, J. *Angew. Chem. Int. Ed.* 2006, 45, 3414-3439.
[11] Xia, Y.; Xiong, Y. J.; Lim, B.; Skrabalak, S. E. *Angew. Chem. Int. Ed.* 2009, 48, 60-103.
[12] Kuo, C. H.; Chiang, T. F.; Chen, L. J.; Huang, M. H. *Langmuir* 2004, 18, 7820-7824.
[13] Im, S. H.; Lee, Y. T.; Wiley, B.; Xia, Y. *Angew. Chem. Int. Ed.* 2005, 44, 2154-2157.
[14] Kim, F.; Connor, S.; Song, H.; Kuykendall, T.; Yang, P. *Angew. Chem. Int. Ed.* 2004, 116, 3759-3763.
[15] Wang, Y.; Chen, P.; Liu, M. *Nanotech.* 2006, 17, 6000-6006.
[16] Murphy, C. J.; Sau, T. K.; Gole, A. M.; Orendorff, C. J.; Gao, J. X.; Gou, L.; Hunyadi, S. E.; Li, T. *J. Phys. Chem. B* 2005, 109, 13857-13870.
[17] Gou, L.; Murphy, C. J. *Chem. Mater.* 2005, 17, 3668-3672.
[18] Fenske, D.; Borchert, H.; Kehres, J.; Kroger, R. Parisi, J.; Kolny-Olesiak, J. *Langmuir* 2008, 24, 9011-9016.
[19] Watt, J.; Cheong, S.; Toney, M. F.; Ingham, B.; Cookson, J.; Bishop, P. T.; Tilley, R. D. *ACS Nano* 2010, 4, 396-402.
[20] Song, Y.; Yang, Y.; Medforth, C. J.; Pereira, E.; Singh, A. K.; Xu, H.; Jiang, Y.; Brinker, C. J.; van Swol, F.; Shelnutt, J. A. *J. Am. Chem. Soc.* 2004, 126, 635-645.
[21] Sau, T. K.; Murphy, C. J. *J. Am. Chem. Soc.* 2004, 126, 8648-8649.
[22] Bakr, O. M.; Wunsch, B. H.; Stellacci, F. *Chem. Mater.* 2006, 18, 3297-3301.
[23] Ren, J.; Tilley, R. D. *Small* 2007, 3, 1508-1512.
[24] Mahmoud, M. A.; Tabor, C. E.; Ding, Y.; Wang, Z. L.; El-Sayed, M. A. *J. Am. Chem. Soc.* 2008, 130, 4590-4591.
[25] Teng, X.; Liang, X.; Maksimuk, S.; Yang, H. *Small* 2006, 2, 249-253.
[26] Zhong, X.; Feng, Y.; Liberwirth, I.; Knoll, W. *Chem. Mater.* 2006, 18, 2468-2471.
[27] Ren, J.; Tilley, R. D. *J. Am. Chem. Soc.* 2007, 129, 3287-3291.
[28] Ramirez, E.; Jansat, S.; Philippot, K.; Lecante, P.; Gomez, M.; Masdeu-Bultó, A. M.; Chaudret, B. *J. Organomet. Chem.* 2004, 689, 4601-4610.
[29] Pan, C.; Pelzer, K.; Philippot, K.; Chaudret, B.; Dassenoy, F.; Lecante, P.; Casanove, M. J. *J. Am. Chem. Soc.* 2001, 123, 7584-7593.
[30] Cordente, N.; Respaud, M.; Senocq, F.; Casanove, M. J.; Amiens, C.; Chaudret, B. *Nanolett.* 2001, 1, 565-568.
[31] Nayral, C.; Viala, E.; Fau, P.; Seoncq, F.; Jumas, J. C.; Maisonnat, A.; Chaudret, B. *Chem. Eur. J.* 2000, 6, 4082-4090.
[32] Cheong, S.; Watt, J.; Ingham, B.; Toney, M. F.; Tilley, R. D. *J. Am. Chem. Soc.* 2009, 131, 14590-14595.
[33] Xia, Y.; Xiong, Y.; Lim, B.; Skrablak, S. E. *Angew. Chem. Int. Ed.* 2009, 48, 60-103.
[34] Yin, Y.; Alivisatos, A. P. *Nature* 2005, 437, 664-670.
[35] Berhault, G.; Bausach, M.; Bisson, L.; Becerra, L.; Thomazeau, C.; Uzio, D. *J. Phys. Chem. C* 2007, 111, 5915-5925.
[36] Ingham, B.; Li, H.; Allen, E. L.; Toney, M. F. arXiv:0901.4782v1, 2009.
[37] Pedersen, J. S. *J. Appl. Cryst.* 1994, 27, 595-608.

[38] Ingham, B.; Toney, M. F.; Hendy, S. C.; Cox, T.; Fong, D. D.; Eastman, J. A.; Fuoss, P. H.; Stevens, K. J.; Lassesson, A.; Brown, S. A.; Ryan, M. P. *Phys. Rev. B* 2008, 78, 245408.

[39] Eastman, J. A.; Thompson, L. J.; Kestel, B. J. *Phys. Rev. B* 1993, 48, 84-92.

[40] Pundt, A.; Kirchheim, R. *Ann. Rev. Mater. Sci.* 2006, 36, 555-608.

[41] Pundt, A.; Dornheim, M.; Guerdane, M.; Teichler, H.; Ehrenberg, H.; Reetz, M. T.; Jisrawi, N. M. *Eur. Phys. J. D* 2002, 19, 333-337.

[42] Suleiman, M. PhD thesis, University Göttingen, 2004.

[43] Watt, J. Ph.D. thesis, Victoria University of Wellington, New Zealand, 2010.

[44] Baranowski, B.; Majchrzak, S.; Flanagan, T. B. *J. Phys. F: Met. Phys.* 1971, 1, 258-261.

[45] Ingham, B.; Hendy, S. C.; Fong, D. D. Fuoss, P. H.; Eastman, J. A.; Lassesson, A.; Tee, K. C.; Convers, P. Y.; Brown, S. A.; Ryan, M. P.; Toney, M. F. *J. Phys. D: Appl. Phys.* 2010, 43, 075301.

[46] van de Sandt, E. J. A. X.; Wiersma, A.; Makkee, M.; van Bekkum, H.; Moulijn, J. A. *Appl. Catal. A* 1997, 155, 59-73.

[47] Ocko, B. M.; Wang, J.; Davenport, A.; Isaacs, H. *Phys. Rev. Lett.* 1990, 65, 1466-1469.

[48] Wang, J.; Ocko, B. M.; Davenport, A.; Isaacs, H. *Phys. Rev. B* 1992, 46, 10321-10338.

[49] Tidswell, I. M.; Marković, N. M.; Lucas, C. A.; Ross, P. N. *Phys. Rev. B* 1993, 47, 16542-16553.

[50] Zegenhagen, J.; Renner, F. U.; Reitzle, A.; Lee, T. L.; Warrne, S.; Stierle, A.; Dosch, H.; Scherb, G.; Fimland, B. O.; Kolb, D. M. *Surf. Sci.* 2004, 573, 67-79.

[51] De Marco, R.; Pejcic, B.; Prince, K.; van Riessen, A. *Analyst* 2003, 128, 742-749.

[52] Herron, M. E.; Doyle, S. E.; Roberts, K. J.; Robinson, J.; Walsh, F. C. *Rev. Sci. Instrum.* 1992, 63, 950-955.

[53] De Marco, R.; Jiang, Z. T.; Pejcic, B., Poinen, E. *J. Electrochem. Soc.* 2005, 152, B389-392.

[54] De Marco, R.; Jiang, Z. T.; John, D.; Sercombe, M.; Kinsella, B. *Electrochim. Acta* 2007, 52, 3746-3750.

[55] Veder, J. P.; Nafady, A.; Clarke, G.; Williams, R. P.; De Marco, R.; Bond, A. M. *Electrochim. Acta* (2010, in press) DOI: 10.1016/j.electacta.2010.09.106

[56] Robinson, K. M.; O'Grady, W. E. *Rev. Sci. Instrum.* 1993, 64, 1061-1065.

[57] Tamura, K.; Ocko, B. M.; Wang, J. X.; Adžić, R. R. *J. Phys. Chem. B* 2002, 106, 3896-3901.

[58] Tamura, K.; Wang, J. X.; Adžić, R. R.; Ocko, B. M. *J. Phys. Chem. B* 2004, 108, 1992-1998.

[59] Ingham, B.; Ko, M.; Kear, G.; Kappen, P.; Laycock, N.; Kimpton, J. A.; Williams, D. E. *Corr. Sci.* 2010, 52, 3052-3061.

[60] Ingham, B.; Illy, B. N.; Toney, M. F.; Howdyshell, M. L.; Ryan, M. P. *J. Phys. Chem. C* 2008, 112, 14863-14866.

[61] Kermani, M. B.; Morshed, A. *Corrosion* 2003, 59, 659-683.

[62] de Waard, C.; Lotz, U. "Prediction of CO_2 Corrosion of Carbon Steel", *Corrosion* 93, Paper 69, NACE International, Houston, TX, 1993.

[63] *http://www.esrf.eu/computing/scientific/FIT2D/*

[64] Ingham, B.; Ko, M.; Kappen, P.; Laycock, N.; Williams, D. E. (unpublished).

[65] Takahashi, K.; Nish, T.; Suzaka, S.; Sigeyama, Y.; Yamaguchi, T.; Nakamura, J.; Murata, K. *Chem. Lett.* 2005, 34, 768-769.
[66] Wang, X. *Phys. Lett. A* 2008, 372, 2900-2903.
[67] Kenanakis, G.; Vernardou, D.; Koudoumas, E.; Kiriakidis, G.; Katsarakis, N. *Sensor Actuator B Chem.* 2007, 124, 187-191.
[68] Pillai, S. C.; Kelly, J. M.; McCormack, D. E.; O'Brien, P.; Ramesh, R. *J. Mater. Chem.* 2004, 13, 2586-2590.
[69] Law, M.; Green, L. E.; Johnson, J. C.; Saykally, R.; Yang, P. D. *Nat. Mater.* 2005, 4, 455-459.
[70] Arnold, M. S.; Avouris, P.; Pan, Z. W.; Wang, Z. L. *J. Phys. Chem. B* 2003, 107, 659-663.
[71] Johnson, J.; Yan, H.; Yang, P.; Saykally, R. *J. Phys. Chem. B* 2003, 107, 8816-8828.
[72] Goux, A.; Pauporté, T.; Chivot, J.; Lincot, D. *Electrochim. Acta* 2005, 50, 2239-2248.
[73] Peulon, S.; Lincot, D. *J. Electrochem. Soc.* 1998, 145, 864-874.
[74] Izaki, M.; Omi, T. *Appl. Phys. Lett.* 1996, 68, 2439-2440.
[75] Gu, Z. H.; Fahidy, T. Z. *J. Electrochem. Soc.* 1999, 146, 156-159.
[76] Mahalingam, T.; John, V. S.; Raja, M.; Su, Y. K.; Sebastian, P. J. *Solar En. Mater. Solar Cells* 2005, 88, 227-235.
[77] Dalchiele, E. A.; Giorgi, P.; Marotti, R. E.; Martin, F.; Ramos-Barrado, J. R.; Ayouci, R.; Leinen, D. *Solar En. Mater. Solar Cells* 2001, 70, 245-254.
[78] Cao, B.; Cai, W.; Sun, F.; Li, Y.; Lei, Y.; Zhang, L. *Chem. Commun.* 2004, 1604-1605.
[79] Pauporté, T.; Cortès, R.; Froment, M.; Beaumont, B.; Lincot, D. *Chem. Mater.* 2002, 14, 4702-4708.
[80] Marotti, R. E.; Guerra, D. N.; Bello, C.; Machado, G.; Dalchiele, E. A. *Solar En. Mater. Solar Cells* 2004, 82, 85-103.
[81] Izaki, M.; Omi, T. *J. Electrochem. Soc.* 1996, 143, L53-L55.
[82] Illy, B. N.; Ingham, B.; Ryan, M. P. *Cryst. Growth Design* 2010, 10, 1189-1193.
[83] Pauporté, T.; Lincot, D.; Viana, B.; Pelle, F. *Appl. Phys. Lett.* 2006, 89, 233112.
[84] Pradhan, D.; Kumar, M.; Ando, Y.; Leung, K. T. *Nanotech.* 2008, 19, 035603.
[85] Elias, J.; Tena-Zaera, R.; Lévy-Clément, C. *J. Electroanal. Chem.* 2008, 621, 171-177.
[86] Pauporté, T.; Lincot, D. *J. Electroanal Chem.* 2001, 517, 54-62.
[87] El Belghiti, H.; Pauporté, T.; Lincot, D. *Phys. Stat. Solidi* 2008, 205, 2360-2364.
[88] Liu, R.; Vertegel, A. A.; Bohannan, E. W.; Sorenson, T. A.; Switzer, J. A. *Chem. Mater.* 2001, 13, 508-512.
[89] Ingham, B.; Illy, B. N.; Ryan, M. P. *J. Phys. Chem. C* 2008, 112, 2820-2824.
[90] Ingham, B.; Ko, M; Kappen, P.; Laycock, N.; Kimpton, J. A.; Williams, D. E. *"In Situ Synchrotron X-ray Diffraction Study of Scale Formation During CO_2 Corrosion of Carbon Steel in Sodium and Magnesium Chloride Brines"*, in preparation, 2010.

In: X-Ray Scattering
Editor: Christopher M. Bauwens
ISBN: 978-1-61324-326-8

Chapter 3

APPLICATIONS OF X-RAY SCATTERING IN EDIBLE LIPID SYSTEMS

Cristián Huck-Iriart[1,2], Noé Javier Morales-Mendoza[1,2], Roberto Jorge Candal[1,3] and María Lidia Herrera[2]

[a]Instituto de Química Inorgánica, Medio Ambiente y Energía (INQUIMAE), Consejo Nacional de Investigaciones Científicas y Técnicas (CONICET), Ciudad Universitaria, Buenos Aires, Argentina

[2]Facultad de Ciencias Exactas y Naturales (FCEN), Universidad de Buenos Aires (UBA), Ciudad Universitaria, Buenos Aires, Argentina

[3]Escuela de Ciencia y Tecnología, Universidad Nacional de San Martín (UNSAM), Campus Miguelete, Provincia de Buenos Aires, Argentina

ABSTRACT

The phase behavior as well as the thermal and structural behavior of many lipid systems such as chocolate, butter, margarine, milk fat and its fractions have been thoroughly studied by traditional X-ray diffraction techniques. The effect of the addition of emulsifiers or additives on crystallization kinetics of fat systems has also been investigated by X-ray. More recently, the polymorphic behavior of a new fat, cupuassu fat, with properties very similar to cocoa butter, has also been described. This behavior is very relevant since it is related to functional properties of final products. Studies with small angle X-ray scattering (SAXS) allowed describing the structural dynamics of several fats such as milk fat fractions in the early stage of crystallization. This early stage of crystallization is very important since it determines the later evolution of the system. A synchrotron source allows diffraction patterns to be acquired during real-time crystallization, and therefore, further and less speculative information about mechanisms of action can be obtained. SAXS was a valuable tool to explain some unexpected behavior in milk fat isothermal crystallization and also to clarify the effect of sucrose ester on fat crystallization. Several reports have dealt with the latter subject, both in bulk and in emulsion systems, but some of these results may be considered contradictory. With the aid of SAXS, it was possible to improve our understanding of the mechanism of action of a palmitic sucrose ester on low *trans* fat blends since it was strongly related to the effects on fat polymorphism.

Introduction

Polymorphism may be defined as the ability of a substance to crystallize in different 3D unit cell structures. Polymorphism results from the different possibilities of lateral packing of the fatty acid chains and of the longitudinal stacking of molecules in lamellae. These two levels of organization are easily identifiable from the short- and long-spacings observed by X-ray diffraction (XRD) at wide and small angles, respectively. The three main organizations frequently observed for the lateral packing of TAG have been related to different subcells that have been described in detail. The lipid long-spacings correspond to the repeat distance in the direction perpendicular to the lamellae. For TAG, long-spacings are commonly double or triple chainlengths (2L or 3L). For a fixed set of conditions (temperature, pressure, composition), only one solid phase will be consistent with a minimum free energy of the system. For any polymorphic system phase transitions toward the stable phase are unavoidable, due to thermodynamic drive to energy minimization. When considering industrial processes, in general, only one phase is required for its technological properties, and hence the crystallization conditions must be controlled to obtain the expected structure only [1]. Polymorphism of triacylglycerols (TAG) has been studied extensively because of its importance in oil chemistry and biophysical sciences. It has long ago been realized that TAG can crystallize in different monotropic modifications; the three main of which are called α, β' and β, in the order of their increasing stability. The α subcell has hexagonal geometry with each chain surrounded by six others at equal distances. The chains have some freedom to move and therefore there is a partial disorder. The β' polymorph shows orthorhombic subcell with a denser and more perfect packing. The β crystals has a tri-clinic subcell with the densest packing of the three subcells. A fourth crystalline structure, often called sub-α, although it contains a β′ subcell and would be less stable than the α-form, is also reported in the literature. Multi-component fats tend to form compound crystals, and this significantly affects polymorphic forms and transitions. In some cases, a group of closely related TAG can almost behave like a single triglyceride, especially in the α and β' forms. One example of this is cocoa butter [2].

Polymorphism of Cocoa Butter

Cocoa butter comprises the major solid fat in chocolate and its crystallization is of substantial importance for the quality of the product, which should be firm and smooth (both visually and in the mouth) and melt rapidly to provide a sensation of coolness. In chocolate manufacturing, careful control of the solidification process is necessary because it significantly influences both rheological properties of chocolate which determine the workability in the production processes, and physical properties of end products such as gloss, snap, texture, heat resistance, and fat bloom stability. The physical properties are related to the polymorphism of cocoa butter, the main structuring material in chocolate and confections [3,4]. Due to the commercial importance of chocolate and chocolate-like products, polymorphism of cocoa butter has been extensively studied, especially in relation to blooming. The need for proper tempering in handling chocolate before enrobing or molding is well known in the confectionary industry. Inevitable during storage at ordinary temperatures the chocolate develops bloom, which appears to be a deposit of small fat crystals on the chocolate surface giving it a white or gray over-all-appearance. It is not completely clear at this time whether the change from form V to VI is the cause of bloom or is just coincidental. There is no report of a bloomed sample that was not form VI, or a bright sample that was not form V. However, it might be that bloom is simply the growth of crystal agglomerates of form VI from submicroscopic to macroscopic size brought about by digestion, migration and resolidification of fat molecules under the influence of small (or large) temperature variations, but blooming is always accompanied by an internal crystal modification which gives rise to the change in X-ray diffraction pattern from form V to VI.

Depending on its origin, cocoa butter has different fatty acids (FA) and triacylglycerols (TAG) compositions. Tables 1 and 2 reported some examples as summarized by Lucas [5].

Table 1. Fatty Acid (FA) Composition of Cocoa Butter from Different Origins

	Country of Origin				
FA	Brazil	Ivory Coast	Malaysia	Ghana	Ecuador
C16:0	25.1	25.8	24.9	25.3	25.6
C18:0	33.3	36.9	37.4	37.6	36.0
C18:1	36.5	32.9	33.5	32.7	34.6
C18:2	3.5	2.8	2.6	2.8	2.6
C18:3	0.2	0.2	0.2	0.2	0.1
C20:0	1.2	1.2	1.2	1.2	1.0
C22:0	0.2	1.0	0.2	0.2	0.1

Table 2. Triacylglycerol (TAG) Composition of Cocoa Butter from Different Origins

	Country of Origin				
TAG	Brazil	Ivory Coast	Malaysia	Ghana	Ecuador
POS	34.6	36.6	36.6	37.3	36.3
SOS	23.7	23.8	28.4	26.8	26.9
POP	14.0	15.9	13.8	15.2	15.3
SOO	8.4	6.0	3.8	4.5	4.8
PLiS	3.4	3.6	2.8	3.2	2.8
POO	5.5	4.4	2.7	2.6	3.5
SOA	1.6	1.6	2.5	2.2	2.1
PLiP	1.7	1.9	1.5	1.9	1.9
SLiS	2.1	1.8	2.0	2.1	1.5

Cocoa butter is composed of a mixture of three classes of TAG (saturated, monounsaturated, and polyunsaturated TAG), the monounsaturated being by far the major component because they represent more than 80% of the total. Moreover, only three TAG 1,3-distearoyl-2-oleoylglycerol (SOS), 1-palmitoyl-2-oleoyl-3-stearoylglycerol (POS) and 1,3-dipalmitoyl-2-oleoylglycerol (POP) account for more than 95% of this fraction. Polyunsaturated and trisaturated TAG correspond to about 13 and 3%, respectively, of the TAG content. This specificity gives cocoa butter a thermal and structural behavior similar to that of a pure compound. POP, POS, and SOS are very similar molecules that form compound crystals even in the β-form. This behavior is nevertheless complex because all major TAG in its composition, like most lipids, show complex polymorphism.

In early studies, Wille and Lutton [6] described six crystalline states, I to VI, in order of increasing melting points, by X-ray diffraction. That work reported in detail long- and short-spacings for all polymorphic forms (Table 3). According to these authors, the normal state of cocoa butter in chocolate is apparently form V. Bloom has not been observed for pure form V and does not exists in the absence of form VI. The change from state V to VI, a real and reproducible change, must nevertheless involve only very minor changes in the crystalline structure of cocoa butter, because there are no changes in long-spacings and only minor changes in short-spacings. The forms I to V can form from the melt and by transition from a less stable polymorph; the latter is the only route to obtain form VI [4]. Other authors have described the polymorphic forms of cocoa butter. Chapman et al. [3] and Adenier et al. [7] have also studied the polymorphism of cocoa butter by X-ray diffraction (Table 3). Although these six forms have been confirmed by other authors, the existence of some of them was debated at that time. According to some authors, Form III may correspond to a mixture of Forms II and IV and is not a separate crystalline variety. In the same way, Form I could be a phase mixture, and Form VI may result from phase separation within a solid solution. More recently, van Malssen [8-10] criticized the designation of solid states of complex mixtures, crystallized as polycrystalline aggregates, as different polymorphs when the chemical compositions of the different crystallites are not identical. According to him, several polymorphs of cocoa butter and at least the phases sub-α, α, and β′ probably correspond to non-homogeneous crystalline states [11]. Today, it is generally accepted that cocoa butter solidifies in six crystalline forms, although obvious discrepancies in melting point and differences in nomenclature continue to exist.

Loisel et al. [11], using simultaneous DSC and XRDT techniques, obtained thermograms and spectra from the same sample by taking advantage of the high-energy flux of a synchrotron. This approach was previously employed for pure TAG and TAG blends. They extended it to a fat such as cocoa butter that was considered to behave as a pure compound and described its complex crystallization and melting behavior. In their work, they characterize the intermediate phase transitions of cocoa butter polymorphs and confirm the existence of the six polymorphic forms (called I to VI) by *in situ* characterization of their formation in the sample holder. The high energy of the synchrotron source allows phase separations to be detected and the competition between the different polymorphic species to be followed quantitatively, even at fast scanning rates, in the short-spacing domain at large angles as well as in the long-spacing domain at very small angles. Their main findings may be summarized as follows: thermal XRD and DSC analysis indicated that phase separations systematically occur during cocoa butter crystallization. The main TAG segregation involves a trisaturated fraction, which partially phase-separates from the mono- and polyunsaturated species by crystallizing, as shown by chemical analysis of the fractions. This behavior may be explained by the low solubility of the trisaturated TAG within the monounsaturated TAG, which represent more than 80% of cocoa butter. However, the phase separation of polyunsaturated TAG, which remain a liquid at low temperature and only crystallize at about 4°C on cooling, should also be considered because it might play an important role in the evolution of unstable cocoa butter polymorphs, as well as in chocolate blooming. This phase separation in the solid state does not affect the validity of the nomenclature proposed by Wille and Lutton [6], especially because the six forms that they described have been re-identified. However, it should serve as a reminder that even for cocoa butter, which is a specific fat mainly composed of three similar monounsaturated TAG, it only describes the behavior of its main fraction and not the whole fat.

Table 3. Long- and short-spacings (Å) of the different polymorphic forms and melting temperatures (°C) of cocoa butter[a]

	Wille and Lutton, 1966 [6]			Chapman, 1971 [3]			Adenier et al., 1975 [7]		
Forms	Long spacing (Å)	Short spacing (Å)	M.T. (°C)	Long spacing (Å)	Short spacing (Å)	M.T. (°C)	Long spacing (Å)	Short spacing (Å)	M.T. (°C)
γ (I)	54 (s), 27 (m)	4.17 (s), 3.87 (m)	---	55.1 (w), 34 (m)	4.19 (vs), 3.70 (s)	17.3	---	---	-----
α (II)	51 (vs), 16.3 (m)	4.20 (vs)	---	49 (vs), 16.3 (s)	4.24 (vs)	23.3	---	---	---
β'$_2$ (III)	51 (vs), 16.4 (m)	4.20 (vs), 3.87 (w)	20.7	49 (vs), 16.35 (s)	4.25 (vs), 3.86 (s)	25.5	---	---	---
β'$_1$ (IV)	49 (vs), 14.8 (w)	4.32 (s), 4.13 (s)	25.6	45 (vs), 14.87 (s)	4.35 (vs), 4.15 (vs)	27.5	45 (s)	4.35 (s), 4.15 (s)	25
β$_1$ (V)	66 (s), 33 (s), 12.8 (m)	4.58 (vs), 3.98 (ms), 3.87 (m), 3.73 (m), 3.65 (ms)	30.8	63.1 (m), 32.2 (vs), 12.8 (s)	5.40 (m), 4.58 (vs), 3.98 (s), 3.87 (m), 3.75 (m), 3.67 (w)	33.8	64 (s)	4.58 (s), 3.98 (s)	30
β$_2$ (VI)	63 (s), 31 (s), 12.7 (mw)	4.53 (vs), 4.01 (w), 3.84 (m), 3.67 (s)	32.3	63.1 (vs), 32.2 (m), 12.76 (s)	5.43 (m), 4.59 (vs), 4.04 (w), 3.86 (m), 3.70 (s)	36.3	63.8 (s)	4.59 (s), 3.7 (s)	33.5

[a] The relative intensity is noted as very strong (vs), strong (s), medium (m), or weak (w).
M.T.: melting temperature.

Very recently, da Silva et al. [12] have also studied the polymorphism of cocoa butter by X-ray diffraction using synchrotron radiation. The synchrotron results are summarized in Table 4.

To fully understand the polymorphism of cocoa butter, Sato et al. have studied symmetric mixed saturated/unsaturated acid triglycerides (TAG), most particularly POP, POS and SOS and their mixture systems, since the three TGs are the major components of cocoa butter and its behavior may be interpreted from the behavior of these mixed model systems. They have observed new findings on the polymorphic modifications of POP and SOS. The diffraction patterns of α, γ, β_2 and β_1 of POP were essentially the same as those of SOS [13]. Among those polymorphic forms, β_2 and β_1 exhibited thermal and structural behavior almost identical to those of form V and form VI of cocoa butter, respectively [14]. Studying seeding effects on the solidification behavior of cocoa butter and dark chocolate, Hachiya et al. (1989) conclude that the crystallization kinetics of cocoa butter and dark chocolate was remarkably accelerated by the seeding of SOS, BOB and cocoa butter form VI, the degree being highly dependent on the molecular and thermodynamic properties of the seed materials. When cocoa butter was seeded with the form VI of cocoa butter or the form β_1 of SOS, the x-ray short-spacing spectra of bloomed samples showed form VI-like patterns, but the non-bloomed sample, which contained the β_2 form of BOB as seeds, showed a form V-like pattern. On the other hand, in one cycle test consisting on keeping the sample at 38°C for 12 h followed by 20°C for 12 h, both the bloomed samples seeded using cocoa butter (Form VI) and SOS (β_1), and the non-bloomed sample using BOB (β_2) showed only one melting peak of form V at 30.7~32.0°C. All of the samples that were cycled revealed the XRD short-spacing spectra of form V [15].

Table 4. Long- and short-spacings (Å) of the different polymorphic forms and melting temperatures (°C) of cocoa butter determined by synchrotron radiation[a]

	Loisel et al. [11]			da Silva, 2009 [12]		
Forms	Long spacing (Å)	Short spacing (Å)	M.T. (°C)	Long spacing (Å)	Short spacing (Å)	M.T. (°C)
γ (I)	112 (w), 52.6 (m), 26.3 (m)	4.19 (s), 3.77 (s)	14-16	55.14 (m), 35.12 (w), 27.41 (m)	4.15 (s), 3.70 (m)	14.2
α (II)	48.5 (s), 36.4 (m)	4.22 (s)	15-16	49.76 (s)	4.21 (s)	22.7
β'_2 (III)	48.5 (s)	3.86 (s)	16-19	---	---	---
β'_1 (IV)	44.9 (s)	4.33 (s)	18-20	45.90 (s)	4.33 (s), 4.16 (s)	28.2
β_1 (V)	66 (s), 44.4 (w), 33 (s)	4.58 (s), 3.98 (s), 3.89 (m), 3.77 (m), 3.67 (m)	34.5	64.50 (vs), 45.94 (vs), 32.88 (vs)	5.43 (m), 4.59 (vs), 3.98 (s), 3.86 (m), 3.75 (m), 3.66 (m)	33.3
β_2 (VI)	63.8 (s), 44.4 (s)	4.59 (s), 3.7 (s)	36.3	71.72 (vs), 46.00 (s), 34.53 (s)	5.45 (m), 4.59 (vs), 4.53 (m), 4.03 (m), 3.92 (m), 3.86 (s), 3.74 (m), 3.70 (s), 3.66 (m)	36.4

[a] The relative intensity is noted as very strong (vs), strong (s), medium (m), or weak (w).
M.T.: melting temperature.

It is well-known that form V is the preferred polymorph for chocolate since it has a melting point closely matched to the human body, thus facilitating optimal dissolution during eating. This phase also gives to good demolding during processing as well as producing high gloss and favorable snap for the resultant product. Usually, problems of poor texture, lack of contraction, and inadequate gloss have been associated with "lower polymorphs". Tempering is an important stage in the manufacturing of chocolate to obtain the desired phase V of cocoa butter. During this processing stage, the completely liquid cocoa butter is first brought to a temperature at which nuclei from both phases IV and V form. The mixture is then heated to a temperature where the seeds of form IV are re-melted. After these two steps, a sufficient number of nuclei of phase V are present in the melt to allow solidification in the desired form during the final cooling in the mold. By this process, the crystallization kinetics of phase V is much faster than for direct crystallization from the melt at the same temperature. Due to the relevance regarding cocoa butter crystallization, other authors have also studied polymorphism of pure TAGs. Rousset and Rappaz [16] in order to improve knowledge of the so-called melt-mediated crystallization studied the behavior of POS performing a series of experiments by differential scanning calorimetry (DSC), coupled with polarized light microscopy (PLM). Starting from a completely liquid state, the melt was first cooled down and maintained at a temperature, $T1$, during a time, $t1$, where the α-phase formed. Then it was heated to a temperature, *T2*, above the melting point of α for isothermal solidification into a solid phase, which was identified as δ. Their main findings may be summarized as follows: the first plateau during α-melt-mediated crystallization of POS accelerates subsequent solidification of δ owing to two main mechanisms: first, the δ-nuclei formed directly from the melt are more numerous and appear more quickly because of the passage at *T1*. Second, if this first plateau is sufficiently long for α to form, it seems that some structure of α remains in the melt at the beginning of the second plateau ("memory effect") and that some α embryos re-transform into δ, which will dramatically accelerate the formation of δ nuclei. Using the same methodology, Rousset et al. [17] determine the phase diagrams of the various polymorphic forms of the POS-SOS binary system (the most metastable sub-α and α, intermediate δ and β', and stable β).

Butter is used in confections primarily for the buttery flavor, although some chocolate manufacturers add 2-3% anhydrous milk fat (AMF) to control hardness of dark chocolate. Milk fat is also widely known to inhibit fat bloom formation in chocolates. Offsetting these advantages is the relatively high cost of milk fat compared to other fats and oils. In addition, the range of melting point and plasticity of milk fat inhibit its use in some confectionery applications. For this reason, fractionation or modification of milk fat has gained increasing popularity in past years. In particular, fractionation of milk fat can produce a range of products with different physical and chemical characteristics which are very interesting in the confectionery industry [18]. Milk chocolate is always formulated with cocoa butter (CB) and milk fat (MF). However, in compound chocolate formulated for coating, cocoa butter substitute (CBS) is the main fat added to reduce cost. The presence of many different types of fats used in the chocolate system could change chocolate's physical properties. For example, such changes will cause chocolate products to soften, bloom, lose gloss and color, snap, and contract. To study the effect of the addition of other fats to cocoa butter, Sabariah et al. [19] determine the melting properties, polymorphic stability, and solid fat content (SFC) changes of binary and ternary blends of CBS, MF, and CB. Two ternary systems of confectionery fats were studied. In the first system, lauric cocoa butter substitutes (CBS), anhydrous milk fat (AMF), and Malaysian cocoa butter (MCB) were blended. In the second system, high-melting fraction of milk fat (HMF42) was used to replace AMF and also was blended with CBS and MCB. CBS contained high concentrations of lauric (C12:0) and myristic (C14:0) acids, whereas palmitic (C16:0), stearic (C18:0), and oleic (C18:1) acid concentrations were higher in MCB. In addition, AMF and HMF42 contained appreciable amounts of short-chain fatty acids. Characteristics of CBS included two strong spacings at 4.20 and 3.8 Å, which corresponded to the β'-form. MCB showed a strong spacing at 4.60 Å and a weak short-spacing at 4.20 Å, characteristics of the β-form. On the other hand, AMF exhibited a very weak short- spacing at 4.60 Å and two strong spacings at 4.20 and 3.8 Å, while HMF42 showed an intermediate short-spacing at 4.60 Å and also two strong short-spacings at 4.20 and 3.8 Å. According to these authors, increasing the concentration of CBS into MCB increased the proportion of the β′-form in the mixture. However, addition of AMF to CBS at any ratio did not change the polymorphic form. Pure HMF42 shows that the β′-form still dominated over the β-form, but the degree of domination was reduced compared to AMF. Surprisingly, the addition of HMF42 to CBS at any ratio showed only a formation of β′ polymorph. For both ternary systems in a ratio of 1:1:1, the predominantly polymorph was the β-form. Smith et al. [20] characterize the chemical composition, polymorphism, and melting behavior of the bloom that develops on bars of compound chocolate prepared using palm kernel stearin and hydrogenated palm kernel stearin. Modified lauric fats such as these mentioned oils, are used extensively in the chocolate confectionery industry as cocoa butter substitutes. In this application, they can be used in place of most of the cocoa butter (CB) that is present in normal chocolate. Compound chocolates based on these fats have the advantage that, unlike normal CB-based chocolate, they do not require tempering. However, a limitation of their use is the fact that they cannot be used in recipes containing more than about 4 or 5% CB on the fat phase. This low level of CB can result in a poor chocolate flavor. However, if this limit is exceeded, the products become prone to phase separation and crystallization of fat at the surface of the compound chocolate. Smith et al. [20] demonstrated that bloom formation in palm kernel stearin and hydrogenated palm kernel stearin compound chocolates was not due simply to trilaurin or CB separation. Instead, the composition of the bloom was temperature dependent, in line with the published phase diagrams. Whereas the TAGs were predominantly from the lauric fat, there was a considerable enrichment in CB TAGs at 15 and 20 °C. It was principally at 25 °C that they observed enrichment in trilaurin. In all cases, bloom was almost fully solid and was sharper melting than the corresponding compound chocolate.

Pore et al. [21] studied the effect of thermal gradients within a sample on the rates of phase transformations in cocoa butter. They performed in-situ X-ray studies of cocoa butter droplets undergoing simulated spray freezing. They used a version of the single droplet freezing apparatus they had described in a previous paper [22] which was fitted inside the X-ray system so that drops could be observed in situ over the freezing process. They showed that phase transformations of cocoa butter in small frozen droplets occur much faster than the equivalent transformations would be expected in bulk samples. In particular, it is possible to achieve Form V (the desirable form for chocolate) much faster in small droplets: this may have practical implications for chocolate manufacturers. Droplets that are frozen at a lower temperature result in more rapid subsequent phase transformations, even after heating to a constant higher temperature. According to these authors, there is clearly a ''memory'' of the freezing conditions retained in the droplet, which affects the subsequent rate of phase transformation. They attributed these observations to a high density of nucleation in a small droplet, due to the even temperature distribution within the droplet; this results in a relatively high density of nuclei for the formation of thermodynamically more stable phases as the droplet is warmed and the less stable phases melt.

MacMillan et al. [23] performed in situ small angle X-ray scattering (SAXS) studies of polymorphism of cocoa butter fat using shearing conditions. The direct crystallization of the phase V polymorphic form is associated with very long induction times rendering it unsuitable for manufacturing environments using continuous processing. Thus, in production environments, the fat is tempered through quenching to the temperature region associated with one of the lower polymorphic phases (forms II/III) and then reheated to phase interconvert the crystals back to the required phase V. Practical experience has shown that the tempering process requires mixing by shearing to produce the required product performance. The exact mechanism involved in this process was poorly understood and was one of the aims of their work. Their main findings may be summarized as follows: the phase behavior under static and sheared conditions has shown that when the sample is not sheared, forms III and IV appear. This contrasts with the sheared case where forms III and V appear, with the absence of form IV. From the data obtained by MacMillan et al., it seems that form III undergoes a polymorphic transformation to the β trilayer packed form V. Shear definitely forces the triglyceride molecules from double to triple packing. Form V formation times reduce as the rate of shear is increased, confirming the significant role it plays in the formation of form V.

Fat bloom continues to be a problem in the confectionery industry as it compromises both the visual and the textural quality of chocolates. This physical defect, which appears during storage as a dull, grayish-white film on the chocolate surface, has been ascribed to the uncontrolled formation of large, light-diffusing TAG fat crystals >5 μm in length. The three root causes of fat bloom are related to composition, processing, and improper storage. Compositionally, when two incompatible fats (e.g., CB and palm kernel oil) are combined, the resulting lower solid fat content (SFC) (below that of either individual fat) enhances liquid fat movement throughout the chocolate matrix. Alternatively, in filled confections, unsaturated TAGs from soft center fillings, which are normally rich in oil (e.g., hazelnut oil), may migrate toward the chocolate surface. Both scenarios promote indiscriminate crystal growth. From a processing standpoint, fat bloom may occur if tempering and/or cooling are poorly controlled, which encourages improper crystallization (i.e., into form IV) or insufficient form V formation and the subsequent form V to VI transition. Post-processing bloom is often related to undesirably high storage temperatures (> 30 °C) and undue temperature fluctuations. In this case, if the chocolate's fat phase is partly or fully liquefied, subsequent cooling leads to unstable polymorph (re)crystallization. Temperature fluctuations during storage, even to a small degree, decrease the induction time for fat bloom and increase its rate of formation [24]. Theories have been propounded for the mechanism of bloom formation, but it is fair to say that the details of the actual mechanism are still not fully understood, although a number of different factors may play a role. Lonchampt and Hartel make exactly this point in their comprehensive review of bloom in chocolate and compound coatings [25]. According to these authors, underlying questions remain: Do the different aspects of chocolate correspond to the same bloom mechanism? And, is the bloom mechanism similar for the different chocolate compositions? Many factors have been found to affect bloom in chocolates and compound coatings. Some tend to accelerate bloom formation, like oil migration from a coated center, the presence of incompatible fats and improper storage temperatures. Some other factors tend to inhibit bloom formation, like the use of milk fat in chocolate and proper storage temperatures.

POLYMORPHISM OF CUPUASSU

The difference between chocolate and compound coatings is based on the cocoa solids. Chocolate must contain not less than 32% total dry cocoa solids, including not less than 18% cocoa butter and not less than 14% of dry nonfat cocoa solids. Compound coating corresponds to product that doesn't match this definition. In most cases, the use of fats other than CB leads to the name compound coating on a product. Other vegetable fats can be used in order to obtain new flavors, to enhance the physico-chemical properties of the product or to reduce production price. Fats used in coatings can be classified according to their compatibility with CB, under three major family names [25]:

- Fats that are totally compatible with CB are called, cocoa butter equivalent, CBE.

- Fats that are partially compatible with CB correspond to cocoa butter replacers, CBR.
- Fats that are incompatible with CB correspond to cocoa butter substitutes, CBS.

Cocoa butter equivalent (CBE) fats should be totally compatible with cocoa butter. Compatibility in this context corresponds to the ability of the TAG of two distinct fats to crystallize together without forming a eutectic, although some CBE do not show total compatibility with CB. CBE are usually issued from some exotic fats (from the equatorial, tropical and sub-tropical countries) such as illipee, borneo tallow, shea fraction or fractionated salfat. They can also correspond to synthesized oils, like coberine. In general, CBE have a similar TAG composition and the same polymorphism as CB, which accounts for their general compatibility with CB.

Fat from cupuassu beans is extensively used in the production of candies and confectionery in the northern and northeastern regions of Brazil, where the Theobroma species was originally found and is considered as an excellent raw material for utilization in the food industry. Besides, cupuassu fat is considered a good candidate to partially substitute cocoa butter in candies and confectionery since the unsaturated fatty acid content is around 47 % and the SOS-type TAGs (POP, POS, SOS, and SOA) are around 55 %. However, these values do not exactly agree with EU standards to qualify cupuassu fat as a cocoa butter equivalent (CBE) material. Cupuassu fat does not reach two of all requirements: to have <45% unsaturated fatty acids (UFAs) content and >65% SOS-type TAGs. Cocoa butter has a unique characteristic profile that makes it such a valuable fat. It shows a sharp decrease in SFC at temperatures slightly lower than the body temperature. A comparison of the SFC curves indicates that cupuassu fat also showed a sharp decrease in solids content between 30 and 35°C. Incorporation of cupuassu fat in up to a 10% addition level yielded SFC curves that were nearly indistinguishable from the results reported for cocoa butter alone. Thus, the addition of cupuassu fat did not produce softening of cocoa butter. This compatibility makes it suitable for use in cocoa butter-based confections. Although SFC values for temperatures below 30°C were lower than the ones for cocoa butter, it is likely that functionality of cupuassu fat for chocolate applications may be improved by fractionation [5]. Most likely, the stearin fraction obtained would meet regulation for CBE. Despite the differences in chemical composition between cocoa butter and cupuassu fat, this material may be considered a good candidate to partially substitute cocoa butter in candies and confectionery [12]. da Silva et al. (2009) provided the first results reported on the polymorphic behavior of Brazilian cupuassu fat. Table 5 shows the long- and short-spacings determined for each phase together with the melting temperatures [12].

That study proved the existence of great thermal and structural similarities between the phases of cupuassu fat and cocoa butter. The polymorphic phases observed for cupuassu fat were identified as γ, α, β', and β in order of increasing melting temperature. Concerning the β phase, two stable thermal states were detected for cupuassu fat, labeled β_2 and β_1 in order of increasing temperature. We note that their melting temperatures are similar to those of the V(β2) and VI(β1) phases of cocoa butter listed in Table 4. On the other hand, the structural parameters of the cupuassu fat for the states labeled β_2 and β_1 are very similar. The X-ray data does not provide structural differences that would clearly identify the β_2 and β_1 phases in cupuassu fat as two different polymorphs. Consequently, there is no complete proof of a $\beta_2 \rightarrow \beta_1$ transition in cupuassu fat. In cupuassu fat, the β_2 phase seems to be more stable, suggesting that the phenomenon of "blooming" may not occur so easily in products manufactured with total or partial substitution of cocoa butter by cupuassu fat, since this effect is related to crystallization of the β_1 phase. Additional studies should be performed to prove that these β states are distinct polymorphic phases. The knowledge of the thermal pathway to reach the β_2 crystallization is very important for chocolate manufacturing. For this reason, the description of the behavior of the β phases is fundamental for the industrial use of cupuassu fat as a partial or total substitute for cocoa butter.

Table 5. Long- and Short- d-Spacings of the Different Crystallographic Phases and Melting Temperatures of Cupuassu Fat [12]

Forms	d-spacings (Å)[a]	Melting temperature (°C)
γ	61.36(w)[b], 57.97(vs)[b], 46.6(m)[b,c], 34.95(diff)[b], 28.83(w)[b], 4.13(vs), 3.68(s)	4.0
α	57.24(s)[b], 28.35(w)[b], 4.16(s)	17.7
β'	36.27(s)[b], 4.56(m), 4.44(m), 4.26(vs), 4.13(vs), 3.83(s)	26.8
β_2	69.38(vs)[b], 47.76(vw)[b], 33.95(vs)[b], 5.44(m), 4.58(vs), 4.00(m), 3.90(m), 3.77(m), 3.68(m), 3.59(w)	32.5
β_1[d]	69.64(vs)[b], 33.93(vs)[b], 5.43(m), 4.59(vs), 3.97(m), 3.89(m), 3.76(m), 3.68(m), 3.59(vw)	36.0

[a] Relative intensity: vw=veryweak,w=weak,mw=medium to weak,m=medium, ms=medium strong, s=strong, vs=very strong, diff=diffuse. [b] Long-spacings. [c] Only observed at -40°C. [d]The melting temperature corresponds to a β_1 phase, but the diffraction data do not show significant structural differences from the β_2 phase.

Effect of Sucrose Esters on Crystallization

Sucrose esters (SE) can be used in foods as emulsifiers because they are non-toxic, tasteless, and odorless and are digested to sucrose and FA in the stomach. They can also be used in pharmaceuticals and cosmetics and in other products where a non-ionic, biodegradable emulsifier is required. Emulsifiers are useful functional additives without which many food products would be impossible to make. Emulsifiers typically act in multi-phase systems in two main ways. The first is as an emulsifying agent to enable two distinct phases to be combined in a stable quasi-homogeneous state for an indefinite length of time. SE have a common feature that makes them suitable as emulsifying agents: they are ambiphilic, processing both lipophilic and hydrophilic properties. The nature of this property is often expressed as the hydrophilic/lipophilic balance (HLB) on a scale of 0 to 20, with low numbers indicative of the oil-like tendency. The second function of an emulsifier is often to modify the behavior of the continuous phase of a product so as to bring about a specific effect or benefit. For example, the use of lecithin in chocolate reduces the viscosity of the product and improves the ease of handling and processability. In addition to their major function of producing and stabilizing emulsions, SE have numerous other functional roles as texturizers, film formers, and modifiers of crystallization. Typical applications are baked goods, fruit coatings, and confectionary.

Several reports have dealt with the effect of SE on the crystallization behavior of fats, both in bulk and in emulsion systems; however, some of these results may be considered contradictory. Yuki et al. [26] studied the crystallization behavior of a fat mixture formulated with 60% hydrogenated soybean oil, 30 wt% palm oil, and 10 wt% rapeseed oil by recording the cooling curve and measuring the latent heat released when crystallization occurred. The process was also followed by Differential Scanning Calorimetry (DSC). They reported that the addition of 0.5 wt% of palmitic SE, waxy form (P-195), and stearic SE, waxy form (S-195), accelerated the crystallization process, whereas lauric SE, waxy form (L-195), retarded nucleation. Oleic SE, liquid form (O-190), however, had no effect on crystallization behavior. Nasir [27] described the isothermal crystallization behavior of a hydrogenated blend of 90% soybean and 10% cottonseed oils crystallized at 17°C by reporting the DSC melting curves. They showed that the addition of 0.5 wt% of palmitic SE, powder form (P-170), reduced the crystallization rate, whereas the addition of stearic SE, powder form (S-170), at the same concentration yielded a higher amount of solid crystals. Herrera and Marquez Rocha [28] studied the nucleation process of hydrogenated sunflower oil (SFO) with laser polarized turbidimetry (LPT). When P-170 was added to the hydrogenated oil at different concentrations, an elongation of the induction times of nucleation and a hindering of $\beta' \rightarrow \beta$ polymorphic transformation were observed. The effect was explained as a co-crystallization mechanism between the fat and P-170. Hodate et al. [29] studied the crystallization behavior of palm oil with and without the addition of P-170 and S-170 by ultrasonic velocity measurements. These authors reported that the addition of S-170 and P-170 at 0.5 wt% retarded the crystallization of palm oil compared with pure palm oil in the bulk system. Martini et al. [30] studied the effect of SE with different HLB values on the nucleation behavior of high-melting milk fat fractions/sunflower oil blends (HMF/SFO). Following the behavior by LPT, they found a slight elongation of the induction times of crystallization for SE with high affinity for the hydrophilic phase (high HLB). P-170 and S-170 (HLB = 1), however, significantly delayed the crystallization process, especially when added at 0.5 wt%. For all samples, both with and without the addition of P-1670 at the 0.1 wt% level and P-170 and S-170 at the 0.1 and 0.5 wt% levels, only one polymorphic form was obtained, the β' form. All patterns had two strong signals at 3.9 and 4.3 Å. The spectra were recorded with conventional X-ray equipment and usually around 20 min were required which means that the early events of crystallization were missing for these samples. The α form had a short life and could not be detected. No signal at 4.6 Å was found indicating that transformation to the β form did not occur during the experiments. Cerdeira et al. [31,32] also analyzed the crystallization behavior of these systems. They conclude that because of their somewhat similar chemical structures, the SE selected for those studies were able to co-crystallize with the fats. However, the structural dissimilarities between triacylglycerols and emulsifiers were responsible for a delay in nucleation and an inhibition of growth. The modification of the nucleation process was then related to molecular structure of SE and, more specifically, to the hydrophilic head. Activation free energies of nucleation were higher with the addition of SE in agreement with the fact that SE impeded nucleation. Chemical similarity of hydrophobic tails had a greater effect on nucleation than tail chain length. SE were also able to modify the polymorphic behavior and crystalline microstructure of the low trans HMF/SFO blends [33]. P-170 and S-170 modified the polymorphic behavior of HMF when it crystallized in the α form or in blends with up to 40% SFO at all crystallization temperatures selected. The addition of P-170 and S-170 favored crystallization in the β' form, and the appearance of the β form was delayed. P-1670 had no effect on polymorphism. When HMF and the blends were crystallized under dynamic conditions, the addition of P-1670 and S-170 markedly decreased crystal sizes. P-1670, however, showed no effect on microstructure. Sonoda et al. [34] studied the crystallization and polymorphic behavior of palm stearin both in a bulk state and in oil-in-water emulsion droplets. By combining differential scanning calorimetry (DSC) and synchrotron radiation X- ray diffraction (SR-XRD) techniques they were able to describe the effect of compartmentalization on fat crystallization. The crystallization process was followed in situ and taken spectra every 10 sec by SR-XRD. The bulk sample suffered the following transformation processes: the low and high melting fractions crystallized in the α form at 3 and 31°C, repectively, and the crystallization of the β' form gave a small exhotermic peak around 21°C; then, the α form corresponding to the low melting fraction melted, there was a melt mediated transformation from α to β' (15-25°C), melting of β' (36°C), and melting of β (53°C) of the high-melting fraction. As for O/W emulsion sample, the DSC and SR-XRD measurements during the cooling and heating processes exhibited basically the same behavior as that of palm stearin in the bulk state, except that β' did not crystallize during the cooling process, and the temperatures of crystallization of α, melt mediated α → β' → β transformation, and melting of β were lower in the emulsion droplets than in the bulk state. Awad and Sato [35] monitored the crystallization behavior of palm mid fraction (PMF) in oil-in-water emulsions by ultrasonic velocity measurements and X-ray diffraction. They studied the effect of three different emulsifiers: a SE containing palmitic acid (P-170), or stearic acid (S-170), and a polyglicerol ester (DAS-750) containing a stearic acid moiety. Their results show that the hydrophobic emulsifiers accelerated the nucleation of PMF in the emulsion system, while retarding the rate of crystal growth. The effects of the three additives were different; the addition of DAS-750 remarkably enhanced crystallization at low concentrations and the extent of β' nucleation was reduced by comparison with that observed with SE. The DSC thermograms and SR-XRD patterns of the PMF emulsions including SE, taken during the heating process, were almost the same as those of the pure PMF emulsions. Therefore the effects of SE did not appear in the transformation during the heating process; instead, they appeared only in the crystallization process as a kinetic effect on nuclei formation [36]. The results were explained in terms of adsorption of the hydrophobic emulsifier additives at the oil-water interface, which provides a template for acceleration of surface heterogeneous nucleation of PMF in the emulsion system. The interfacial heterogeneous nucleation was most enhanced, as observed in the rise of the crystallization temperature of the oil phase, when the emulsifier additive became more hydrophobic through a higher degree of esterification. Thus, it is reasonable to assume that the increased rate of nucleation of the PMF crystals by the P-170 additive caused the formation of small PMF crystals of α form, which transformed to the more stable polymorph of β' during isothermal treatment. These authors also studied the acceleration of crystallization of palm kernel oil (PKO) in oil-in-water emulsions by the hydrophobic emulsifier additives S-170 and P-170 [37]. They proposed two mechanisms of interfacial heterogeneous nucleation. The first mechanism of the heterogeneous nucleation of PKO crystals assumes an acceleration effect that is caused by molecular interactions between Tween 20 and the SE additives occurring at the oil-water interface. The second mechanism of the heterogeneous nucleation assumes reversed micelles in the interior oil phase. The crystallized SE molecules, either in an adsorbed layer form or in a reversed micelle form, can catalyze the nucleation of PKO through a role of template for nucleation. X-ray spectra and DSC thermopeaks showed that the emulsion without the additives crystallized in the α polymorph, whereas β' was crystallized with the addition of S-170. According to these authors, the occurrence of the X-ray long spacing spectrum before the short-spacing spectra may be indicative of the presence of the template formed by the S-170 additive. The template films may first organize the lamella formation of the crystals that is reflected on the long-spacing spectra.

For a selected fat in bulk, none of these authors have reported both effects: acceleration or delay with supercooling. Moreover, differences in behavior between fat in bulk and in emulsion were not explained by the mechanisms proposed in those articles. Some authors used techniques that do not allow studying in detail the early stage of crystallization, which can lead to misleading interpretations of the phenomenon. This early stage of crystallization is very important since it determines the later evolution of the system. Others used powerful sources such as synchrotron radiation, which allows diffraction patterns to be acquired during real-time crystallization, and therefore, further and less speculative information about mechanisms of action can be obtained. However, these studies were usually performed as a function of temperature and were not performed isothermally.

Huck Iriart et al. [38] reported that according to the SFC with time experiments performed at different temperatures, addition of P-170 to a low *trans* blend formulated with HMF and SFO had different effects on isothermal crystallization in bulk depending on crystallization temperature: acceleration or delay. As an example of each behavior, Figure 1 shows the increase in SFC with time for a blend of 60% HMF and 40% SFO crystallized in bulk, both with and without the addition of P-170.

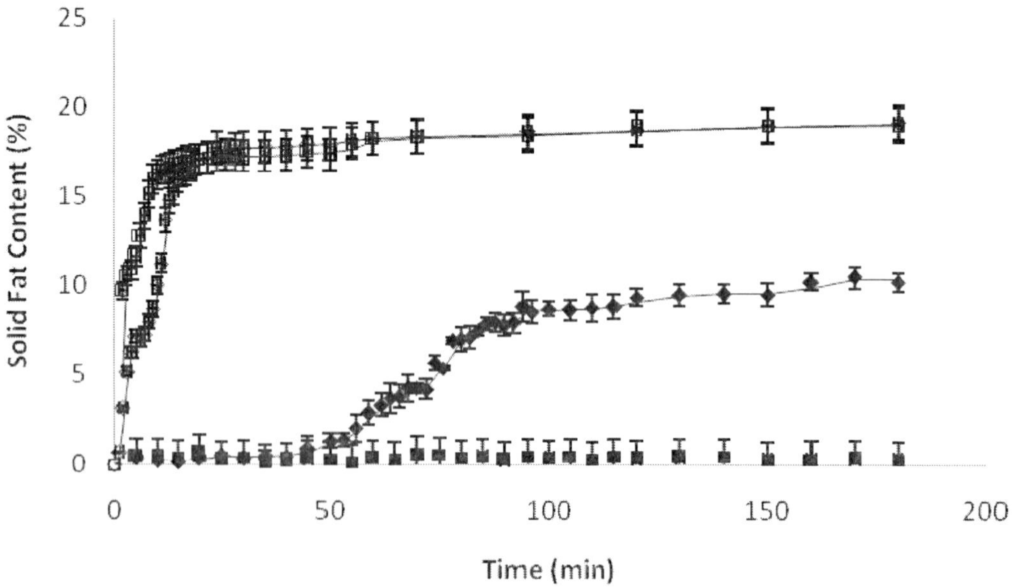

Figure 1. Solid Fat Content (%) vs. time for a 60% HMF/40% SFO mixture isothermally crystallized at 30°C , without (filled diamonds) and with addition of P-170 (filled squares), and at 20°C, without (empty diamonds) and with addition of P-170 (empty squares).

As may be noticed from Figure 1, P-170 strongly delayed the crystallization process at 30°C while it accelerated crystallization at 20°C. There was a slightly significant acceleration or no significant effect on crystallization below 26.0 ± 0.2°C. Above this temperature, P-170 significantly delayed crystallization. The resulting effect, that is, inhibition or acceleration, strongly depended on supercooling (defined as the difference between melting temperature base on Mettler dropping point and crystallization temperature). For HMF/SFO blends, a supercooling lower than 11°C led to inhibition of crystallization, that is, crystallization started later.

Figure 2 reports SAXS patterns of a blend of 80% HMF and 20 % SFO crystallized isothermally in bulk with and without addition of P-170 at 26 and 28°C.

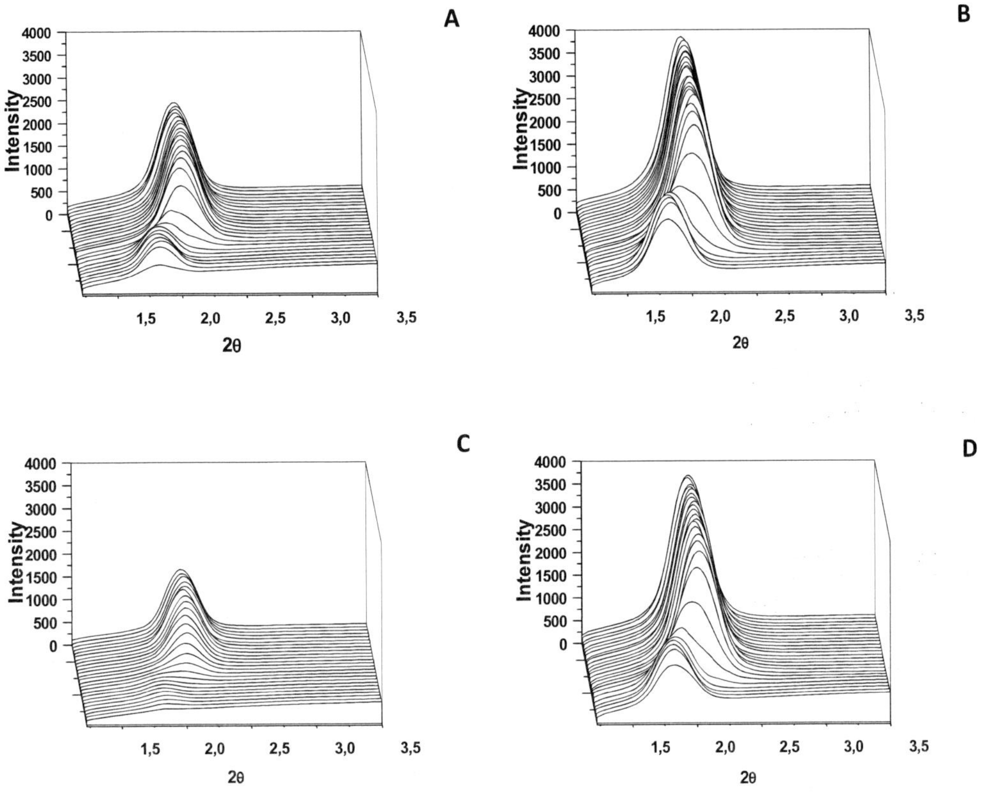

Figure 2. Three-dimensional plots of SAXS vs. the diffraction angle (2θ) recorded as a function of time for bulk 80% high-melting fraction (HMF)/20% sunflower oil (SFO) blend isothermally crystallized at 26 °C, a without and b with addition of P-170, and at 28 °C, c without and d with addition of P-170. SAXS patterns were recorded as a function of time with frames of 30 s.

After 30 sec at 26 °C (a), the SAXS spectrum of 80% HMF/20% SFO blend displayed a single peak in the 0.1° to 3.5° region (2θ) with a d of 47.14 Å. According to the literature [39-41], these X-ray diffraction patterns are consistent with a bilayer (2L001) lamellar packing arrangement and a hexagonal subcell or α polymorph. The intensity of the diffraction pattern of the blend increased until 3.5 min of isothermal crystallization. After this time, the initial diffraction peak began to move to higher 2θ values. The second diffraction peak with a d of 40.60 Å corresponded to the 2L001 packing of the orthorhombic subcell (β′ polymorph), [39-41]. After 5 min at 26 °C, the α peak disappeared. The polymorphic transition α to β′ lasted 1.5 min. When P-170 was added to the 80% HMF/20% SFO blend (b), the α form appeared after 30 s at crystallization temperature. A second signal corresponding to the β′ form occurred after 2.5 min. P-170 accelerated the appearance of the β′ form. Both forms, α and β′, coexisted for 1 min. After 3 min only, the β′ signal was present in the X-ray diffraction pattern. NMR measurements showed that P-170 accelerated crystallization at temperatures below 26 °C if it is likely to have this effect. At 28 °C (c), the α-form was almost not detectable. The β′ signal appeared after 5 min of isothermal crystallization at 28 °C and was the only polymorphic form found even until 30 min. When P-170 was added to the 80% HMF/20% SFO blend (d), the α form appeared after 30 s at crystallization temperature. P-170 favored the crystallization in the α form and delayed the formation of the β′ polymorph. A second signal corresponding to the β′ form was found after 3 min. Both forms, α and β′, coexisted for 30 s. According to NMR measurements, P-170 delayed crystallization at this temperature. As a summary, when the effect was acceleration of crystallization, P-170 expanded the time interval during which the α and β′ forms coexisted. In the case of HMF/SFO blends crystallized at 28 °C (effect of retardation of crystallization), the control samples (without emulsifier) showed one signal corresponding to the β′ signal. Thus, the α form was only noticeable with the addition of P-170. In addition, α and β′ forms coexisted for shorter times than at lower temperatures. When dissolved in SFO, P-170 crystallizes in the α form in the conditions selected in this study. No other polymorphic forms were present after 1 h at crystallization temperature. However, when added to these samples, the signal of the α form, if present, disappeared after a short time of isothermal crystallization. This seems to indicate that P-170 is incorporated in these blends crystals. It was reported that sucrose esters tend to form molecular aggregates such as inverse micelles even at low concentrations [42, 43], which may act as template accelerating crystallization of fats [35]. This mechanism occurred with high supercooling when P-170 elongated the time of coexistence of the α and β′ forms of the fat. Under high supercooling, the fat crystallizes in the α form [39], as is the case of P-170. This phenomenon explains the acceleratory effect of P-170, which can act as seed for fat crystallization in the α form. For temperatures close to the melting point, the effect of P-170 on crystallization was the one reported for palm mid-fraction and palm kernel oil, a delay of crystallization. Fats and sucrose esters are able to co-crystallize because of their somewhat similar chemical structure. At high temperature, the α form was either not present or the intensity of the α-form diffraction line was weak, so P-170 was not able to accelerate crystallization and was incorporated in β′ fat crystals. However, the structural dissimilarities between triacylglycerols and emulsifiers caused delayed crystallization. Time-resolved in situ synchrotron X-ray scattering is a more sensitive technique than NMR to monitor isothermal crystallization. Zero time in Figures 1 and 2 is the moment at which the samples in the NMR tubes reached crystallization temperature. The first pattern in X-ray figures corresponded to 30 s at crystallization temperature. The NMR registers crystal appearance when SFC starts increasing. Typical detectable levels of SFC are 0.5% to 1.0%. An SFC of 0.1%, as found in several samples in the first 5 min of crystallization, is below the detection threshold of a pNMR machine. However, X-ray patterns clearly showed early crystallization. Since the effects of P-170 were determined for the events that took place in those early steps of crystallization, they could only be understood by synchrotron X-ray scattering. Conventional X-ray techniques do not allow these studies since they are not performed in real time. By analyzing polymorphism in real time, with the aid of synchrotron X-ray scattering, Huck Iriart et al. [38] established that the different effects caused by P-170 during isothermal crystallization of HMF or its blends with SFO, as described by SFC curves, were strongly related to the effects of P-170 on fat polymorphism, especially to the value of the time interval of coexistence of the α and β′ forms.

CONCLUSIONS

Triacylglycerols {TAGs) are major components of naturally occurring fats and oils. TAGs are utilized in edible fats, such as margarine, chocolate, and in cosmetics and pharmaceuticals. They are usually employed in mixed systems. The polymorphism of mixed saturated-unsaturated TAGs has been extensively studied in view of their industrial applications, in particular those related to the confectionery and food industry. In many production lines, quality of a final product like gross, snap, texture and shelf life is decided by the polymorphic structure of the main fat in formulation. X-ray is the preferred technique for polymorphic studies since the two levels of organization, lamellar stacking and lateral chain packing, are identified univocally from the Bragg reflections using X-ray diffraction at wide and small angles (WAXS, SAXS), respectively, allowing to assign the polymorphic forms present in a fat. Due to the relevance of polymorphism in chocolate properties, it is one of the most studied fat systems. Tempering is an important stage in its manufacturing to obtain the desired phase V of cocoa butter. During this processing stage, the completely liquid cocoa butter is first brought to a temperature at which nuclei from both phases IV and V form. The mixture is then heated to a temperature where the seeds of form IV are re-melted. After these two steps, a sufficient number of nuclei of phase V are present in the melt to allow solidification in the desired form during the final cooling in the mold. X-ray also was very useful to understand early events in crystallization. With the aid of synchrotron X-ray scattering, it was possible to establish that the different effects caused by some hydrophobic emulsifiers such as P-170 during isothermal crystallization of fat blends were strongly related to the effects of P-170 on fat polymorphism.

REFERENCES

[1] Aquilano, D., Sgualdino, G. Fundamental aspects of equilibrium and crystallization kinetics. In: *Crystallization Processes in Fats and Lipid Systems*. Garti, N., Sato, K., edts. Marcel Dekker, Inc., New York, USA, 2001, pp. 1-52.

[2] Walstra, P. *Crystallization. In: Physical Chemistry of Foods*. Marcel Dekker, Inc. New York, USA, 2003, pp. 583-649.

[3] Chapman, G. M., Akehurst, E. E., Wright, W.B. *J. Am. Oil Chem. Soc.* 1971, 48,824-830.

[4] Marangoni, A. G., McGauley, S. E. *Crystal Growth Design*. 2003, 3, 95-108.

[5] Luccas, V. In Fracionamento termico e obtençao de gorduras de cupuaçu alternativas a manteiga de cacau para uso na fabricaçao de chocolate, Ph.D. thesis, Universidade Estadual de Campinas, Campinas, São Paulo, Brasil, 2001, *http://libdigi.unicamp.br/document/?code=vtls000235715*.

[6] Wille, R. L.; Lutton, E. S. *J. Am. Oil Chem. Soc.* 1966, 43, 491–496.

[7] Adenier, H., Ollivon, M., Perron, R., Chaveron, H. *Chocolaterie Confiserie de France*. 1975, 315, 7–14.
[8] van Malssen, K.; Peschar, R.; Schenk, H. *J. Am. Oil Chem. Soc*. 1996, 73, 1209–1215.
[9] van Malssen, K.; Peschar, R.; Brito, C.; Schenk, H. *J. Am. Oil Chem. Soc*. 1996, 73, 1225–1230.
[10] van Malssen, K.; van Langevelde, A.; Peschar, R.; Schenk, H. *J. Am. Oil Chem.* Soc.1999, 76, 669–676.
[11] Loisel, C., Keller, G., Lecq, G., Bourgaux, C., Ollivon, M. *J. Am. Oil Chem. Soc.* 1998, 75, 425-439.
[12] da Silva, J., Plivelic, T.S., Herrera, M.L., Ruscheinsky, N., Kieckbusch, T.G., Luccas, V., Torriani, I.L. *Crystal Growth Design* 2009, 9, 5155-5163.
[13] Sato, K., Arishima, T., Wang, Z. H., Ojima, K., Sagi, N., Mori, H. *J. Am. Oil Chem. Soc*. 1989, 66, 664-674.
[14] Koyano, T., Hachiya, I., Arishima, T., Sato, K., Sagi, N. *J. Am. Oil Chem. Soc.* 1989, 675-679.
[15] Hachiya, I., Koyano, T., Sato, K. *J. Am. Oil Chem. Soc*. 1989, 66, 1763-1770.
[16] Rousset, Ph., Rappaz, M. *J. Am. Oil Chem. Soc*. 1997, 74, 693–697.
[17] Rousset, Ph., Rappaz, M., Minner, E. *J. Am. Oil Chem. Soc*. 1998, 75, 857–864.
[18] Hartel, R.W. *J. Am. Oil Chem. Soc*. 1996, 73, 945-953.
[19] Sabariah, S., Ali, A.R. Md., Chong, C.L. *J. Am. Oil Chem. Soc*. 1998, 75, 905-910.
[20] Smith, K.W., Cain, F.W., Talbot, G. *J. Agric. Food Chem.* 2004, 52, 5539-5544.
[21] Pore, M., Seah, H.H., Glover, J.W.H., Holmes, D.J., Johns, M.L., Wilson, D.I., Moggridge, G.D. *J. Am. Oil Chem. Soc*. 2009, 86, 215-225.
[22] Gwie, C.G., Griffiths, R.J., Cooney, D.J., Johns, M.L., Wilson, D.I. *J. Am. Oil Chem. Soc.* 2006, 83, 1053–1062.
[23] Mac Millan, S.D., Roberts, K.J., Rossi, A., Wells, M.A., Polgreen, M.C., Smith, I.H. *Crystal Growth Design* 2002, 221-226.
[24] Sonwai, S., Rousseau, D. *Crystal Growth Design*. 2008, 9, 3165-3174.
[25] Lonchampt, P.; Hartel, R. W. *Eur. J. Lipid Sci. Technol*. 2004, 106, 241-274.
[26] Yuki, A., Matsuda, K., Nishimura, A. *J. Jpn, Oil Chem. Soc*. 1990, 39, 24–32.
[27] Nasir, M.I. *Effect of sucrose polyesters and sucrose polyester–lecithins on crystallization rate of vegetable ghee, in Crystallization and solidification properties*, ed. by N. Widlak, R.W. Hartel, S.S. Narine (AOCS, Champaign, 2001), pp. 87–95.
[28] Herrera, M.L., Marquez Rocha, F.J. *J. Am. Oil Chem. Soc*. 1996, 73, 321–326.
[29] Hodate, Y., Ueno, S., Yano, J., Katsuragi, T., Tezuka, Y., Tagawa, T., Yoshimoto, N., Sato, K. *Colloids Surf. A Physicochem. Eng. Asp*. 1997, 128, 217–224.
[30] Martini, S., Herrera, M.L., Hartel, R.W. *J. Agric. Food. Chem*. 2001, 49, 3223–3229.
[31] Cerdeira, M., Martini, S., Hartel, R.W., Herrera, M.L. *J. Agric. Food Chem*. 2003, 51, 6550-6557.
[32] Cerdeira, M., Pastore, V., Vera, L.V., Martini, S., Candal, R.J., Herrera, M.L. *Eur. J. Lipid Sci. Technol.* 2005,107,877-885.
[33] Cerdeira, M., Martini, S., Candal, R.J., Herrera, M.L. *J. Am. Oil Chem. Soc*. 2006, 83, 489-496.
[34] Sonoda, T., Takata, Y., Ueno, S., Sato, K. *J. Am. Oil Chem. Soc*. 2004, 81, 365-373.
[35] Awad, T., Sato, K. *J. Am. Oil Chem. Soc*. 2001, 78, 837-842.

[36] Arima, S., Ueji, T., Ueno, S., Ogawa, A., Sato, K. *Colloids Surf. B Biointerfaces* 2007, 55, 98-106.
[37] Awad, T., Sato, K. *Colloids Surf. B Biointerfaces* 2002, 25, 45–53.
[38] Huck Iriart, C., Candal, R.J., Herrera, M.L. *Food Biophysics* 2009, 4, 158–166.
[39] Cisneros, A., Mazzanti, G., Campos, R., Marangoni, A.G. *J. Agric. Food Chem.* 2006, 54, 6030–6033.
[40] López, C., Lavigne, F., Lesieur, P., Bourgaux, C., Ollivon, M., *Dairy Sci.* 2001, 84, 756–766.
[41] Mazzanti, G., Guthrie, S.E., Sirota, E.B., Marangoni, A.G., Idziak, S.H.J. *Cryst. Growth Des.* 2004, 4, 1303–1309.
[42] Kunieda, H., Kanei, N., Tobita, I., Kihara, K., Yuki, A. *Colloid Polym. Sci.* 1995, 273, 584–589.
[43] Kunieda, H., Ogawa, E., Kihara, K., Tagawa, T. *Prog. Colloid Polym. Sci.* 1997, 105, 237–243.

In: X-Ray Scattering
Editor: Christopher M. Bauwens

ISBN: 978-1-61324-326-8

Chapter 4

SMALL ANGLE X-RAY SCATTERING ANALYSIS OF NANOMATERIALS FOR ULTRA LARGE SCALE INTEGRATED CIRCUITS

T. K. S. Wong and T. K. Goh
Division of Microelectronics, School of Electrical and Electronic Engineering,
Nanyang Technological University, Singapore

ABSTRACT

Small angle X-ray scattering is a non-destructive versatile characterization technique that is becoming increasingly important in nanomaterial analysis. Its applications include nanolithography, quantum dots, quantum wires and nanoporous materials. In this chapter, we focus on the determination of the properties of nanoporous dielectrics by small angle X-ray scattering. After an overview and the historical background, the basic theory of small angle X-ray scattering is discussed. This theory is then applied to the characterization of two types of solution processed nanoporous dielectric films: (i) sol-gel derived silica and (ii) porous organic polymer films used for reducing interconnect parasitic capacitance in ultra large scale integrated circuits. For each type of sample, the small angle X-ray scattering pattern is collected in a transmission geometry and a parametric model is developed to fit to the measured data. By means of model fitting, the characteristic pore size and other critical parameters can be deduced. This information is useful to the development of nanomaterials for advanced integrated circuits.

1. INTRODUCTION

Small angle X-ray scattering (SAXS) is an old technique that was first applied to the study of natural fibres such as collagen and colloidal coal in the late 1920s and 1930s [1]. In a typical SAXS experiment, the specimen or sample is probed by a collimated beam of hard X-rays. The electromagnetic field of the incident X-ray photons interacts with the electrons within the sample. When the excited electrons relax, X-ray photons are emitted in a different direction and is said to be elastically scattered. The superposition of the coherently scattered

X-rays by all electrons in the sample results in a scattering pattern in the far field. By analyzing the pattern of scattered X-rays as recorded by a position sensitive detector, much useful information about the sample can be deduced through a process of model fitting.

SAXS characterization can be performed in either a transmission geometry or a grazing incidence geometry. In the former, the X-ray beam from the source penetrates the substrate of the sample or the solution medium. When the structures of interest are located only at the surface, the grazing incidence geometry is used instead. In this chapter, we restrict the discussion to transmission SAXS. Reviews of grazing incidence SAXS can be found in references [2, 3].

The main advantages of SAXS are that the technique is non-destructive and minimal sample preparation is required. Unlike transmission electron microscopy (TEM), SAXS can yield statistically significant information because a larger ensemble of particles is probed by the X-ray beam. Especially important is the fact that the technique becomes more accurate as the characteristic feature size of the sample approaches 1nm [4]. This distinct property enables the SAXS technique to an excellent complement to conventional techniques such as scanning electron microscopy (SEM) and scanning probe microscopy (SPM). A new nanometrology technique that illustrates this property is the critical dimension CD-SAXS technique [5-8]. Due to the scaling of device features below 45nm, it is becoming difficult to perform CD metrology on nanoscale devices by CD-SEM. However, this feature size is accessible by CD-SAXS and in prelimimary studies, CD measurements had been performed and validated for regular lines and space arrays patterned by extreme ultra-violet lithography.

Another important application of SAXS in nanotechnology is the characterization of the size of nanoparticles or quantum dots. These zero dimensional nanostructures already have wide applications in medicine as marker labels as well as single electron devices and quantum dot lasers [9]. In addition, they are also increasingly being explored in third generation photovoltaic devices for boosting the absorption and enhancing power conversion efficiency beyond the Shockley-Quiesser limit [10-12]. The key advantage of SAXS over alternative techniques such as transmission electron microscopy (TEM) is that the entire ensemble of quantum dots can be characterized simultaneously and the obtained size distribution will thus be more representative [4]. This is in contrast to measurements based on TEM image analysis where only a small ensemble can be studied. SAXS has been successfully applied to the characterization of gold nanoparticles chemisorbed with organic ligands [13] and quantum dots grown epitaxially on a semiconductor substrate [14-20].

This chapter is concerned with a third application of SAXS, namely the characterization of mesoporous thin films in microelectronic interconnects. The semiconductor industry introduced porous dielectric films such as carbon doped silicon oxide (SiOCH) in ~2005 in order to mitigate the signal propagation delay and crosstalk in interconnects of ultra large scale integrated circuits [21]. The porosity of these dielectric films reduces the effective mass density and dielectric constant (k) of the layer and result in k values below that of the host matrix. In order to control the value of the dielectric constant, it is important to correlate the typical pore size and pore size distribution with the deposition and curing conditions during process development [22]. Existing techniques include gas adsorption porosimetry (Barrat-Joyner-Halenda (BJH) method) [22], ellipsometric porosimetry (EP) [23,24] and positiron annihilation lifetime spectroscopy (PALS) [25]. However, mercury probe porosimetry is only suitable for bulk porous solids. EP can be applied to thin films but the technique requires specially configured ellipsometers and only open pores that are connected to the surface can be probed. PALS can probe both open and closed pores but since radioactive gamma ray sources are required, the technique may not be readily used by industry. As a result of these constraints, there has been interest in applying SAXS to the analysis of porous low dielectric constant insulating films [26-33]. In the remainder of this chapter, we will describe a pore analysis methodology based on transmission SAXS. This approach complements the GISAXS approach reported by Omote et.al. [33]. The chapter is organized as follows. In section 2, we first briefly discuss the background of the SAXS technique and present the detailed theory of SAXS in section 3. In section 4, the experimental set-up and preparation of sol-gel derived silica and porous polymer films are described. In section 5, the obtained SAXS data and the development of parametric models to numerically fit the experimental data are detailed. Some of the other results in this study had been published elsewhere [34]. It is worth pointing out that the SAXS technique described in this chapter should also be useful to the characterization of porous electrodes in electrical energy storage devices in future [35].

2. Historical Background

The structure of materials had been investigated extensively using X-ray techniques during the last 30 years [36]. This is perhaps due to the development and maturity of X-ray scattering techniques (both small and wide angle) and the application of porous and polymer materials in thin films for microelectronics. As a result of this, a better understanding of the polymer or porous materials had been achieved. X-rays were discovered in 1895 by W. C. Roentgen, who called them X- rays because their nature was at first unknown [37]. They are sometimes also called Röentgen, or Röntgen, rays [38]. The small angle scattering of X-rays was discovered in 1930 by Krishnamurti [39]. Its basic principle forms the basis of the concepts that are normally associated with the theory of X-ray diffraction. However, the experimental aspect is quite different. It is performed by focusing a low divergence X-ray beam onto a sample and observing a coherent scattering pattern that arises from electron density inhomogeneities within the sample. Since the dimensions typically analyzed are much larger than the wavelength of the typical X-ray used (0.154 nm, for Cu K_α), dimensions from tens to thousands of angstroms can be analyzed within a narrow angular scattering range. This angular range or pattern is analyzed using the inverse relationship between particle size and scattering angle to distinguish the characteristic shape and size features within a given sample.

X-ray scattering experiments are usually performed either by X-rays from conventional X-ray tubes or synchrotron sources. Conventional X-ray tubes can be easily installed within a laboratory without taking up much space. On the contrary, synchrotron sources require a large dedicated space. However, the intensity of X-rays emitted from tubes is smaller than that from synchrotron sources. In addition, the spectra emitted from conventional X-ray tubes include only one useful wavelength (characteristic wavelength of anode material) and hence the sample size range accessible to study is limited. The spectrum of synchrotron source is continuous and intense over a wide range from infrared to hard X-ray wavelengths. Thus, measurements involving different beam wavelengths can be easily carried out. High intensity synchrotron radiation only required a very short measurement time (less than 30 minutes per sample) and structural changes are expected to be observed more clearly compared with conventional X-ray tube sources. Hence, in-situ kinetic studies are possible with this facility.

Bragg found in the early 20th century the relationship between the wavelength, λ, lattice spacing, d and scattering angle, θ of X-rays scattered by a crystalline solid [38]:

$$2d\sin\theta = \lambda \tag{1}$$

The Bragg law shows that larger a lattice spacing will cause reflections to smaller scattering angles and vice versa. Hence, X-ray scattering methods can be differentiated into two categories. The first is wide angle X-ray scattering (WAXS), which covers measurements made at scattering angle larger than 2°. The second category, small angle X-ray scattering (SAXS) is usually performed at scattering angles less than 2°. One can conclude from equation 1 that WAXS is more sensitive in resolving microstructure (atomic order) information. SAXS, on the other hand, probes the macroscopic structure of a material, such as pores.

3. Theory Of Small Angle X-Ray Scattering

Small angle scattering (SAS) experiments are commonly performed in the configuration shown in Figure 1 [40]. They generally follow this procedure. The sample is irradiated with some type of radiation such as x-rays, neutrons or light. The resulting scattering pattern is measured and then the structure that caused the observed pattern is deduced. Scattering of x-rays is caused by differences in electron density in the sample and the superposition of X-ray wavefronts emitted by excited electron upon de-excitation.

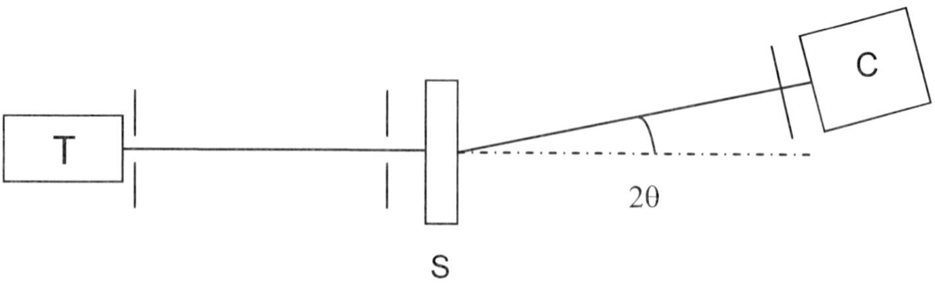

Figure 1. Schematic illustration of small-angle scattering experimental setup: Source *T*, Sample *s*, slits used to define the incident and scattered beams and the detector *C*. The scattering angle is 2θ and θ is the angle used to calculate the scattering vector.

SAS data is usually measured with a position sensitive proportional counter. The distance from the sample to the detector determines the scattering vector q. The magnitude of the scattering vector is given as:

$$q = \frac{4\pi}{\lambda}\sin\theta \tag{2}$$

where θ is one half of the scattering angle. X-rays scattered by the electrons in the sample are mainly coherent. Although incoherent (Compton) scattering can also occur, it can be neglected at the small angles that are involved in structure analysis [41]. In the present context, coherence means the wave amplitudes can be added and the intensity is given by the absolute square of the resultant amplitude. As in any scattering process, X-rays that are scattered by different parts of the sample will propagate in various directions and distance from source to detector. Waves from different parts of the sample thus arrive at the detector with different phases. The resultant amplitude of the scattered wave from the sample is the sum of the amplitude of the waves from all parts of the sample, with allowance for these phase differences. Thus, the scattered intensity is proportional to the product of the resultant amplitude and its complex conjugate. The amplitude $A(q)$ of small angle scattering can be written as:

$$A(q) = A_e F(q) \tag{3}$$

where A_e is the scattering amplitude from one electron. Thus $F(q)$ can be called the scattering amplitude in electron units. Intensity $I(q)$ is then given by the following equation:

$$I(q) = A(q)A^*(q) = I_e|F(q)|^2 \tag{4}$$

where I_e is the electron scattering intensity. The scattered intensity is determined by the structure of the sample and information about this structure can be derived from an analysis of the scattered intensity. The fluctuations in the mass density of the sample actually lead to the fluctuations of the electron density that is the source of small angle X-ray scattering. Since λ is usually about 10^{-10} m for X-rays used for small angle scattering, this phase difference is small compared to 1 for distances d of the order of the diameters of atoms or small molecules. Consequently, the rays scattered from different parts of an atom or small molecules arrive at the detector almost completely in phase. Hence, the phase differences $q.r$ of the waves from two points in a scatterer will be appreciably different from zero only when the condition, given as:

$$q.r >> 1 \tag{5}$$

is satisfied. In this inequality, r is the vector from one point in a scatterer to another point in the same scatterer. Also:

$$q = \frac{2\pi}{\lambda}(S - S_0) \tag{6}$$

S and S_0 are unit vectors in the direction of the scattered and incident beams, respectively, as shown in Figure 2. Little or no information about the structure can be obtained from scattering measurements when the inequality 5 is not satisfied. Further derivation will give equation 2.

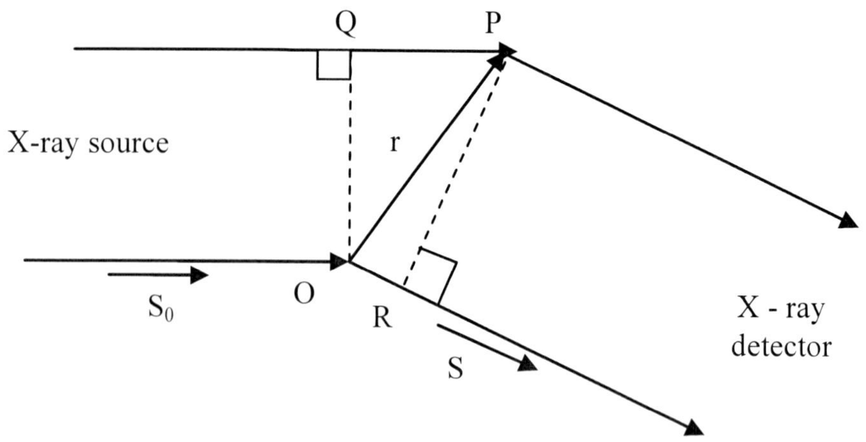

Figure 2. Geometry of the path length difference.

The scattered wave amplitude $F(q)$ of a system of n scatterers with amplitude $f_j(q)$ can be written as:

$$F(q) = \sum_{j=1}^{n} f_j(q) e^{iq.r} \tag{7}$$

The amplitudes $f_j(q)$ are proportional to the volumes V_j of the scatterers. When q and the diameters l_j of the scatterers are so small that $ql_j << 1$, the $f_j(q)$ are very nearly equal to the $f_j(0)$ and therefore are almost completely independent of q. It is important to note that different symbols can be found in the literature corresponding to the same definition of scattering vector: q, h, Q, $2\pi s$ are identical [42]. The total amplitude at q is sum of the waves scattered by all atoms in the sample. Since small angle scattering is related to distances that are large compared to interatomic distances, it is not possible to separate the contributions of individual atoms to the scattering and the sum over discrete atoms can be replaced by an integral if one introduces a scattering density [42]:

$$\rho(r) = \sum \frac{b_i}{V_{res}} \tag{8}$$

where the scattering amplitude b_i are summed over a volume V_{res} of the size of a resolution element. Then, equation 3 can be expressed as:

$$A(q) = \iiint \rho(r) e^{-iq.r} dr \tag{9}$$

This equation is the 3-dimensional Fourier transform of $\rho(r)$. The scattered intensity is the product of $A(q)$ and its complex conjugate, as described by equation 4. Hence, it can be re-expressed in:

$$I(q) = A(q)A^*(q)$$
$$= \iiint dr_1 \iiint \rho(r_1)\rho(r_2)e^{-iq.(r_1-r_2)}dr_2 \quad (10)$$
$$= V \iiint \gamma(r)e^{-iq.r}\, dr$$

with

$$\gamma(r) = \frac{1}{V}\iiint \rho(r_0)\rho(r_0+r)dr_0 \quad (11)$$
$$=< \rho(r_0)\rho(r_0+r) >$$

The intensity distribution in q or reciprocal space is uniquely determined by the structure in real space. $\gamma(r)$ is the autocorrelation function. It expresses the correlation between the densities measured at two points separated by a vector r, averaged over the irradiated volume V. It can be related to the structure of the scattering object. Figure 3 gives a simple illustration about the relationship between intensity, amplitude and autocorrelation function.

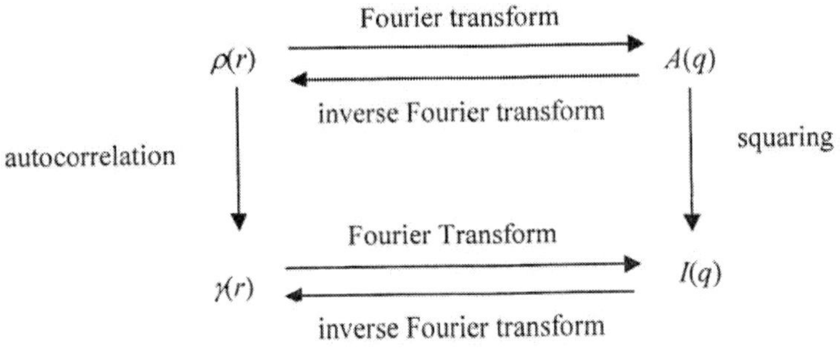

Figure 3. Illustration of the relationship between intensity, amplitude and autocorrelation function.

In the typical domain of small angle scattering, one would introduce two restrictions which are met in the majority of cases, in order to simplify the problem. The first one is that the system being investigated is considered to be statistically isotropic. So, there is no difference whether it is a property of the structure itself or a consequence of some changes in time. In addition, there is no long range order or no correlation between two points separated widely enough. The first restriction that the system follows the distribution in C-space depends only on the magnitude r of the distance, though this will not be true for $\rho(r)$ in ordinary space. So the phase factor e^{-iqr} can be replaced by its average, taken over all directions of r. This is expressed by the Debye formula [43,44]:

$$\left\langle e^{-iqr} \right\rangle = \frac{\sin qr}{qr} \quad (12)$$

Combining equations 10 and 12 results in equation 13:

$$I(q) = \int 4\pi r^2 dr \cdot \gamma(r) \frac{\sin(qr)}{qr} \tag{13}$$

3.1. Form Factor and Structure Factor

Although the analysis of microscope images is straightforward and independent of the size distribution of pore or particle, the images may not be representative of the whole sample. This is because only a very small portion of sample is studied. By contrast, the small angle x-ray scattering method takes into account the degree of polydispersity and packing. The scattering intensity from a collection of pores can be written as [45, 46]:

$$I(q) = NP(q)S(q) \tag{14}$$

where N is the number of scattering objects. The form factor, $P(q)$ describes the electron density distribution of an individual scatterer and $S(q)$, the structure factor, describes how the objects are arranged in space. On length scales smaller than the size of the primary scatterer, the structure factor is constant and hence the scattered intensity is proportional to the form factor $P(q)$. The influence of particle shape and size in small angle scattering must be seriously taken into account. The influence of polydispersity is accounted for in equation 14 by including a suitable particle size distribution $n(R)$ such as:

$$n(R) = \int_0^{\infty} \frac{1}{r_0\sqrt{2\pi\sigma^2}\exp(0.5\sigma^2)} \exp(-\frac{\ln(R/r_0)^2}{2\sigma^2})dR \tag{15}$$

where R is the radius of the pore, σ is the standard deviation, r_0 is the radius at peak of distribution. The scattering of identical spherical pores distributed randomly with no interference between particles can be written as follows:

$$P(q) = \left(3\frac{\sin(qR) - qR\cos(qR)}{(qR)^3}\right)^2 \tag{16}$$

For systems comprising spherical colloidal particles, a structure factor that is often used is the Percus-Yevick factor:

$$\frac{1}{S(q)} = 1 + \frac{24\phi}{u^3}\left\{a(\sin u - u\cos u) + b\left[(\frac{2}{u^2} - 1)u\cos u + 2\sin u - \frac{2}{u}\right] + \frac{\phi a}{2}\left[\frac{24}{u^3} + 4(1 - \frac{6}{u^2})\sin u - (1 - \frac{12}{u^2} + \frac{24}{u^4})u\cos u\right]\right\} \tag{17}$$

where ϕ is the packing density of the pores, $a = (1+\phi)^2(1-\phi)^{-4}$, $b = -1.5(1+\phi)^2(1+\phi)^2$ and $u = 2qR$. For some sol-gel samples, the SAXS intensity are modeled by the Debye-Bueche form factor and the Frelfort structure factor given as follows [47, 48]:

$$P(q) = \frac{1}{\left(1 + a_{cor}^{2} q^{2}\right)} \tag{18}$$

$$S(q) = 1 + \frac{B}{\left(1 + q^{2}\xi^{2}\right)^{(d_f - 1)/2}} \cdot \frac{\sin\left[(d_f - 1)\arctan(q\xi)\right]}{(d_f - 1)(q\xi)} \tag{19}$$

where a_{cor} is the correlation length, B is a constant, d_f is the fractal dimension and ξ is the cut off length.

4. EXPERIMENT

The SAXS measurements were carried out using a Bruker Nanostar SAXS system. A simplified schematic of the experimental setup is shown in Figure 4. The measurements were performed with a Cu X-ray tube operated at 1.5 kW mounted in the point focus position and the beam was monochromatized by forming an image of the entrance slit in the exit plane at the wavelengths present in the source. Data was measured with a position sensitive proportional counter. The distance from the sample to the detector was 106 cm. Thus, the magnitudes of the accessible scattering vectors are from 0.05 nm^{-1} to 2 nm^{-1}.

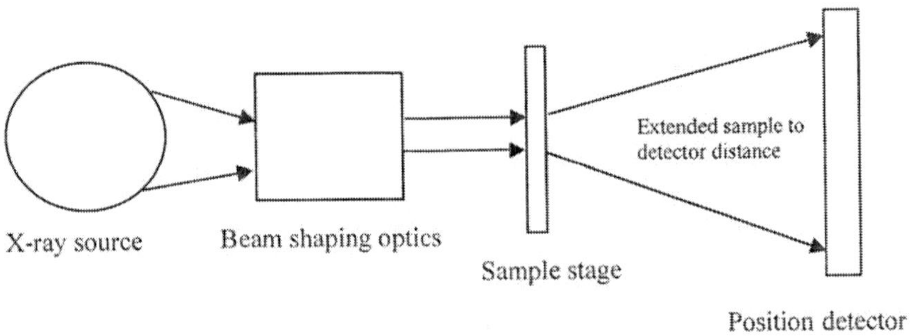

Figure 4. Experimental setup of SAXS.

In addition, this setup also comprises of cross-coupled Gobel mirrors to achieve an intense and parallel output X-ray beam, a 0.3 mm pin hole collimator, a high resolution chamber evacuated < 10^{-4} bar, and a two-dimensional gas filled detector to reduce scattering by air.

The operating procedure of the SAXS instrument is briefly described as follows. First, the sample chamber was opened to remove the sample holder from its original position on the setup. This sample holder has six 4 mm diameter pinholes and each has a different position as defined in the Bruker SAXS 4.0 NT version software. The experiment samples, preferably with size less than 15 mm x 15 mm are then placed on the sample holder. Next, the sample holder was attached into its original position inside the chamber. After the chamber has been pumped down by the turbomolecular pump, the X-ray source was turned on to start the experiments. The background scattering of air and the substrate material are required for transmissivity calibration and background subtraction. The transmissivity of the sample was obtained by performing a short scattering of 120 s duration with a glassy carbon standard plus air and the combination of sample and standard. The transmissivity of the sample was obtained by inputting the scattering obtained from standard plus air and the combination of sample and standard into the SAXS software. A detailed explanation of the transmissivity measurement will be given later in this section. Finally, the scattering experiment was started with the collection of scattering data by the proportional counter. After completion of the scattering experiment (typically 50,000 s per sample), the scattered intensity versus scattering vector q was computed through a Chi integration of the scattering pattern with the SAXS software provided by Bruker AXS. An example of a scattering curve is shown in Figure 5.

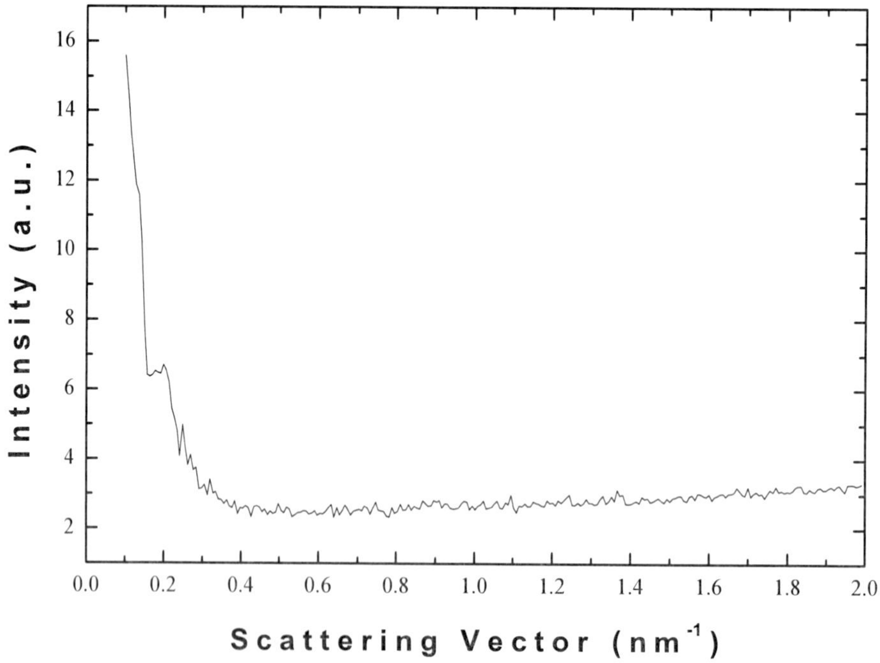

Figure 5. Scattering curve obtained through a Chi integration on the scattering pattern of sol-gel prepared silica.

Cross sectional transmission electron microscope (TEM) images of porous SiLK and LKD samples were prepared to obtain structural information of the film. Preliminary evaluation on the substrate material was required for the transmission geometry SAXS as the scattering intensity of low-k film on a thick Si substrate was otherwise rather weak. This evaluation aims to identify the proper substrate materials for SAXS experiments. Silicon wafers with different thickness ranging from 650 μm to 100 μm were prepared and evaluated by both theoretical calculation and measurement. In addition, transmissivity data of mica and glass were collected. The theoretical calculation of the absorbance of silicon with different thickness was performed using Beers' Law [49], absorption coefficient and density. The ratio of transmitted and incident intensity is as follows:

$$I = I_0 \exp(-\mu \rho d) \tag{20}$$

where ρ is the density, μ is the linear absorption coefficient and d is the thickness of the sample. Absorbance is also related to the transmittance T through the following equation:

$$A = \log_{10} \frac{1}{T} \tag{21}$$

where transmittance is the ratio of transmitted and incident intensity. Table 4.1 presents the transmissivity of substrate materials with different thickness investigated with Nanostar system for 120 s. The SAXS transmissivity was obtained through the transmissivity calibration.

The computation of the transmissivity of substrate materials was performed with known values of the energy of x-ray, mass absorption data and substrate density [50]. It was found that a Si substrate with a thickness of 100 μm and a thin mica substrate with a thickness of 100 μm and below have the appropriate thickness for high transmissivity, and hence a stronger x-ray signal in the transmission mode of x-ray scattering. In view of this, porous film deposited on the normal silicon wafers of 650 μm thick were sent for backgrinding to reduce the substrate thickness to 100 μm. This was necessary in order to increase the scattering signal that are rather weak on normal Si substrate.

Table 1. Transmissivity of substrate materials with different thickness

Materials	Thickness (μm)	Transmissivity	
		Theoretical calculation	SAXS reading
Si	650	0.0005144	0.0064
	200	0.0937	0.056
	100	0.312	0.234
	80	0.3938	-
Mica	150	-	0.133
	50	-	0.532
Glass	100	-	0.319

Porous silica films synthesized by a sol-gel process were spin coated onto mica substrate [60] in an ambient environment. Two types of sols were prepared, one to form the matrix and the other is to provide a bridging between the molecules. Sol A was prepared by adding tetraethyl orthosilicate (TEOS) in ethanol with HCl as a catalyst. Sol B was made by dispersing methyltriethoxysilane (MTES) in ethanol with NH_4OH as a catalyst. Both sols were allowed to hydrolyze for about 24 hours before mixing them together with mole ratio of sol A/sol B = 1:1. Following spin coating, the films were thermally treated in air up to 450

°C, which is the maximum temperature for both Al and Cu metallization processes. The above-mentioned preparations were all carried out in a class 100 clean room.

Polymeric dielectric porous SiLK™ and SiLK™ were spin-coated onto 200mm silicon p-type wafers at 2500rpm for 45s and underwent normal baking in hot plate at 325°C to remove the solvent and curing in a horizontal tube furnace at 425°C for 30 minutes. The dielectric constants at 0.1MHz of the cured and porous and non-porous films are about 2.4 and 2.65 respectively.

5. Results and Discussion

5.1. Sol-Gel Prepared Silica

The fractal geometry concept introduced by Mandelbrot was applied for the description of the structure of complex, disordered, porous materials [51]. The main concept in fractal geometry is the property of self-similarity in which objects appear invariant under transformation of the scale of observation. A fractal model is commonly used to describe porous materials made of aggregates. The SAXS intensity from a system of scatterers which exhibit the property of self-similarity can be described by [52]:

$$I(q) = I_0 q^{-\alpha} \tag{22}$$

where I_0 and α are constants. For mass fractals, $\alpha = D_m$ and surface fractals, $\alpha = 6 - D_s$. D_m and D_s are the mass and surface fractal dimension respectively. For mass fractals, $1 < \alpha < 3$ since $1 < D_m < 3$, while for surface fractals, $3 < \alpha < 4$ since $2 < D_s < 3$. Mass fractals have a non-uniform density distribution while surface fractals consist of two non-fractal regions separated by a fractal surface. A surface fractal is an indication of the roughness of the surface, where in porous materials larger pores may have on their surface smaller pores which in turn exhibit similar irregularities, for example giving a 'rough surface' down to atomic scale. A mass fractal is one where the mass within a spherical volume increases as a non-integer power of the radius. Many particle aggregates show this behavior and it is possible that either the network of pores running through a material or the matrix material itself may resemble such a structure. In the present work, the scattered intensity of the sol-gel silica samples has been first analyzed by the power law to determine its fractal dimension. Next, the data was fitted by the Debye-Anderson-Brumburger fractal model, which is commonly adapted to describe porous materials made of aggregates [53]. This model was developed for materials consisting of two strongly segregated, interpenetrating phases and it assumes a random distribution of pores both in size and shape at any scale [53].

The scattering curves of the sol-gel silica samples were fitted according to equation 22 by a least square fit (Figure 6). The α values obtained for all of the samples lie between 1.157 and 2.163, indicating that the pore structure in the film is of the mass fractal type. This type of fractal represents materials with porosity, usually much greater than 50% by volume [54]. The deviations from the power-law behaviour at high scattering vector region is the result of scattering structure at larger length scales compared to the size of the porous structure and might be indicative of cavities in the porous structure. In such cases, the solid phase is very often an aggregate of primary particles formed during aggregation or gelation process. The aggregate of mass objects can be described by a mass fractal. The mass fractal dimension D_m increases with treatment temperature and shows that annealing the film at higher temperature leads to the formation of a more branched, polymeric type structure.

A rather low fractal dimension was observed when the film was annealed at 300 °C, having a fractal dimension of 1.44. Hence, the system may consist of a network of air-filled channels in a solid matrix with an apparent density larger than unity [55].

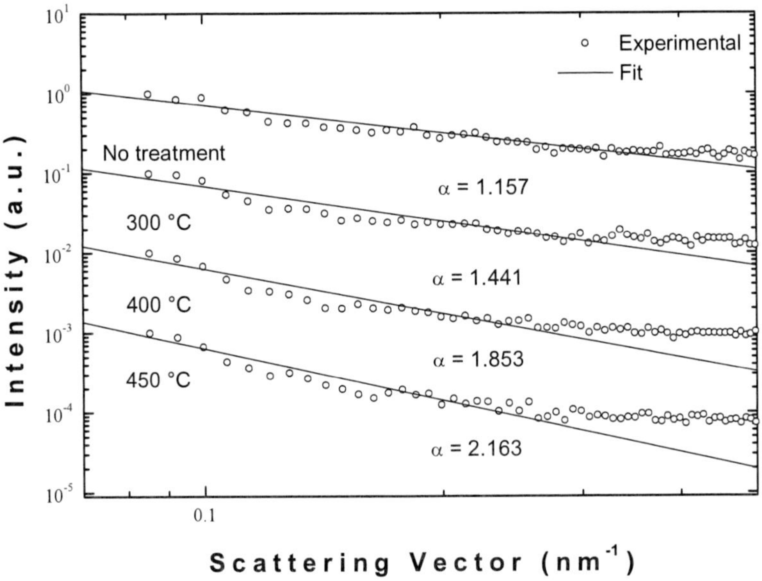

Figure 6. Sol-gel samples and modeling fits.

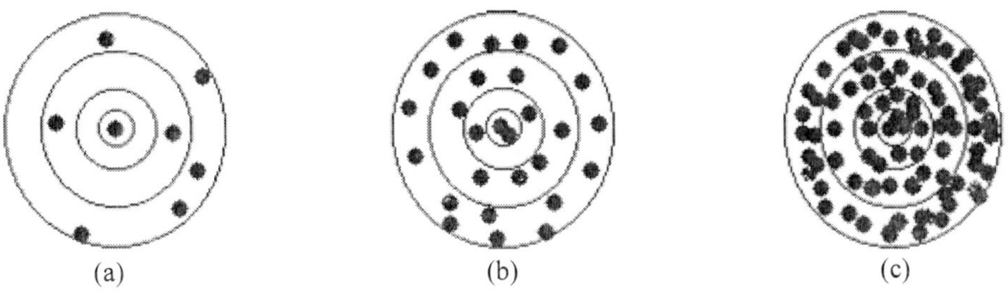

Figure 7. Two dimensional schematic representation of the distribution of pores as a function of the radius for various fractal dimension of: (a) 1, (b) 1.5and (c) 2.

The low fractal dimension of scattering is caused by scattering of electrons by pores and thus at these length scales the scattering data provide information about the pore structure in the film. Figure 7 illustrates the development of the numbers of pores as a function of the particle radius for three different fractal dimensions [54]. Clearly, number of pores increases with the fractal dimension. Each black dot represents a pore and the circles indicate different radii for each fractal dimension [54].

The scattering curves did not show a shoulder, which is an indication of scattering by micropores [56-58]. Hence, to determine the intensity scattered by the pore interface, the intensity can be fitted by the Debye-Anderson-Brumburger equation [58]:

$$I(q) = \frac{A}{\left(1 + q^2 a_c^{\ 2}\right)^2} \tag{23}$$

where A is a constant and a_c is the correlation length. The values found from linear regression are given in Table 4.2 together with I_0 and α. The numerical fits to the experimental data are given in Figure 8.

The correlation length is the distance from a point beyond which there is no further correlation of a physical property associated with that point. Values for a given property at distances beyond the correlation length can be considered purely random. On the other hand, the greater the fractal dimension, the farther a particular volume separates between two points. This is because the fractal dimension increases toward the dimensionality of Euclidean space.

Table 2. Summary of fit parameters deduced from analysis of SAXS curves

Calcination temperature	I_0	α	A	a_c (nm)
No treatment	0.00345	1.157	0.770798	3.35431
300 °C	0.000894	1.441	1.07941	5.25021
400 °C	0.000123	1.853	2.28605	9.28465
450 °C	0.0000299	2.163	3.62238	11.5811

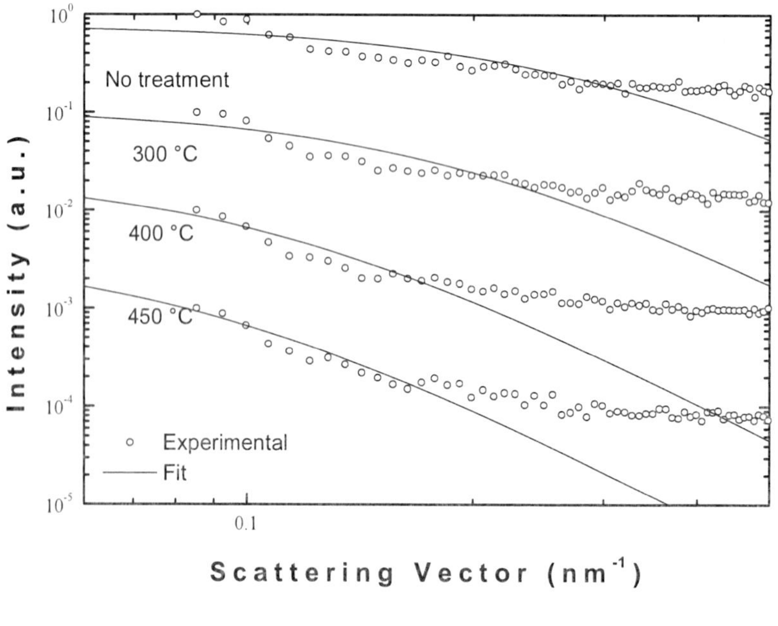

Figure 8. Fits to Debye-Anderson-Brumburger equation to determine correlation length.

The increase of both the fractal dimension D_m and correlation length a_c signaled an increase of pore dimensions and volumes upon annealing from 300 °C to 450 °C. The fractal dimension of films annealed at 450 °C has a value of 2.163, which corresponds to more compact aggregates formed. It is expected for an aggregated system that the fractal dimension increases toward dimensionality of Euclidean space, $d = 3$, when the aggregates become more and more entangled upon annealing at higher temperature.

Based on the Debye model, several other characteristics of the sol-gel silica film can be readily obtained. The specific surface area S can be calculated from from the porosity P and correlation length a_c through [53,59]:

$$S = 4P(1 - P)/a_c \tag{24}$$

and the average size of the aggregates l_c is obtained from:

$$\begin{aligned} l_c &= 2\int_0^\infty \gamma(r)dr \\ &= 2a_c \end{aligned} \tag{25}$$

where $\gamma(r)$ is the characteristic function of a fractal object. The estimated porosity from separate X-ray reflectivity data reported in [60] was about 0.57. With this value, the specific surface and average size of aggregates are listed in Table 3.

Table 3. The specific surface and average size of aggregates deduced for the film

Sample	Average size of aggregates (nm)	Specific surface/volume (nm^{-1})
No treatment	6.70862	0.29228
300 °C	10.50042	0.18674
400 °C	18.5693	0.10559
450 °C	23.1622	0.08466

5.2. Porous Silk

The porous SiLK samples showed circulary symmetric small angle scattering pattern. Hence, circular averaging could be performed on the scattering pattern to obtain the scattering intensity, plotted as a function of the momentum transfer q. Figure 9 shows the scattering curve of a porous SiLK film. A dense SiLK scattering curve is also given in Figure 10 for comparison. The curve was obtained by subtraction of the scattering curve of a bare silicon substrate of the same thickness from the raw scattering data.

A scattering experiment involves a few precise stages in acquisition and treatment of data which are essential. First, the scattering data below $q = 0.05$ nm^{-1} cannot be obtained due to the beam stop on the detector. The scattering intensity decreases with q and peaks can be seen at certain values of q. A peak in a scattering profile indicates the presence of a two phase structure.

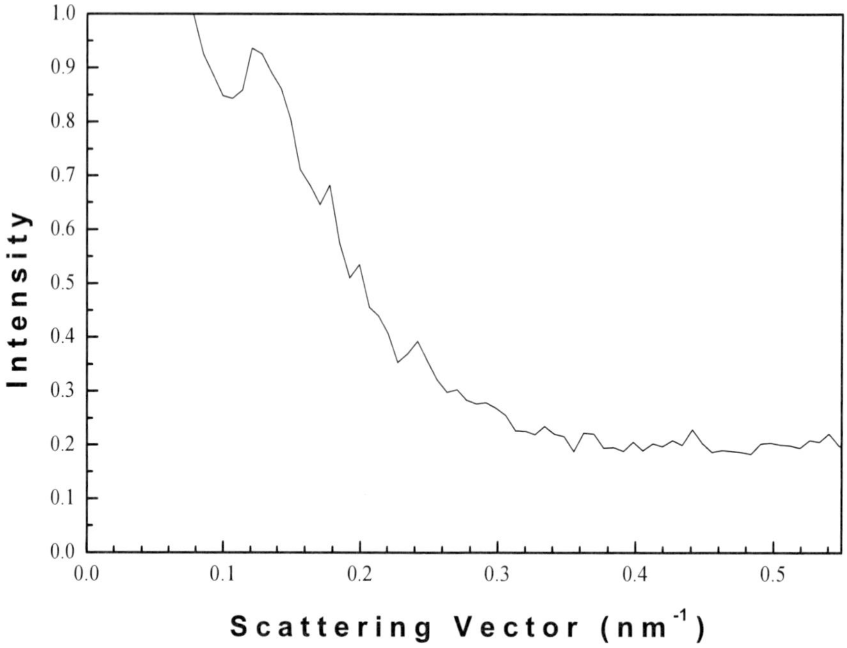

Figure 9. SAXS scattering curve of Porous SiLK.

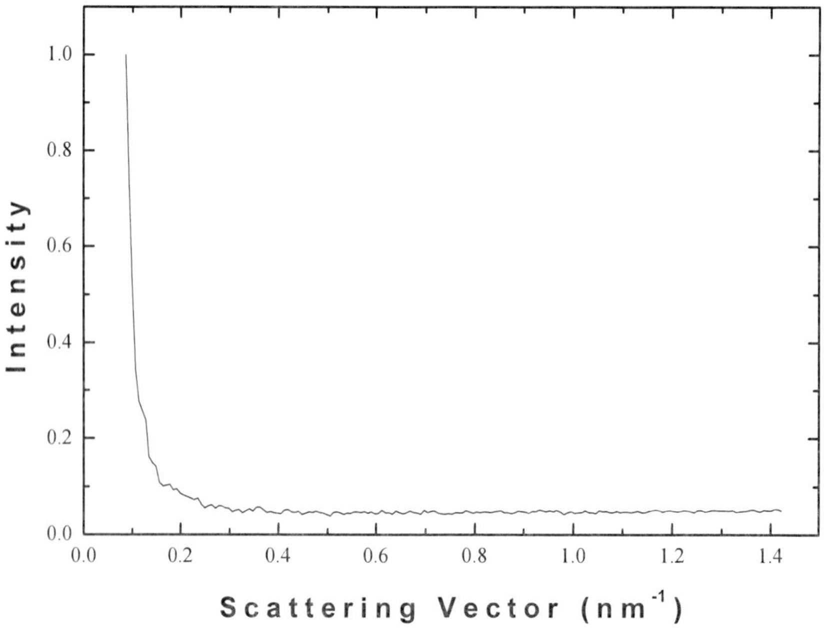

Figure 10. SAXS scattering curve of Dense SiLK.

Hence, the existence of peaks can be attributed to the voids present within the matrix of the film, as shown in the bright field cross sectional transmission electron micrograph (TEM) of porous SiLK in Figure 11. The brighter areas in this image correspond to regions of lower electron density, which is the pore. The typical pore size is found to be in the range of 15 nm to 25 nm. With the aid of this image, a structural model based on spherical pores has been developed to fit the scattering curve. The flow chart of the fitting is given in Figure 12.

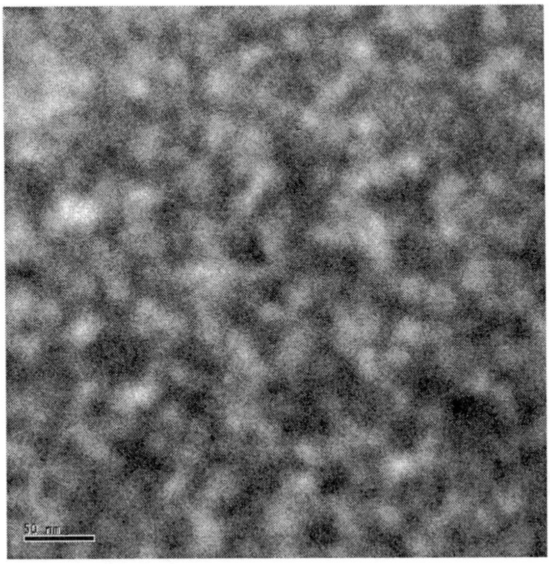

Figure 11. TEM image of porous SiLK.

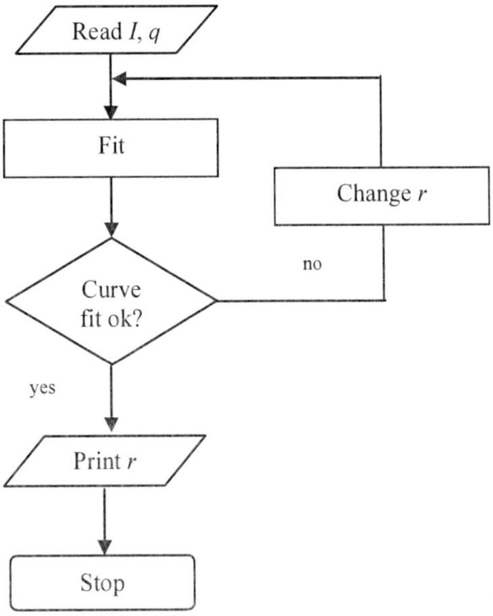

Figure 12. Flow chart of how fitting was performed. r is the average radius of the pores.

5.2.1. Development of Porous Silktm Structural Model

From the TEM image, the most appropriate scattering model to start with is the spherical model in which most of the pores are considered to have spherical shape. The curve was fitted with this equation:

$$I(q) = Nv^2\left(3\frac{\sin(qR) - qR\cos(qR)}{(qR)^3}\right)^2 \tag{26}$$

where v is the volume of the sphere and N is the number of scatterers. The fitted and experimental scattering curves are shown in Figure 13.

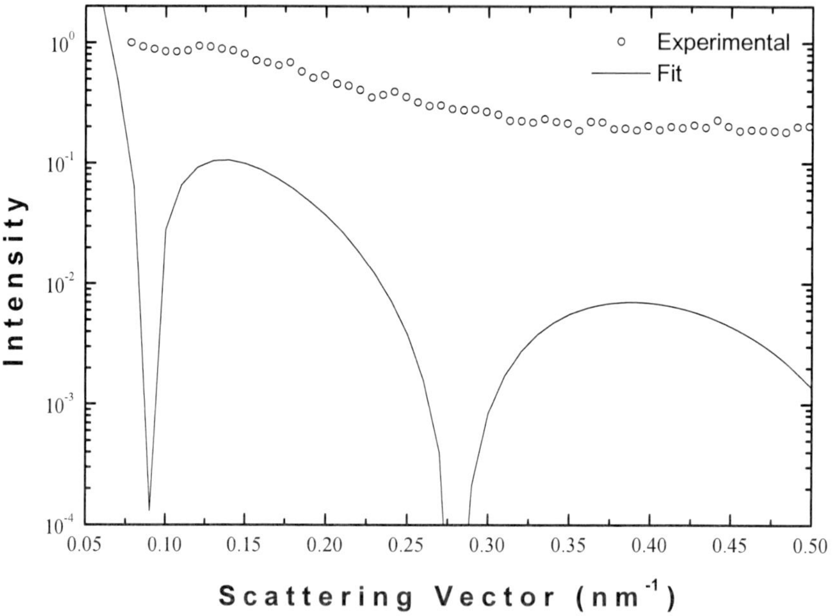

Figure 13. Fitted scattering curve according to spherical form factor.

The fitted value for Nv^2 and R are 2.48×10^{-11} cm^{-1} and 11.18 nm respectively. The main peak at 0.121 nm^{-1} was matched quite satisfactorily. However, the overall curve is not very well matched since this model only considered an average pore size for all of the pores. The scattering curve did not show oscillations after 0.25 nm^{-1} and suggests the possibility of a size distribution in the system [61]. In addition, a certain size distribution is evident from the TEM image in Figure 11. Influence of the shape and size distribution may be accounted by combining equations 15 and 16, as:

$$I(q) = a\int_0^\infty\left(3\frac{\sin(qR) - qR\cos(qR)}{(qR)^3}\right)^2 \frac{1}{r_0\sqrt{2\pi\sigma^2}\exp(0.5\sigma^2)}\exp\left(-\frac{\ln(R/r_0)^2}{2\sigma^2}\right)dR \tag{27}$$

where a is a constant to replace Nv^2. The fitted curve is given in Figure 14. The fitted value for a, r_0 and σ are $2.46\text{x}10^{-9}$ cm^{-1}, 5.77 nm and 0.19 nm respectively. The fitted curve does not exhibit the expected oscillations in the SAXS data, at both the high q and low q region, even at a low polydispersity level (defined as σ/r_0) of 3.5. This observation showed that the scattering pattern of this porous film cannot be due to a collection of one single size scatterer with its own distribution, but contributions from two significant or average sizes of scatterers. In fact, in one prior work by IBM on a similar type of low-k film, it was shown that there exists defect or killer pores with size in the range of 50 nm to 70 nm [62]. Using the above information, we modified the scattering intensity formula to take into account of killer and normal pores:

$$I(q) = a\left(3\frac{\sin(qR_1) - qR\cos(qR_1)}{(qR_1)^3}\right)^2 + b\left(3\frac{\sin(qR_2) - qR\cos(qR_2)}{(qR_2)^3}\right)^2 + c \quad (28)$$

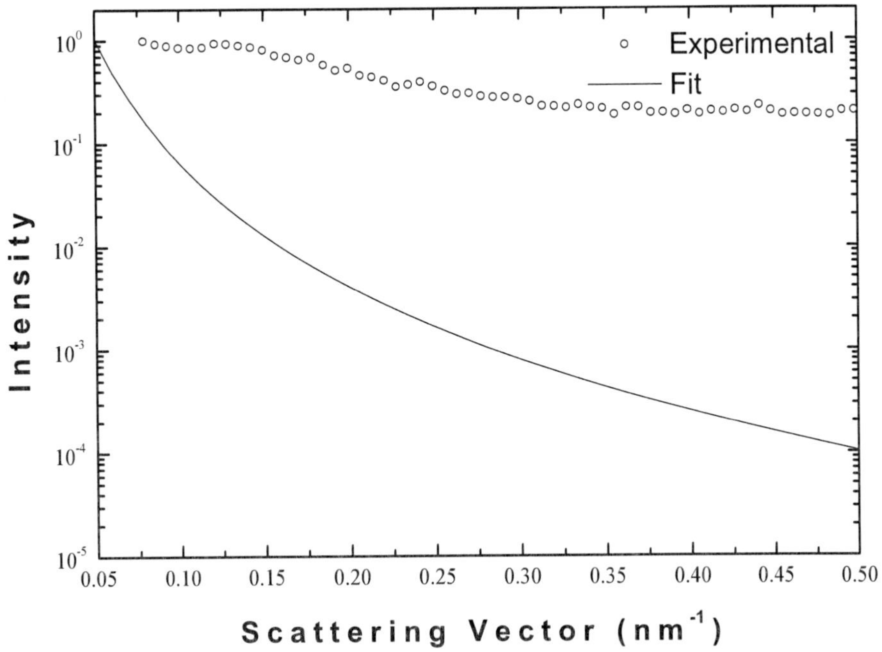

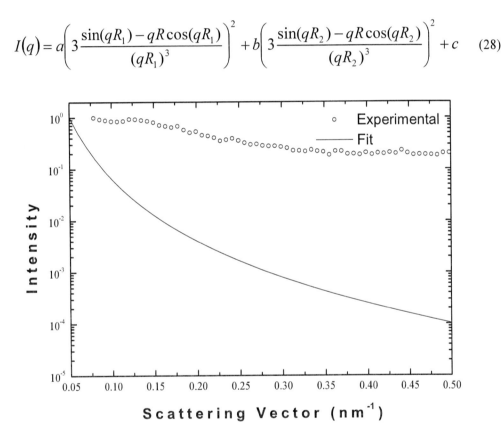

Figure 14. Fitted scattering curve according to spherical form factor and a size distribution.

where a is the constant to replace Nv^2 for one collection of average pore size, b is the constant to replace Nv^2 for another collection of average pore size, for easier identification. R_1 and R_2 are the average radii of the collections of killer and normal pores, respectively. c is the total scattering contributions from other sizes of pores and imperfections of background subtraction. This model comprises of two form factors (bi-form factor) to account for the scattering contribution of killer pores (larger pores) and average pores. On the other hand, constant a and b give us information on population of pores in the film. The fitted results are given in Figure 15. The fitted value for a, R_1, b, R_2 and c are 0.00089 cm^{-1}, 25.07 nm, 2.91 cm^{-1}, 11.8 nm and 0.231 cm^{-1} respectively.

Figure 15 shows a satisfactory fit for the scattering curve compared to previous models. The first peak at $q = 0.12$ nm^{-1} was matched successfully and the fitted curve resembles the pattern of the scattering curve. This was accomplished in a nonlinear least square iterative by normalizing the scattering curve and adjust the mean (average) pore size of the calculated curves to align with the first maximum of the scattering curve.

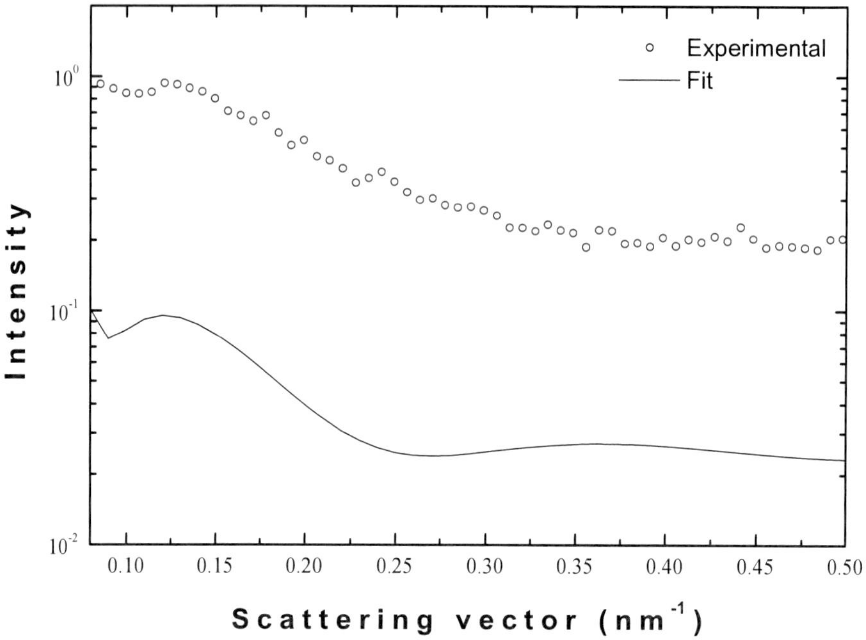

Figure 15. Fitted scattering curve according to two sphere form factors.

The constant *a* has a lower value compared to the constant *b* indicating that the volume and number of killer pores in the film are much lower than the normal pores. Due to the transmission geometry of SAXS and relatively weaker Cu K_α source used in this experiment, the scattering in high *q* region was limited for our sample. If a stronger source such as synchrotron radiation facility can be employed, the quality of our scattering data will definitely be better and easier for interpretation. Nevertheless, the main peak position is reasonably well matched, except for some peaks due to intensity fluctuations and imperfection in background subtraction. The results obtained are different from the work of Huang *et al.* on methyl silsesquioxane films [63]. First, the fringes observed in our scattering curve prove that there is a relatively small and narrow pore size distribution. In fact, when the pore size distribution function was included by numerical integration and fitted against the experimental data [64], the scattering oscillation disappeared, at polydispersity levels as low as 3.5%, as demonstrated in Figure 14. With respect to this observation, the constants *a* and *b*, which are representative of the "population" of pores in the film can be used to estimate the fraction of killer pores in the film. The *b*:*a* ratio was found to be 3226:1, indicating the population of killer pores is very low in the film. Secondly, Huang *et al.* fitted their scattering curves in the *q* region near the drop off in intensity, typically at $q = 0.3$ nm^{-1} to $q = 10$ nm^{-1} [63]. No attempt of fit was performed below this *q* value. This is because their facility had a much wider q detection range to detect smaller pores while the maximum pore size of their sample was typically in the range of 4 nm to 10 nm. However the size of the pores created in our film were generally larger than 12 nm and previous study showed that detection of larger pores is important in low q region ($q < 0.3$ nm^{-1}) [65]. The detection ability of a scattering system can be deduced from $\pi / 2R$ (denoted as q^*). From the observation of TEM image and computation of q^*, this scattering vector in the q range of 0.126 nm^{-1} is of most relevance to our sample. Our fitting and scattering curve (Figure 15) proved the validity of this assumption. Hence, it is important to fit our data in the low q region rather than the high q region. Thirdly, fringes in the scattering curve that are inherent in the spherical form factor are an indication that a system is a well-defined one. Some peaks at q values greater than 0.3 nm^{-1} cannot be well matched and subsequently the curve appear relatively flat. We attribute these discrepancies to the following factors: (1) the pore shape is not perfectly spherical and (2) lack of well defined scattering data at $q > 0.4$ nm^{-1}, due to the low intensity tube source. The quality of data will be improved if a stronger source is used. This can also reduce the scattering time required. Thirdly, transmission SAXS is usually performed on bulk homogeneous samples. For thin film deposited on silicon substrate, the films must first be backgrinded to reduce the attenuations and a discrepancy may arise during the background subtraction. In view of this, the constant c was added to account for these imperfections in our fitting to the scattering experiments. The challenges highlighted above had been pointed out by Omote *et al.* [33], where they measured their silsesquioxane sample with reflection geometry mode of SAXS. He and co-workers mentioned that using conventional SAXS with transmission geometry is difficult, for thin films on substrates due to weak intensities. When a film is very thin, the resultant scattering intensity is very weak. In addition, a very high fraction of the scattering X-rays can be absorbed by a semiconducting substrate. Nevertheless, this method is feasible by reducing the substrate thickness with the knowledge of transmissivity.

The effect of interactions between pores is included next in the model. Normally in the case of a two-phase system, where pores are dispersed in a matrix of different density, the intensity is given by the product of the contributions from the pores and the interaction term. The pore can be assumed to have a uniform radius R and interact with neighnours with a 'hard-sphere' interaction. In this model, the radius in the Percus-Yevick structure factor (refer to equation 17) is assumed not to be identical with radius in the spherical the form factor (refer to equation 16) [66]. This difference is interpreted by existence of a shell around the spherical pores made by molecules of polymer present in the system [67,68]. Hence in the following, the hard sphere interaction radius and spherical pore radius will be treated as different parameters denoted as R_{HS} and R, respectively, where $R_{HS} > R$. The formula related to the model described above is given as follows:

$$I(q, R_1, R_{HS1}, R_2, R_{HS2}, \phi_1, \phi_2) = aP(q, R_1)S(q, R_{HS1}, \phi_1) + bP(q, R_2)S(q, R_{HS2}, \phi_2) + c \tag{29}$$

For comparison, the model with $R = R_{HS}$ is written as:

$$I(q, R_1, R_2, \phi_1, \phi_2) = aP(q, R_1)S(q, R_1, \phi_1) + bP(q, R_2)S(q, R_2, \phi_2) + c \tag{30}$$

The fitting results are summarized in Table 4 and the relevant fits are given in Figure 16.

Table 4. Fitted parameters of hard-sphere model

Model	a ($\times 10^{-4}$ cm^{-1})	b (cm^{-1})	R_1 (nm)	R_2 (nm)	R_{HS1} (nm)	R_{HS2} (nm)	ϕ_1	ϕ_2 ($\times 10^{-4}$)	C (cm^{-1})
$R = R_{HS}$	7.5	2.63	26.27	11.60	-	-	0.326	1.04	0.26
$R < R_{HS}$	1.77	1.06	26.15	11.57	27.08	12.58	0.323	1.01	0.27

The fits in the vicinity of the scattering peak are satisfactory. The R and R_{HS} values obtained agree with the values found in the TEM images. The packing density of the voids or the overall pore volume fraction ϕ, was found to be ~ 32% for both models. The values obtained are lower than the annealed samples in the X-ray reflectivity experiment but are in reasonable agreement for the as-cured porous SiLK sample.

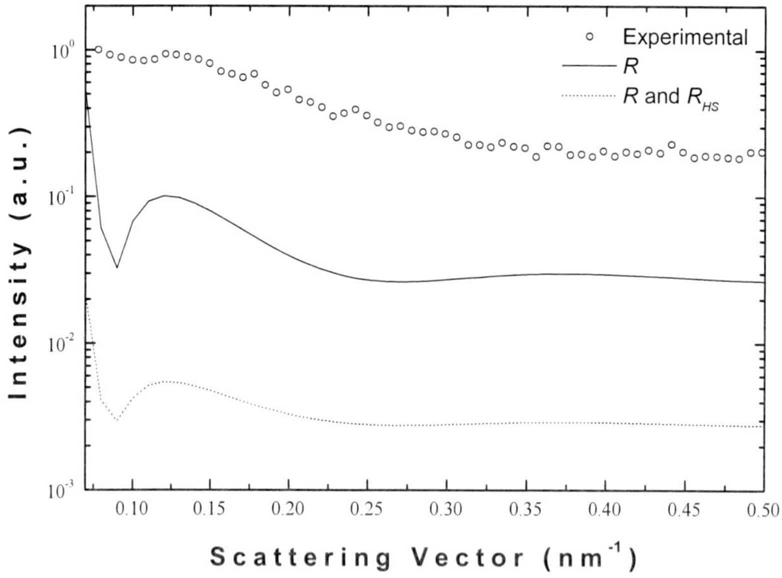

Figure 16. Fitted scattering curve according to hard sphere models.

CONCLUSION

SAXS is feasible to be used for pore characterization of porous dielectric materials if suitable structural models were developed. The pore size and volume fraction of the pores can be determined from suitable structural model. Structural models can only be developed with knowledge or understanding of the sample through some other complementary background data obtained TEM or SXR. TEM gives us the real image of the sample while SXR provides information about volume fraction. However, this method is limited by long scattering time, scattering vector range and sample nature. A more efficient way is to employ synchrotron radiation sources with an adjustable sample-to-detector distance setup to solve the first two limitations. The thick silicon substrate is not suitable for transmission and can only be performed in reflection geometry [66, 69]. Employing a thin silicon substrate (~ 80 μm to 100 μm) for the transmission scattering experiment [63] is not the solution for a wafer foundry because of the possibility of wafer breaking.

For the sol-gel prepared silica, the fractal size and average size of aggregates increase with calcination temperature. Various models were employed to fit the scattering data of porous SiLK. The pore radius and packing density data deduced were in agreement with other supporting characterization techniques such as TEM and SXR. However, the proposed model is material or pore type sensitive and cannot be applied for all samples. The TEM image shows an average pore size of 22 nm, while fitting of scattering curve employing several models yield value of 22.2 nm to 25 nm. The bi-form factor model take into the consideration of killer pores exist in the film. The ratio of killer pore to normal pore can be found from the constant a and b in the model. A constant c was included in the model for scattering contributed by other means. The structural factor takes into account of the interference effects between particle, the fitted packing density parameter is found to be reasonable with reference to separate specular X-ray reflectivity experiments.

Finally, we would like to point out that the versatile SAXS technique is applicable not just to back end of line processes. In reference [70], Stemmer et.al. had demonstrated the use of SAXS to elucidate a phase separation phenomenon in hafnium silicate which is used as a high permittivty gate dielectric in CMOS transistors.

REFERENCES

[1] Glatter, O; Kratky, O. *Small angle x-ray scattering*, Academic Press: New York, NY, 1982; pp 1-13.

[2] Fratzel, P. *J. Appl. Cryst.* 2003, 36, 397-404.

[3] Metzger, T.H.; Kegel, I.; Paniago, R.; Lorke, A.; Peisl J.; Schulze, J.; Eisele, I.; Schittenhelm, P.; Abstreiter, G. *Thin Solid Films* 1999, 336, 1-8.

[4] Sasaki, A. Rigaku J. 2005, 22, 31-38.

[5] Wang, C.; Choi, K.W.; Fu, W.E.; Ho, D.L.; Jones, R.L.; Soles, C.; Lin, E.K.; Wu, W.L.; Clarke, J.S.; Bunday, B. *Proc. SPIE* 2007, 6922, 69222E1-7.

[6] Wang, C.; Choi, K.W.; Jones, R.L.; Soles, C.; Lin, E.K.; Li, W.L.; Clarke, J.S.; Villarrubia, J.S.; Bunday, B. *Proc. SPIE* 2007, 6922, 69221Z1-8.

[7] Wang, C.; Jones, R.L.; Lin, E.K.; Wu, W.L.; Leu, J. *Appl. Phys. Lett.* 2007, 90, 193122-1-8.

[8] Ito, Y.; Inaba, K.; Omote, K.; Wada, Y.; Ikeda, S. J. *J. Appl. Phys.* 2007, 46, L773-L775.

[9] Boxberg, F.; Tulkki, J. In *The Handbook of Nanotechnology, Nanometer Structures, Theory, Modeling and Simulation;* Lakhtakia, A.; Ed.; SPIE Press: Bellingham, WA, 2004; pp 107-143.

[10] Park, S.; Cho, E.; Song, D.; Conibeer, G.; Green, M.A. *Sol. Energy Mater. Sol. Cells.* 2007, 93, 684-689.

[11] Konig, D.; Rudd, J.; Green, M.A.; Conibeer, G. *Sol. Energy Mater. Sol. Cells.* 2007, 93, 753-758.

[12] Luque, A.; Marti, A.; Nozik, A.J. *MRS Bull.* 2007, 32, 236-241.

[13] Nagao, O.; Harada, H.; Sugawara, T.; Sasaki, A.; Ito, Y. J. *J. Appl. Phys.* 2004, 43, 7742-7746.
[14] Zolotaryov, A.; Schramm, A.; Heyn, Ch.; Zozulya, A.; Hansen, W. *Mater. Sci. Semicond. Proc.* 2009, 12, 75-81.
[15] Hanke, M.; Dubslatt, M.; Schmidbauer, M.; Wang, Zn.M.; Mazur, Yu. I.; Lytvyn, P.M.; Lee, J.H.; Salamo, G.J. *Appl. Phys. Lett.* 2009, 95, 023103-1-3.
[16] Holy, V.; Strangl, J.; Lechner, R.T.; Springholz, G. *J. Phys. Condens. Matter* 2008, 20, 454215.
[17] Sun, W.C.; Chang, H.C.; Wu, B.K.; Chen, Y.R.; Chu, C.H.; Chang, S.L.; Hong, M.; Tang, M.T.; Stetsko, Yu. P. *Appl. Phys. Lett.* 2006, 89, 0919151-1-3.
[18] Dujardin, R.; Poydenot, V.; Uchulli, T.U.; Renaud, G.; Ulrich, O.; Barski, A.; Derivaz, M.; Colonna, S.; Metzger, T. *J. Appl. Phys.* 2006, 99, 063510-1-7.
[19] Flege, J.I.; Schmidt, Th.; Aleksandrovic, V.; Alexe, G.; Clausen, T.; Gehl, B.; Kornnowski, A.; Bernstorff, S.; Weller, H.; Falta, J. *Nucl. Instru. Meth. Phys. Res. B* 2006, 246, 25-29.
[20] Leroy, F.; Eymery, J.; Buttard, D.; Renaud, G.; Lazzari, R. *J. Cryst. Growth* 2005, 275, e2195-e2200.
[21] Grill, A. *Annu. Rev. Mater. Res.* 2009, 39, 49-70.
[22] Maex, K.; Baklanov, M.R.; Shamiryan, D.; Iacopi, F.; Brongersma, S.H.; Yanovitskaya, Z.S. *J. Appl. Phys.* 2003, 93, 8793-8841.
[23] Dulstev, F.N.; Baklanov, M.R. *Electrochem. Solid-State Lett.* 1999, 2 192-194.
[24] Baklanov, M.R.; Mogilnikov, K.P.; Polovinkin, V.G.; Dulstev, F.N. J. *Vac. Sci. Technol.* 2000, B18, 1385-1391.
[25] Sun, J.N.; Gidley, D.W.; Dull, T.L.; Frieze, W.F.; Yee, A.F.; Ryan, E.T.; Lin, S.; Wetzel, J. *J. Appl. Phys.* 2001, 89, 5138-5144.
[26] Chen, H.J.; Li, S.Y.; Liu, X.J.; Li, R.P.; Smilgies, D.M.; Wu, Z.H.; Li, Z. *J. Phys. Chem. B* 2009, 113, 12623-12627.
[27] Jousseaume, V.; Gourhant, O.; Zenasni, Z.; Maret, M.; Simon, J.P. *Appl. Phys. Lett.* 2009, 95, 022901-1-3.
[28] Jin, K.S.; Heo, H.; Oh, W.; Yoon, J.; Lee, B.; Hwang, Y.; Kim, K.S.; Park, Y.H.; Kim, K.W.; Kim, J.; Chang, T.; Ree, M. *J. Appl. Cryst.* 2007, 40, s631-s636.
[29] Simon, J.P.; Jousseaume, V.; Rolland, G. *J. Appl. Cryst.* 2007, 40, s363-s366.
[30] Suzuki, T.; Omote, K.; Ito, Y.; Hirosawa, I.; Nakata, Y.; Sigura, I.; Shimizu, N.; Nakamura, T. *Thin Solid Films,* 2006, 515, 2410-2414.
[31] Dourdain, S.; Bardeau, J.F.; Colas, M.; Smarsly, B.; Mehdi, A.; Ocko, B.M.; Gibaud, A. *Appl. Phys. Lett.* 2005, 86, 113108-1-3.
[32] Hata, N.; Negoro, C.; Yamada, K.; Kikkawa, T. J. *J. Appl. Phys.* 2004, 43, 1323-1326.
[33] Omote, K.; Ito, Y.; Kawamura, S. *Appl. Phys. Lett.* 2003, 82, 544-546.
[34] Goh, T.K.; Wong, T.K.S. *Microelect. Eng.,* 2004, 75, 330-343.
[35] Whittingham, M.S. *MRS Bull.* 2008, 33, 411-420.
[36] Bowen, D.K.; Tanner, B.T. *High-resolution x-ray diffractometry and topography*; Taylor and Francis: London, 1998; pp 1-252.
[37] Cao, G. *Nanostructures & Nanomaterials synthesis, properties & applications;* Imperial College Press: London, 2004, pp 329-332.
[38] Cullity, B.D.; *Stock, S.R. Elements of x-ray diffraction, 3rd Edition*, Prentice Hall: Upper Saddle River, NJ, 2001, pp 1-664.

[39] Krishnamurti, P. *Indian J. Phys*. 1930, 5, 473-500.

[40] Guinier, A. *X-ray diffraction in crystals, imperfect crystals and amorphous bodies,* Dover: Mineola, NY, 1994; pp 319-350.

[41] Glatter, O. *Modern aspect of small-angle scattering*, Kluwer Academic Publishers: Boston, MA, 1995; pp 108.

[42] William, C.E.; Roland, P.M.; Guinier, A. *In X-ray characterization of materials,* Lifshin, E.; Ed. Wiley-VCH, New York, NY, 1999; pp 211-254.

[43] Debye, P. *Ann. Physik* 1915, 351, 809-823.

[44] Debye, P. *J. Phys. Colloid Chem*. 1947, 51, 18-32.

[45] Hasmy, A.; Foret, M.; Pelous. J.; Jullien, R. *Phys. Rev. B*. 1993, 48, 9345-9353.

[46] Megens, M.; van Kats, C.M.; Bosecke, P.; Vos, W.L.; *Langmuir* 1997, 13, 6120-6129.

[47] Debye, P.; Bueche, R.M. *J. Appl. Phys* 1949, 20, 518-525.

[48] Freltoft, T.; Kjems, J.K.; Sinha, S.K. *Phys. Rev. B* 1986, 33, 269-275.

[49] Thompson, A.; Attwood, D.; Gulikson, E.; Howells, M.; Kim, K.J.; Kirz, J.; Kortright, J.; Lindau, I.; Pianetta, F.; Robinson, A.; Schofield, J.; Underwood, J.; Vaughan, D.; Williams, G. *X-ray data booklet*, Lawrence Berkeley National Laboratory: Berkeley, CA. 2001; 1-38-1-43.

[50] D.K. Schroder, *Semiconductor material and device characterization*, Wiley Interscience: New York, NY, 2006; pp 585-591.

[51] Mandelbrot, B.B. *The Fractal Geometry of Nature*, W.H. Freeman: New York, NY, 1983; pp 1-468.

[52] Martin, J.E.; Hurd, A.J. *J. Appl. Cryst*. 1987, 20, 61-78.

[53] P. Debye, H.R. Anderson, Jr. and H. Brumberger, *J. Appl. Phys*. 1957, 28, 679-683.

[54] Dokter, W.H.; Beelen, T.P.M.; van Garderen, H.F.; van Santen, R.A. In *Characterization of Porous Solid III*; Rouquerol, J.; Rodriguez, F.; Sing, K.S.W.; Unger, K.K. Eds.; Elsevier: Amsterdam, 1994; pp 725-734.

[55] Zarzycki, J.; L.L. Hench and J.K. Wiley, Chem. Process. Adv. Mater. 1992. 77-92.

[56] Reichenauer, G.; Emmerling, A.; Fricke, J.; Pekala, R.W. *J. Non-Crsyt. Solids* 1998, 225, 210-214.

[57] Petricevic, R.; Reichenauer, G.; Bock, V.; Emmerling, A.; Fricke, J. *J. Non-Crsyt. Solids* 1998, 225, 41-45.

[58] Bock, V.; Emmerling, A.; Fricke, J. *J. Non-Crsyt. Solids* 1998, 225, 69-73.

[59] Valiev E.Z.; Bogdanov, S.G.; Pirogov, A.N.; Sharygin, L.M.; Barybin, V.I.; *J. Exp. Theo. Phys*. 1993, 76, 111.

[60] Yu, S.; Wong, T.K.S.; Pita, K.; Hu, X. *J. Electrochem. Soc*. 2003, 150, F116-F121.

[61] Rieker, T.; Hanprasopwattana, A.; Datye, A.; Hubbard, P. *Langmuir* 1999, 15, 638-641.

[62] Muraka, S.P.; Eizenberg, M.; Sinha, A.K. *Interlayer dielectrics for semiconductor technologies*, Elsevier: Amsterdam, 2003, pp. 37-76.

[63] Huang, E.; Toney, M.F.; Volksen, W.; Mecerreyes, D.; Brock, P.; Kim, H.C.; Hawker, C.J.; Hendrick, J.L.; Lee, V.Y.; Magbitang, T.; Milller, R.D.; Lurio, L.B. *Appl. Phys. Lett*. 2002, 81, 2232-2234.

[64] Press, W.H.; Teukolsky, S.A.; Vetterling, W.T.; Flannery, B.P. *Numerical recipes,* 2nd edition, Cambridge University Press: Cambridge, 1992, pp 656-706.

[65] Goh, T.K.; Yu, S.; Wong, T.K.S.; He, C. *Proc. ECS 202nd meeting*, 2002, EC PV2002-22, 176-186.

[66] Krakovsky, I.; Bubenikova, A.; Urakawa, H.; Kajiwara, K. *Polymer* 1997, 38, 3637-3643.

[67] Fournet, G. *Acta Crystall.* 1951, 4, 293-301.

[68] Kinning, D.J.; E.L. Thomas, E.L. *Macromol.* 1984, 17, 1712-1718.

[69] Hsu, C.H.; Lee, S.Y.; Liang, K.S.; Jeng, U.S.; Windover, D.; Lu, T.M.; Jin, C.; *Mat. Res. Soc. Symp. Proc.* 2000, 612, D.5.23.1- D.5.23.6.

[70] Stemmer, S.; Li, Y.; Foran, B.; Lysaght, P.S.; Streifer, S.K.; Fuoss, P.; Seifert, S. *Appl. Phys. Lett.* 2003, 83, 3141.

In: X-Ray Scattering
Editor: Christopher M. Bauwens
ISBN: 978-1-61324-326-8

Chapter 5

X-RAY SCATTERING OF BACTERIAL CELL WALL COMPOUNDS AND THEIR NEUTRALIZATION

***Michael Rappolt*[1]*, Manfred Rössle*[2]*, Yani Kaconis*[3]*, Jörg Howe*[3]*, Jörg Andrä*[3]*, Thomas Gutsmann*[3]*, and Klaus Brandenburg*[3]**

[1]Institute of Biophysics and Nanosystems Research, Austrian Academy of Sciences, c/o Sincrotrone Trieste, 34149 Basovizza, Italy

[2]European Molecular Biology Laboratory, Notkestr. 52, D-22603 Hamburg

[3]Forschungszentrum Borstel, Leibniz-Zentrum für Medizin und Biowissenschaften, Borstel, Germany

ABSTRACT

Bacterial infections are still a major concern of human health. Despite the existence of antibiotics (AB), the increasing occurrence of resistant strains and the inability of the AB to neutralize bacterial pathogenicity factors (PF), which are released from the bacterial cells, are responsible for the high death rate on intensive care units. For the major pathogenicity factors of Gram-negatives, endotoxins (lipopolysaccharides, LPS), and of Gram-positives, lipoproteins (LP), which may cause in patients the life-threatening septic syndrome, there is still no effective therapy available.

A new therapeutic approach to neutralize the PF is the use of suitable binding proteins from human or animal origin or peptides derived thereof (defense structures). In this way, the binding of the PF to receptors of the human immune system such as CD14 and the Toll-like receptors (TLR2 and 4), which represent the initial events of the inflammation reaction, may be inhibited competitively. The characterization of the binding process comprises, among others, the supramolecular structures of the PF. In the absence of the defense structures, cubic aggregates have been shown to be the active principle. In the presence of the defense structures, the PF aggregates are converted into a multi-lamellar organization which inhibit the binding to the cell receptors. The effect of a variety of human or animal proteins on the aggregate structures of the PF has been characterized. To this class belong albumin, hemoglobin, lactoferrin, Nk-lysin, and *Limulus* anti-LPS factor. It can be shown that the influence of these proteins on the aggregate structures of the PF is protein-specific: whereas albumin or hemoglobin do not change the aggregate structures or even convert them into a different type of cubic organization, the other proteins and the peptides derived from them lead to a multilamellar aggregate structure of the PF. These results directly correlate with the biological data, leading to no change or even to an increase in bioactivity for the former, but to a decrease for the latter. By constructing suitable peptides with high binding affinity to the PF, it is possible to obtain defense structures which are able to completely down-regulate the bioactivity of the PF, making them suitable effective compounds to combat bacterial infections in general and sepsis in particular.

Keywords: SAXS, endotoxins, antimicrobial peptides AMP, bacterial outer membrane.

INTRODUCTION

Infectious diseases are still a major threat tohuman health worldwide, and are the most frequent causes of death in the 'third world' states. Most of the infections are elicited by micro-organisms, bacteria and viruses, most of them being ubiquitous in the environment. When a viral is followed by a bacterial infection, one speaks of superinfection. To this category belongs the 'Spanish flu' from the 1920's with at least 30 million deaths (21). Interestingly, in this global pandemic infection, elicited by the H1N1 virus, the bacterial infection was the main reason of lethality, because the ill people died from septic shock rather than from the primary viral infection.

Sepsis belongs to the most severe infectious diseases and are elicited on most cases not from the bacteria but from their released pathogenicity factors (PF). The PF are liberated from the bacteria due to the action of the immune system, by antibiotics, or solely due to cell division. Regarding the chemical structures, they are lipopolysaccharides (LPS, endotoxins) in the case of Gram-negative and lipoproteins for Gram-positive bacteria (8, 9). The PF interact with the immune system by binding to serum- or membrane proteins such as lipopolysaccharide-binding protein (LBP) or CD14, and with the TLR-4/MD2 receptor system finally eliciting the inflammation reaction (13),(22).

To understand these mechanisms, it is of utmost importance to understand their molecular conformation responsible for the biological response. One part of this overview deals with the structural prerequisites for biological activity, mainly focused on bacterial LPS.

Since bacterial infections still represent a major problem of human health, the inactivation of the PF by suitable agents remains an important task. One promising approach is the use of antimicrobial peptides (AMP) in most cases based on LPS-binding sequences of human or animal binding proteins such as lactoferrin, NK-lysin, or *Limulus* anti-LPS factor (1-4, 23). Some progress was achieved, however, most investigations were directed to construct AMP effective against the bacteria, and not against the PF. To overcome this problem, in recent investigations, the design and synthesis of suitable anti-LPS peptides became more and more relevant. Therefore, the second part of this overview deals with the structural prerequisites of the inactivation of LPS by these peptides.

X-RAY SCATTERING EXPERIMENTS AND ANALYSIS

Small angle X-ray scattering measurements of endotoxins and endotoxin:peptide mixtures were performed at the European Molecular Biology Laboratory (EMBL) outstation at the Hamburg synchrotron radiation facility HASYLAB using the double-focusing monochromator-mirror camera X33 (11). Diffraction patterns in the range of the scattering vector $0.3 < s < 0.90$ nm^{-1} ($s = 2 \sin \theta/\lambda$, 2θ scattering angle and applying a wavelength $\lambda = 0.15$ nm) were recorded at different temperatures with exposure times of 1 min using an image plate detector with online readout (MAR345, MarResearch, Norderstedt/Germany) (19). The s-axis was calibrated with Ag-behenate, which has a periodicity of 5.84 nm. The diffraction patterns were evaluated assigning the spacing ratios of the main scattering maxima to defined three-dimensional structures. The lamellar and inverted hexagonal structures H_{II} structures are the most relevant in the present study. Their reflection laws are the following, see e.g., (14):

(1) L $(d/d_h) = 1, 2, 3, 4, 5...$ d is the lattice repeat distance and $d_{hkl} = 1/s_{hkl}$
(2) H_{II} $(\sqrt{3}a/2d_{hk}))^2 = 1,3,4,7,9,12,13...$

Electron density profiles of the L_β - phases as well as 2D maps of the H_{II} - phase were derived from the small-angle X-ray diffraction pattern by standard procedures (for details see (12), (16), (17)). The electron density contrast was calculated by the Fourier synthesis

$$\tilde{\rho}(\vec{r}) = \sum_{h,k \neq (0,0)}^{h,k\,\max} \alpha_{h,k} \cdot F_{h,k} \cdot \cos\left(2\pi\, \vec{s}_{h,k}\, \vec{r}\right), \quad (1)$$

where $F(h,k)$ is the amplitude of the peak at the position $\vec{s}(h,k)$, h, k are the Miller indices and □(h,k) is its corresponding phase. For centrosymmetric structures, as in this study, the phases for each diffraction order are either '+1' or '–1'. The phase combination (- - + - -) for the lamellar gel phase was taken from literature (Levine 1973), and for the inverted hexagonal phase concerning the (1,0), (1,1), (2,0), (3,0), (2,2), (3,1) and (4,0) reflections the phase combination (+ - - - - + +) was determined. Here, the phases of the first three orders were taken from literature (12), while all of the phase combinations of the weak reflections were checked. The electron density maps presented show the smallest radial deviations in the position of the phosphate groups, the smoothest maximum electron density distribution, and further, the void regions coincide with the corners of the Wigner-Seitz cell.

The most trustworthy information from electron density maps of lipid/water systems is the location of the molecular groups with highest atomic number, i.e. the position of the phosphate groups. Thus, in this overview, the bilayer thickness is simply given as the phosphate to phosphate group distance d_{PP}, and consequently, the water layer thickness, d_W, can be defined as

$$d_W = d - d_{PP}, \quad (2)$$

where d is the repeat distance of the lamellar lattice. Corresponding structural parameters are given also for the inverse hexagonal lattice:

$$d_L = a - D_{PP}, \quad (3$$

where d_L is the minimum thickness of opposed monolayers, a is the unit cell parameter of the inverse hexagonal cell, and D_{PP} is the water core diameter defined as two times the mean distance from the center of the rod to the phosphate groups.

Structural Prerequisites of Endotoxin Activity

Since the 1980's, some work has been done to characterize the physicochemical properties of bacterial lipopolysaccharides. It was shown that lipid A, the molecular part of LPS anchoring it into the outer membrane of the bacteria, represents its 'endotoxic principle'. The chemical structure(s) of typical enterobacterial LPS, in this case, those from *Salmonella minnesota*, is shown in Figure 1. Its essential part is the hexaacylated bisphosphorypated diglucosamine backbone, to which some non-stoichiometric groups are linked (25). The lipid A structures of other Enterobacteriaceae, such as that from *Escherichia coli,* are more or less identical. It was shown that the three-dimensional supramolecular structure of lipid A/LPS (= endotoxins) is a determinant for its biological activity, i.e., its ability to induce cytokines such as tumor-necrosis-factor α or interleukins in human mononuclear cells (20). Thus, the biological activity is high for endotoxins which adopt an inverted cubic or inverted hexagonal (non-lamellar) aggregate structure, is lower for samples adopting mixed lamellar/non-lamellar structures, and absent for samples adopting multi-lamellar structures (8).

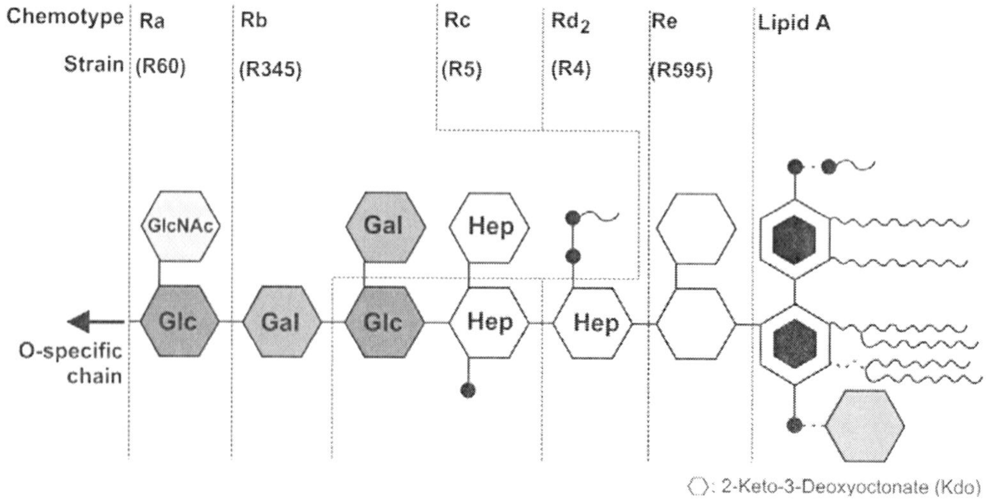

Figure 1. Chemical structures of typical enterobacterial LPS from Salmonella minnesota.

In Figure 2, structural preferences for some lipid A samples are presented, complete heterogeneous lipid A from *E. coli*, its purified hexaacyl and pentaacyl part, and synthetic tetraacyl compound '406' (7). As can be seen from the small-angle X-ray scattering (SAXS) patterns, there are considerable differences. Most probably these differences are mainly due to different hydration levels of the samples. The water content is definitely the lowest in the lipid A *E.-coli* sample, since here a rhombohedral R$\underline{3}$m phase with the lattice parameters a = 13.2 nm and c = 12.3 nm is apparent. For indexing of the Bragg reflections, refer to the inset. This phase has been well-described elsewhere (15), (24).

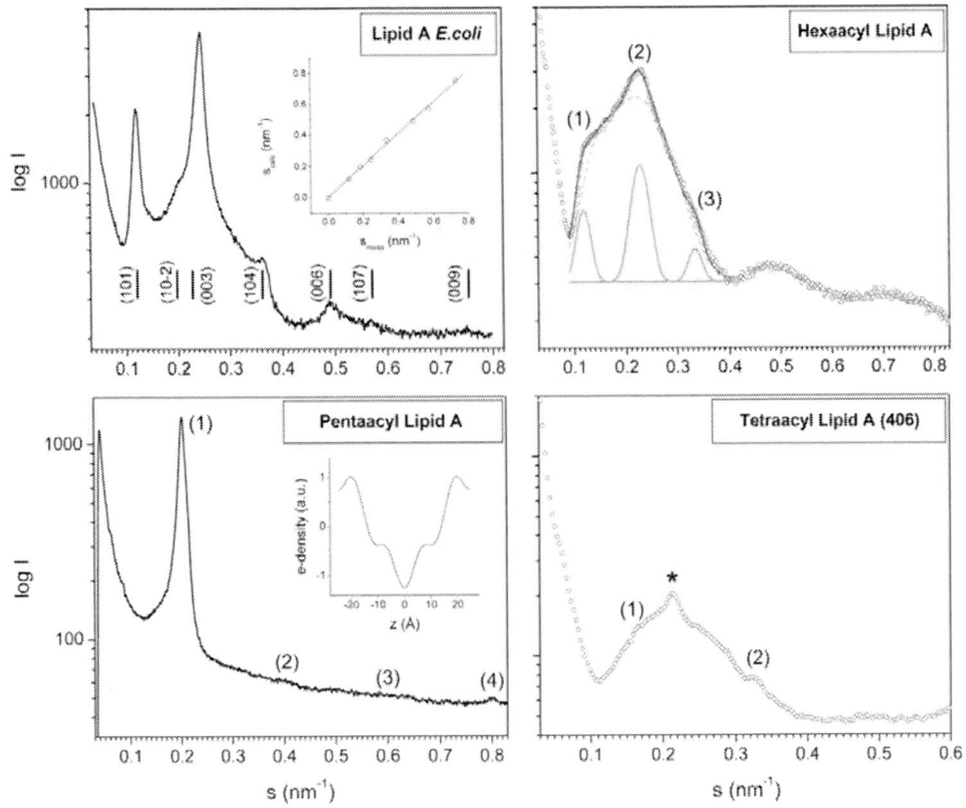

Figure 2. SAXS diffraction patterns indicating the structural polymorphism seen in different lipid A samples.

Briefly, its structural organization can be described as a stack of bilayers, in which neighboring bilayers form in plane hexagonally ordered stalk-like defects. The stalk to stalk distance is given by the lattice parameter a, while the bilayer repeat distance is equal to $c/3$ = 4.09 nm. In comparison, the sample pentaacyl lipid A is more hydrated and forms multi-lamellar vesicles free from defects with a d-spacing value of 4.97 nm, i.e. its interlamellar water layer is 0.9 nm thicker. The diffraction peaks were recorded up to the 4th order and the following amplitudes were determined: F_1 = -1.00, F_2 = F_3 =0 and F_4 = -0.25. The corresponding electron density profile is given in the inset. Note that the bilayer thickness, d_{PP} = 4.0 nm, comes very close to the bilayer repeat distance in the strongly dehydrated lipid A *E. coli* sample. In contrast, the hexaacyl lipid A sample is strongly hydrated. The scattering pattern is governed by the diffuse scattering arising from the bilayer form factor. Three lobes are visible, centered roughly at s = 0.22, 0.49 and 0.73 nm^{-1} and the corresponding form factor minima are found at s = 0.10, 0.40 and 0.64 nm^{-1}, respectively. However, not all membranes are spatially uncorrelated. As can be seen in the region of the first lobe (green dashed line), additionally, the first three diffraction peaks fitted by Gaussian distributions (solid green lines) of a lamellar phase with a *d*-spacing of 8.7 nm are observed. Remarkably, the water spacing, $d_W = d - d_{PP}$, is in the order of 4.7 nm, i.e. $d_{PP} \sim d_W$ in this case. In other words, the main electron density contrast is given for every *d*/2, which explains why the second order peak has the highest intensity (18). The scattering pattern of the tetraacyl lipid A sample seems similar to the latter. Also, here the scattering pattern is mainly governed by diffuse scattering arising from the form factor. Unfortunately, a unique phase assignment can not be made. However, comparing this pattern to the SAXS data of Lipid A *S. minnesota* (Fig. 4A and C), it would make sense to index the observed three Bragg peaks has a lamellar phase with *d*-spacings of 6.1 nm (see first and second order) coexisting with traces of an inverse hexagonal phase with unit cell parameter of 5.4 nm (asterisk).

This data gives a good impression of the complexity of the structural polyphorphism of lipid A samples, in particular, governed by different hydration conditions. As shown earlier, when lowering even slightly the water content of hexaacyl samples down to 75 to 85 5, there is a strong tendency for the adoption of bicontinous phases such that from room group Pn3m (Q^{224}, (5)). A similar behavior was observed for LPS Re (from *S. minnesota* R595, (6)).

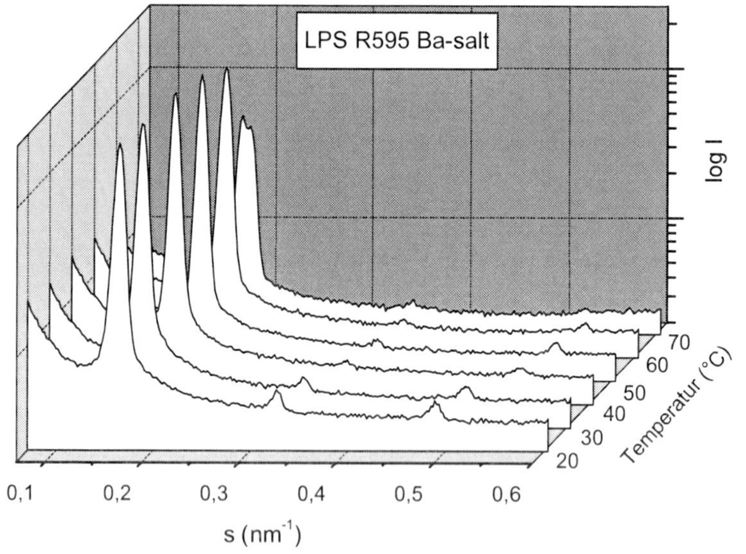

Figure 3. SAXS patterns of the Ba2+ salt form of LPS R595 in the temperature range 20 to 70 °C.

It has also been published that there is a dependence of the aggregate structures, and with it, on the bioactivity of endotoxins, on their salt forms (10). The aggregate structures of the different divalent salt forms of LPS Re (strain R595 from *S. minnesota*) are converted from the original cubic aggregate structure (for monovalent salts such as Na^+ and K^+) into a more and more multi-lamellar structure with an increase in molecular weight of the cation ($Mg^{2+} \rightarrow Ca^{2+} \rightarrow Ba^{2+}$). The SAXS pattern of the Ba^{2+} salt form of LPS R595 is presented in Figure 3, in the temperature range 20 to 70 °C, with the occurrence of the first to third order reflections characteristic for stacks with high quasi long range order. This property is connected with the absence of any bioactivity, which was confirmed in concomitant biological experiments.

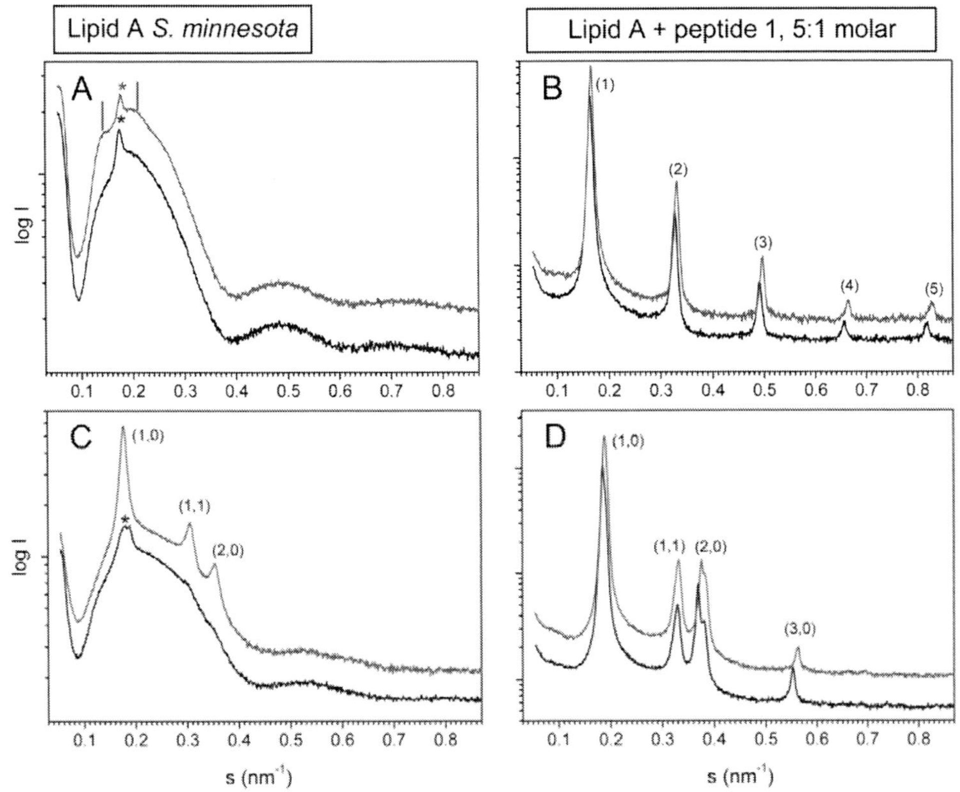

Figure 4. Small angle X-ray scattering patterns of Lipid A and Lipid A plus peptide 1 samples are depicted. Samples were recorded at 20 and 40 °C (A and B) as well as at 60 and 70 °C (C and D). Both the Miller indices of the Lβ and HII phase are given. Traces of the HII phase are marked with asterisks (* in A and C), while traces of an highly disordered lamellar phase is marked with vertical lines (| in A). In each panel the higher temperature recordings have been shifted for clarity (red solid lines).

AGGREGATE STRUCTURES OF ENDOTOXIN:PEPTIDE COMPLEXES

From the foregoing it might become clear that the aggregate structures of endotoxins are determinants of their ability to induce cytokines in human immune cells. Therefore, it was very interesting to find out whether the change of the aggregate structure of LPS in the presence of peptides correlates with the effectivity of the peptides to block the LPS-induced bioactivity. We have investigated some structurally closely related peptides (10-mers, see Table 1) to change the aggregate structures of lipid A. As can be taken from Figure 4A, lipid A adopts at 20 and 40 °C more or less uncorrelated lipid bilayers. At higher temperatures, the uncorrelated bilayers convert into an inverted hexagonal phase H_{II} at 70 °C, the latter deduced from the occurrence of the three reflections at 1/5.65, 1/3.26, and 1/2.83 nm^{-1} corresponding to the (1,0),(1,1) and (2,0) planes (see Figure 4C). In the presence of peptide 1, the scattering patterns at all temperatures change drastically: at 20 and 40 °C, five diffraction peaks are observed which are due to the scattering from multi-lamellar stacks of bilayers with the *d*-spacing of 6.10 and 6.02 nm, respectively (Fig. 4B). It should be noted that the value of the periodicity (~ 6 nm) is much higher than those values found for highly condensed lipid A samples alone (4.09 to 4.90 nm), i.e., at low water content or high Mg^{2+} concentrations (5). At 60 and 70 °C, lipid A in the presence of peptide 1 clearly adopts the H_{II} phase (Figure 4D), which is much better ordered than that found at 70 °C in the absence of the peptide.

Table 1. Structurally related 10-mer and 19/20mer peptides.

Pep 1:	FWQRNIRKVR
Pep 31:	FWQRNIRKWR
Pep 32:	FWQRNIRKYR
Pep19-2.5:	GCKKYRRFRWKFKGKFWFWG
Pep19-8:	GRRYKKFRWKFKGRWFWFG

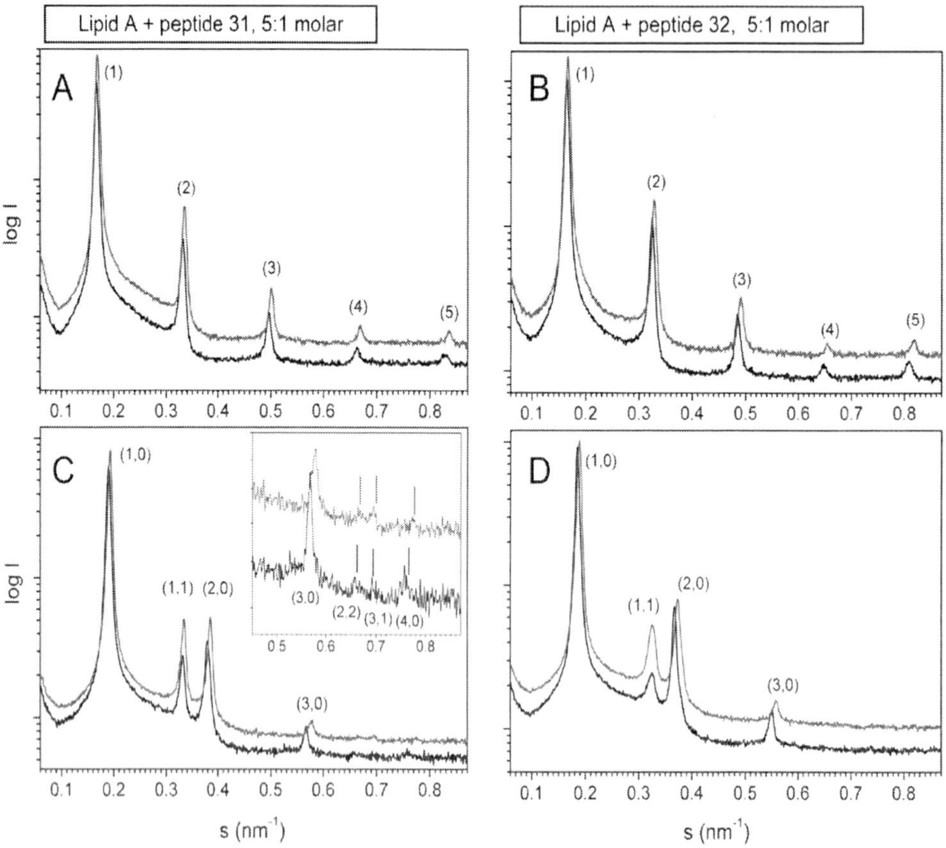

Figure 5. Small angle X-ray scattering patterns of Lipid A plus peptide 31and Lipid A plus peptide 32 samples are depicted. Samples were recorded at 20 and 40 °C (A and B) as well as at 60 and 70 °C (C and D), and for each reflection the Miller indices are given. Higher temperature recordings have been shifted for clarity (red solid lines).

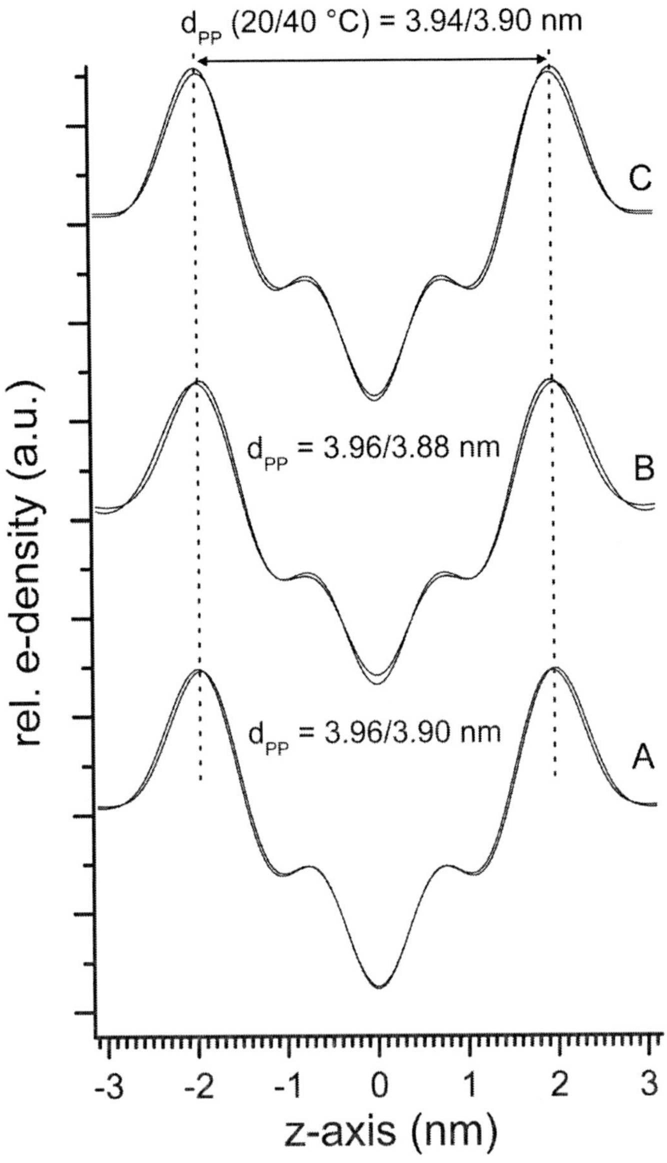

Figure 6. Calculated electron density profiles of the Lβ phase concerning the Lipid A plus peptide 1 (A), the Lipid A plus peptide 31 (B), and the Lipid A plus peptide 32 (C) samples. The amplitudes with their respective phases applied were -1.00/-1.00, -0.60/-0.61, +0.39/+0.37, -0.24/-0.23, -0.30/-0.30 (A: 20/40 °C); -1.00/-1.00, -0.52/-0.55, +0.39/+0.37, -0.24/-0.23, -0.22/-0.27 (B: 20/40 °C); -1.00/-1.00, -0.65/-0.64, +0.42/+0.40, -0.21/-0.18, -0.36/-0.33 (C: 20/ 40 °C). The electron density profiles at 20 and 40 °C of each respective sample are superimposed. Electron density profiles of different samples are shifted for clarity.

Similar experiments were performed with lipid A samples in the presence of peptides 31 and 32 (sequences see Table 1). As can be seen in Figure 5, in the gel-phase regime (20 and 40 °C) multi-lamellar vesicles are induced, and in the fluid phase regime (60 and 70 °C), the inverse hexagonal phase forms. All basic structural parameters referring to Figure 4 and Figure 5 are given in Table 2.

Table 2. Structural parameters of the lamellar L□ and the inverted hexagonal HII phase of different Lipid A samples. The lamellar d-spacings and unit cell parameter, a, of the HII phase are given. Further, the membrane thickness, dPP, of L□ phase, and the rod-diameter, DPP, of the HII phase are listed

	20 °C	40 °C	60 °C	70 °C
Lipid A	†	$d \sim 14$ nm†	†	
S. minnesota	$a = 6.61$ nm*	$a = 6.56$ nm*	$a \sim 6.3$ nm*	$a = 6.52$ nm
Lipid A +	$d = 6.10$ nm	$d = 6.02$ nm	$a = 6.20$ nm	$a = 6.10$ nm
peptide 1	$d_{PP} = 3.96$ nm	$d_{PP} = 3.96$ nm		
Lipid A +	$d = 6.00$ nm	$d = 5.95$ nm	$a = 6.06$ nm	$a = 5.97$ nm
peptide 31	$d_{PP} = 3.96$ nm	$d_{PP} = 3.88$ nm	$D_{PP} = 2.7$ nm	$D_{PP} = 2.7$ nm
Lipid A +	$d = 6.16$ nm	$d = 6.09$ nm	$a = 6.25$ nm	$a = 6.17$ nm
peptide 32	$d_{PP} = 3.94$ nm	$d_{PP} = 3.90$ nm		

†Most of the membranes are spatially uncorrelated; *only traces of the H_{II} phase were observed. All data refer to the scattering patterns presented in Figure 4 and Figure 5.

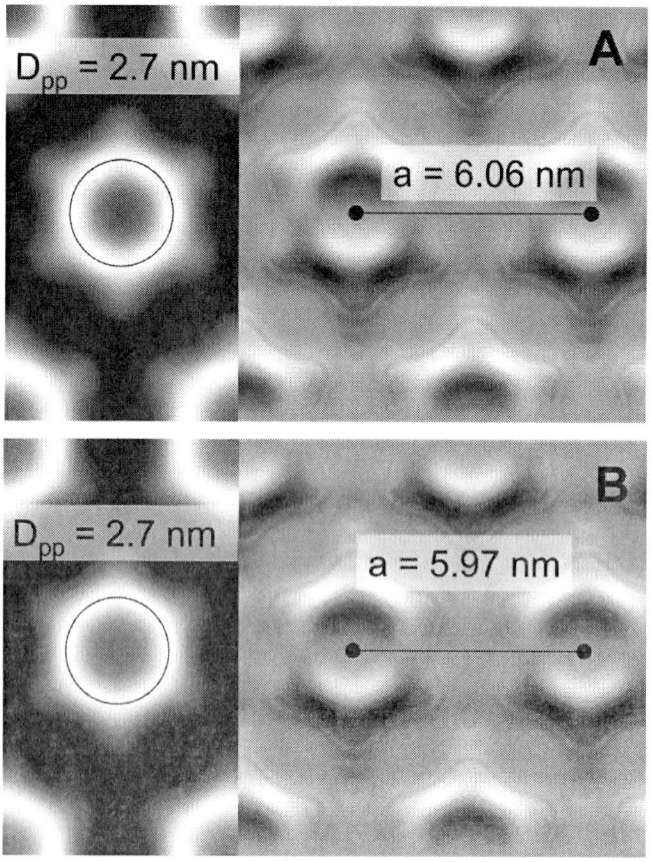

Figure 7. Electron density maps of the HII phase of the Lipid A sample under the influence of peptide 32 at 60 °C (A) and 70 °C (B) are presented. The form factors applied were F10 = +1.00/+1.00; F11 = -0.33/-0.43; F20 = -0.48/-0.50; F21 was not observed; F30 = -0.24/-0.20; F22 = -0.08/-0.00; F31 = +0.00/+0.07, and F40 = +0.15/+0.11 for 60/70 °C. Maximum electron densities are highlighted in white (left) or as protuberance in the relief image (right).

In Figure 6, all of the electron density profiles of the L_β phase induced by the peptides 1, 31 and 32 are illustrated. Obviously, all peptides have about the same capacity to condense formerly uncorrelated membranes as can be seen in the low variation of *d*-spacings (5.95 to 6.16 nm from 20 to 40 °C; see Table 2). The deduced bilayer thickness, d_{PP}, shows an even smaller variation: 3.88 to 3.96 nm (Table 2). Last, temperature has only a very minor effect on the electron density profiles as can be judged from the superpositions of the 20 and 40 °C data.

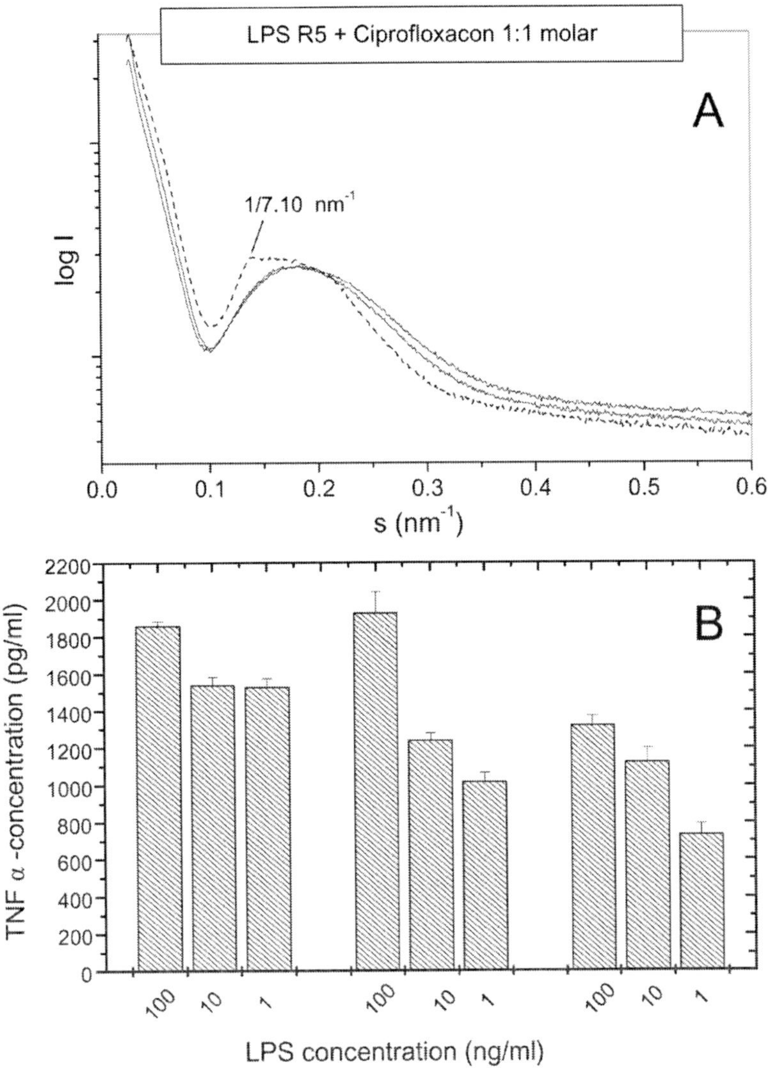

Figure 8. (A) SAXS pattern of LPS from S. minnesota strain R5 to which ciprofloxacin was added. (B) LPS-induced production of TNFα in human mononuclear cells at different concentrations of ciprofloxacin. Left: pure LPS; middle: LPS:Ciprofloxacin 1:1 weight%; right: LPS:Ciprofloxacin 1:3 weight%.

As stated above, in the fluid phase of all three peptides induce the inverse hexagonal phase. Assuming a low monolayer thickness variation, also at 60 and 70 °C, respectively, one can tell from the absolute values of the unit cell parameters, *a*, that the curvature induction is the strongest with peptide 31, followed by peptide 1, and weakest for peptide 32 (Table 2). Again, the differences in the unit cell parameter are not big, but the lipid A/peptide31 complex displays the highest quasi long range order (Fig. 5C), which allowed us to record reflections up to the (4,0)-reflection. In Figure 7, the corresponding electron density maps of this sample are displayed. The water core diameter, D_{PP}, of the inverse hexagonal phase is at both temperatures 2.7 nm. While the polar region of the nanotubes are similar to those found in phosphatidylethanolamine/water H_{II} phases (16),(17), interestingly, the hydrocarbon chain region is not homogeneous at all. It is tempting to believe that this is caused by the extraordinary lipid chain diversity of lipid A.

One important point is the observation, that antibiotics may kill the bacteria, but usually are not able to neutralize the released pathogenicity factors such as LPS (8). For a better understanding of these processes, we have studied the interaction of ciprofloxacin, a very effective antibiotic against a variety of bacteria, with LPS. A SAXS pattern of LPS from *S. minnesota* strain R5 to which ciprofloxacin was added, is shown in Figure8A The spectra may be due to the form factor arising from lipid bilayers, but a sponge phase cannot be excluded, since for a pure lipid bilayers at least two form factor minima can be expected. Interestingly, the addition of ciprofloxacin apparently does not induce changes. Parallel to this, in Figure 8B the results from a biological experiment are shown: The addition of ciprofloxacin to an LPS dispersion does not influence its ability to induce the tumor-necrosis-factor α, a cytokine important for the inflammation reaction.

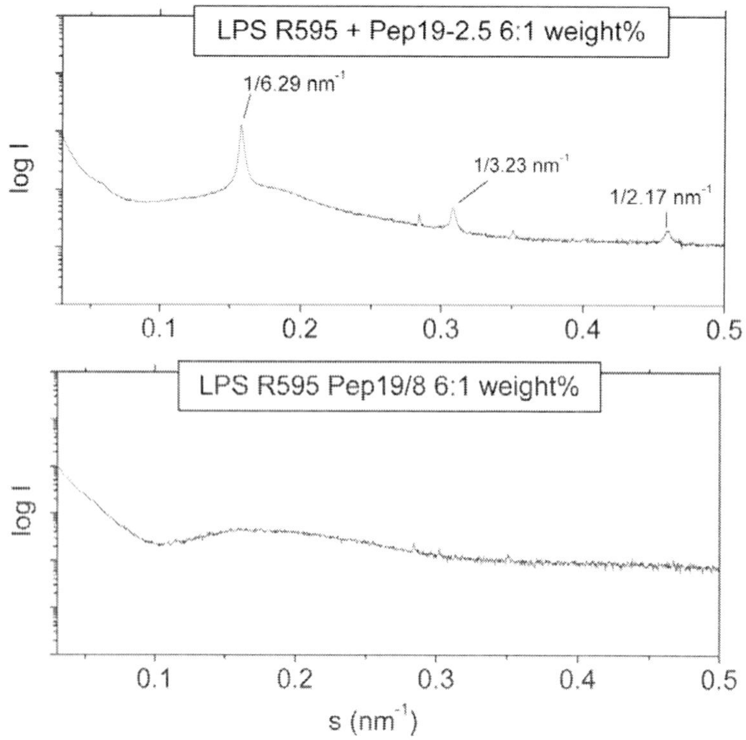

Figure 9. SAXS pattern of LPS from S. minnesota strain R595 in the presence of Pep19-2.5 and Pep19-8.

In contrast to the antibiotic, particularly designed synthetic anti-LPS peptides (SALP) induce changes in the aggregate structures of LPS which are correlated with their efficiency to neutralize endotoxin. This is shown exemplarily in Figure 9, in which the influence of the peptides Pep19-8 and 19-2.5 on the diffraction pattern of LPS from *S. minnesota* strain R595 is shown: As can be seen, the compound Pep19-2.5, which is strongly able to neutralize LPS even if added at low molar excess, induces much more changes in the pattern - conversion into a multi-lamellar aggregate - as compared to compound Pep19-8, which causes only binding to LPS, if added in a high molar excess (11).

A further important point to be considered a suitable therapeutic is the absence of any cytotoxic activity against body-own cells. It has been shown that cytotoxic effects for the Pep19-20'mers can only be seen at concentrations > 20 µg/ml (11). To confirm and extend these findings, SAXS experiments were performed with a characteristic mammalian phospholipid, phosphatidylinositol (PI). The addition of the peptide Pep19-2.5 to a PI bilayer does not change the lipid periodicity of the PI bilayer essentially (from 5.0 to 5.1 nm; Figure 10), but leads to a strong broadening of the first order diffraction peak indicating a drastic decrease in the spatial membrane correlations (severe stacking disorder). From this, it can be deduced that such change may be responsible for the adverse effects of peptide drugs seen at higher concentrations.

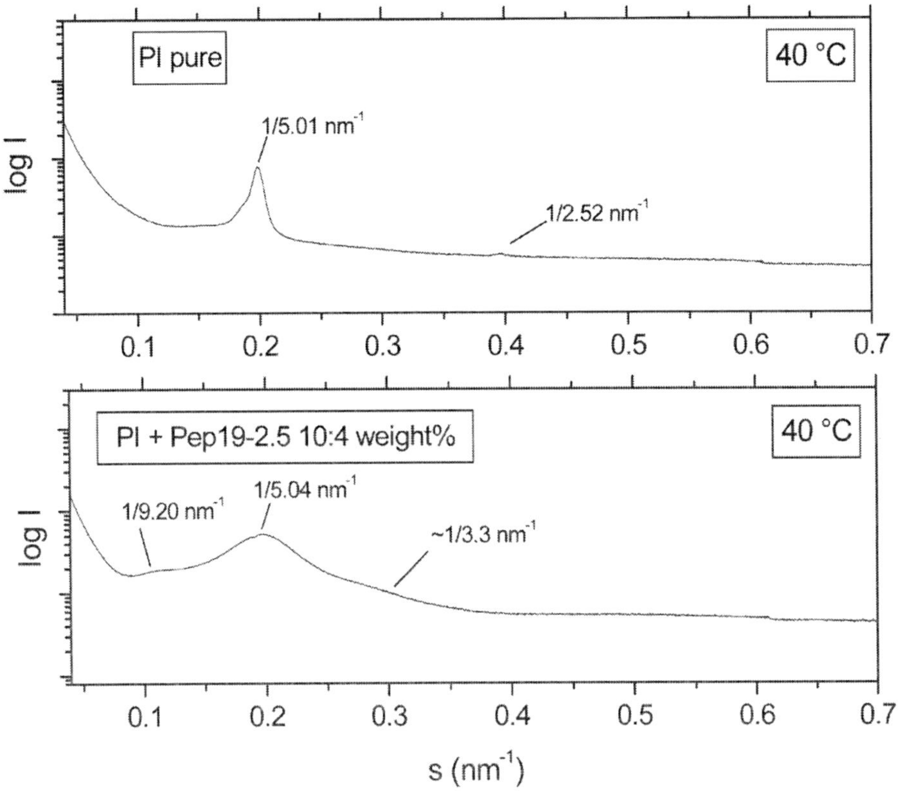

Figure 10. SAXS pattern of a mammalian phospholipid, phosphatidylinositol, in the absence and presence of Pep19-2.5.

Acknowledgments

The authors are indebted to the German ministry BMBF for financial help in the frame of a pre-clinical study 'Therapy of infectious diseases with special regards to bacterial sepsis (project: 01GU0824)'.

References

[1] Andersson, M., Gunne, H., Agerberth, B., Boman, A., Bergman, T., Olsson, B., Dagerlind, A., Wigzell, H., Boman, H. G., and Gudmundsson, G. H. (1996). NK-lysin, structure and function of a novel effector molecule of porcine T and NK cells. *Vet. Immunol. Immunopathol.* 54:123-126.

[2] Andrä, J., Gutsmann, T., Garidel, P., and Brandenburg, K. (2006). Mechanisms of endotoxin neutralization by synthetic cationic compounds. *J. Endotoxin. Res.* 12:261-277.

[3] Andrä, J., Howe, J., Garidel, P., Rössle, M., Richter, W., Leiva-Leon, J., Moriyon, I., Bartels, R., Gutsmann, T., and Brandenburg, K. (2007). Mechanism of interaction of optimized Limulus-derived cyclic peptides with endotoxins: thermodynamic, biophysical and microbiological analysis. *Biochem. J.* 406:297-307.

[4] Baveye, S., Elass, E., Mazurier, J., Spik, G., and Legrand, D. (1999). Lactoferrin: a multi-functional glycoprotein involved in the modulation of the inflammatory process. *Clin. Chem. Lab. Med.* 37:281-286.

[5] Brandenburg, K., Koch, M. H. J., and Seydel, U. (1992). Phase diagram of lipid A from *Salmonella minnesota* and *Escherichia coli* rough mutant lipopolysaccharide. *J. Struct. Biol.* 105:11-21.

[6] Brandenburg, K., Koch, M. H. J., and Seydel, U. (1992). Phase diagramm of deep rough mutant lipopolysaccharide from *Salmonella minnesota* R595. *J. Struct. Biol.* 108:93-106.

[7] Brandenburg, K., Kusumoto, S., and Seydel, U. (1997). Conformational studies of synthetic lipid A analogues and partial structures by infrared spectroscopy. *Biochim. Biophys. Acta* 1329:193-201.

[8] Brandenburg, K., Schromm, A. B., and Gutsmann, T. (2010). Endotoxins: relationship between structure, function, and activity. *Subcell. Biochem.* 53:53-67.

[9] Cross, A. S. and Opal, S. M. (1995). Endotoxin's role in Gram-negative bacterial infection. *Curr. Opin. Infect. Dis.* 8:156-163.

[10] Garidel, P., Rappolt, M., Schromm, A. B., Howe, J., Lohner, K., Andrä, J., Koch, M. H. J., and Brandenburg, K. (2005). Divalent cations affect chain mobility and aggregate structure of lipopolysaccharide from Salmonella minnesota reflected in a decrease of its biological activity. *Biochim. Biophys. Acta* 1715:122-131.

[11] Gutsmann, T., Razquin-Olazaran, I., Kowalski, I., Kaconis, Y., Howe, J., Bartels, R., Hornef, M., Schürholz, T., Rössle, M., Sanchez-Gomez, S., Moriyon, I., Martinez, de Tejada, G., and Brandenburg, K. (2010). New antiseptic peptides to protect against endotoxin-mediated shock. *Antimicrob. Agents Chemother.* 54:3817-3824.

[12] Harper, P. E., Mannock, D. A., Lewis, R. N., McElhaney, R. N., and Gruner, S. M. (2001). X-ray diffraction structures of some phosphatidylethanolamine lamellar and inverted hexagonal phases. *Biophys. J.* 81:2693-2706.

[13] Kirkland, T. N., Finley, F., Leturcq, D., Moriarty, A., Lee, J. D., Ulevitch, R. J., and Tobias, P. S. (1993). Analysis of lipopolysaccharide binding by CD14. *J. Biol. Chem* 268:24818-24823.

[14] Rappolt, M. 2006. (2006). The biologically relevant lipid mesophases as "seen" by X-rays. *In* Advances in Planar Lipid Bilayers and Liposomes. A.Leitmanova-Liu, editor. Elsevier, Amsterdam. 253-83.

[15] Rappolt, M., Amenitsch, H., Strancar, J., Teixeira, C. V., Kriechbaum, M., Pabst, G., Majerowicz, M., and Laggner, P. (2004). Phospholipid mesophases at solid interfaces: in-situ X-ray diffraction and spin-label studies. *Adv. Colloid Interface Sci.* 111:63-77.

[16] Rappolt, M., Hickel, A., Bringezu, F., and Lohner, K. (2003). Mechanism of the lamellar/inverse hexagonal phase transition examined by high resolution x-ray diffraction. *Biophys. J.* 84:3111-3122.

[17] Rappolt, M., Hodzic, A., Sartori, B., Ollivon, M., and Laggner, P. (2008). Conformational and hydrational properties during the L(beta)- to L(alpha)- and L(alpha)- to H(II)-phase transition in phosphatidylethanolamine. *Chem. Phys. Lipids* 154:46-55.

[18] Rappolt, M, (2010) Bilayer thickness estimations with "poor" diffraction data, *J. Appl. Phys.* 107:084701-1-084701-7.

[19] Roessle, M., Klaering, R., Ristau, U., Robrahn,B., Jahn,D., Gehrmann, T., Konarev,P., Round, A., Fiedler, S., Hermes, C., Svergun, D. (2007). Upgrade of the small-angle X-ray scattering beamline at the European Molecular Biology Laboratory, Hamburg. *J. Appl. Cryst.* 40:190-194.

[20] Schromm, A. B., Brandenburg, K., Loppnow, H., Zähringer, U., Rietschel, E. Th., Carroll, S. F., Koch, M. H. J., Kusumoto, S., and Seydel, U. (1998). The charge of endotoxin molecules influences their conformation and interleukin-6 inducing capacity. *J. Immunol.* 161:5464-5471.

[21] Springer, J., Safley, M., Huber, S., Troxclair, D., Craver, R., Newman, W. P., and McGoey, R. R. (2010). Histopathological findings in fatal novel H1N1: an autopsy case series from September-November 2009 in New Orleans, Louisiana. *J. La State Med. Soc.* 162:88-91.

[22] Tapping, R. I., Gegner, J. A., Kravchenko, V. V., and Tobias, P. S. (1998). Roles for LBP and soluble CD14 in cellular uptake of LPS. *Prog. Clin. Biol. Res.* 397:73-78.

[23] Vallespi, M. G., Glaria, L. A., Reyes, O., Garay, H. E., Ferrero, J., and Arana, M. J. (2000). A *Limulus* anti-lipopolysaccharide factor-derived peptide exhibits a new immunological activity with potential applicability in infectious diseases. *Clin. Diagn. Lab Immunol.* 7:669-675.

[24] Yang, L. and Huang, H. W. (2002). Observation of a membrane fusion intermediate structure. *Science* 297:1877-1879.

[25] Zähringer,U., B.Lindner, and E.T.Rietschel. 1999. Chemical structure of lipid A. Recent advances in structrural analysis of biologically active molecules. *In Endotoxin in Health and Disease.* H.Brade, S.M.Opal, S.N.Vogel, and D.C.Morrison, editors. Marcel Dekker, New York. 93-114.

In: X-Ray Scattering
Editor: Christopher M. Bauwens
ISBN: 978-1-61324-326-8

Chapter 6

DECOMPOSITION OF WAXS DIFFRACTOGRAMS OF SEMICRYSTALLINE POLYMERS BY SIMULATED ANNEALING

Gopinath Subramanian **and Rahmi Ozisik***
Rensselaer Polytechnic Institute, Troy, NY, USA

Abstract

A Monte–Carlo analysis technique is presented as a general tool for decomposing experimental WAXS diffractograms of semicrystalline polymers. The technique of simulated annealing used here is an adaptation of the Metropolis–Hastings algorithm for global optimization, and is intended as an analysis technique for polymers whose ideal crystalline and amorphous structures are well known. It is capable of simultaneously extracting from a single WAXS diffractogram (a) the degree of crystallinity (b) relative content of each phase, and (c) most probable lamellar thickness of the individual crystalline phases. It is applicable to specimens containing a single type of polymer capable of crystallizing in multiple phases (such as i-PP), or blends of different polymers. It is also capable of finding multiple solutions when more than one combination of the constituents is consistent with the experimental WAXS diffractogram. The lamellar thickness predicted by the technique contains contributions from lattice strain, and thus, the deformation history of the specimen is required for a proper interpretation of the results. Prior knowledge of the different species that make up the specimen is assumed.

1. Motivation

Polymers have rapidly gained acceptance in modern society as a versatile material for multiple applications. Polymers are used in many forms, such as fibers, coatings, films, and bulk. The wide spectrum of possible uses for polymers arises in part due to the complex structural organization of the chains, which can result in a rich variety of microstructures. The study of processes to control the microstructure is of great interest, as it allows tuning of the bulk material properties of the finished product.

*E-mail address: gsub@scorec.rpi.edu

Many polymers are capable of forming semi–crystalline structures, which can result in a wealth of interesting phenomena. The overall crystallinity can directly affect the structural and electrical properties of such polymers. For example, semi–crystalline polymers are both denser, and tougher than their amorphous counterparts, making them attractive for a variety of structural applications. For other applications, such as polymer electrolytes in lithium ion batteries, it is desirable to have a highly amorphous microstructure, as the ionic mobility within the electrolyte is hindered by the presence of crystalline domains. From a more fundamental standpoint, crystallinity measurements are used to validate theories of polymer crystallization kinetics. [1–4] Some polymers, such as polyethylene oxide are capable of forming only one type of crystalline phase [5,6]. Other polymers, such as isotactic polypropylene, [7–9] and polyethylene [10], are capable of crystallizing as more than one phase. The different shapes and sizes of the unit cells can lead to different structural and transport properties, and thus, it can be useful to measure the relative content of each crystalline phase in a polymer specimen.

In addition to measuring the overall crystallinity, and the relative content of each crystalline phase, the lamellar thickness of the crystallites is of great interest. Polymer crystals grown from dilute solution are are relatively unaffected by entanglement effects, and are thus larger, with a smaller number of defects. In melt–grown crystals, on the other hand, the long time scales associated with disentanglement of polymer chains hinders the migration of chains to the crystal growth fronts, and can result in the formation of complex sheaf–like or spherical lamellar aggregates. [11] The nature of these aggregates depends on lamellar size, and can have far–reaching impacts on the mechanical and electrical properties: the amorphous region between spherulites is less dense than the crystalline domains, and thus serves as sites for crack propagation [12–17], and dielectric breakdown [18–22].

Accurate measurement of the aforementioned aspects of semi–crystalline polymers is also crucial for the continued development of fundamental theories of crystallization. These measurements are also vital for developing new processing methodologies, and quality control. A wealth of microanalysis techniques, with varying degrees of accuracy and ease–of–use exist. The simplest of these microanalysis techniques is density measurement, that is used to estimate the crystallinity of the specimen. Differential Scanning Calorimetry (DSC) is also routinely used in both academia and industry to measure crystallinity with a higher degree of accuracy. Wide angle X–ray Scattering (WAXS) is another technique that is capable of probing microstructure deeper than DSC, and can be used to qualitatively identify the different phases. Small angle X-ray scattering (SAXS), [23–26] Transmission electron microscopy (TEM), [23,27,28] and Atomic force microscopy (AFM) [29–31] are still more sophisticated techniques, that are used to measure lamellar thickness.

The output of most of these techniques is a composite of contributions from the individual components that make up the polymer specimen being examined. In order to obtain useful mirostructural information, both qualitative and quantitative, deconvolution of the experimental signal is necessary. Density measurements are nearly impossible to deconvolve into contributions from the individual components. The experimental signal from DSC can, in principle, be decomposed into contributions from the individual phases, but the decomposition itself can be erroneous, due to limits of experimental accuracy. For example, DSC measurements used to estimate lamellar thickness have been found to be extremely sensitive to experimental parameters, such as heating rate, and constants in the

Gibbs-Thomson equation, thereby rendering DSC unsuitable as a tool for routine quantitative analysis. [32] WAXS is relatively easier to deconvolve. The contributions from the individual components can be clearly distinguished [33,34], and form the basis of qualitative phase identification. Quantitative information can also be extracted from WAXS patterns. For example, the degree of crystallinity can be obtained by deconvolving [2,35–38] the halo obtained from a purely amorphous specimen [39]. The relative heights of peaks generated by the individual phases has been used to determine the relative content of the individual phases using the Turner–Jones method. [40] The most probable lamellar thickness can then be obtained by using SAXS data in conjunction with crystallinity measurements obtained from DSC or WAXS. [23,24,32] Both SAXS, and other techniques used to measure lamellar thickness, such as AFM and TEM, while sensitive to the lamellar thickness, provide little or no information about the phase of the lamellae.

As can be seen from the above discussion, many of the modern microanalysis techniques, while advantageous, are not without their shortcomings. It is the aim of this chapter to elucidate a method that can obtain *simultaneously* the degree of crystallinity, relative phase content, and most probable lamellar thickness of the individual phases. To this end, a general–purpose algorithm that is capable of extracting a large amount of information from a single WAXS pattern is presented. Limitations and shortcomings of the method are discussed.

2. Method

The experimental WAXS pattern (Intensity vs. 2θ) to be analyzed is denoted as $\Psi(2\theta)$. It is assumed to be composed of contributions from the individual phases. Furthermore, the contributions from each phase are assumed to be independent. Under these assumptions, the experimental WAXS pattern can be decomposed as:

$$\Psi(2\theta) = \sum_{i=1}^{n} \phi_i \psi_i(2\theta, L_i) \tag{1}$$

where

$\Psi(2\theta)$ = Integrated experimental WAXS pattern
n = Total number of possible phases
ϕ_i = Volume fraction of phase i
$\psi_i(2\theta, L_i)$ = Ideal WAXS pattern of phase i
L_i = Lamellar thickness of phase i
θ = Bragg angle

For crystalline phases, the ideal WAXS intensity is a function of scattering angle and lamellar thickness, L_i. For amorphous phases, on the other hand, the concept of lamellar thickness is meaningless. For the rest of this chapter, it is understood that lamellar thickness, and other resulting properties, such as peak broadening, are applicable only to the crystalline phases. In equation 1, the weighting factors ϕ_i are chosen to be the volume fractions instead of weight fractions because in an X-ray scattering experiment, the probability with which the beam strikes a given phase is assumed to be the volume fraction of that phase. The ideal WAXS patterns, ψ_i, and the experimental WAXS pattern, Ψ, are treated as probability density functions, and are thus subjected to the normalization condition

$$\int_{2\theta=0}^{\pi} \psi_i(2\theta, L_i)\,d(2\theta) = 1 \; ; \; i = 1 \ldots n \tag{2}$$

$$\int_{2\theta=0}^{\pi} \Psi(2\theta)\,d(2\theta) = 1 \tag{3}$$

WAXS patterns of the amorphous phases can be obtained by fitting a suitable function to scattering data obtained from purely elastomeric specimens. For the crystalline phases, the peak locations and relative intensities can be obtained by using the unit cell properties, or controlled experiments. The effect of finite lamellar thickness can be incorporated by treating each peak as a Gaussian function with full width at half maximum (FWHM) determined using the Scherrer equation [41,42] as:

$$w = \frac{K\lambda}{L_i \cos\theta} \tag{4}$$

where

w_i = full width at half maximum (FWHM) of a peak
K = Scherrer constant, taken to be 0.93 [41]
L_i = lamellar thickness of phase i [43,44]
λ = X-ray wavelength
θ = Bragg angle

More complex functions, such as pseudo-Voigt, Pearson VII and Voigt [45,46] have been used to account for peak broadening of WAXS data. The Gaussian is much simpler, as it is a two-parameter function, and greatly reduces computational time. It is also a reasonable approximation to the WAXS peaks. Furthermore, the accuracy that is gained by using more complex functions may not be necessary, and as shown previously [47], the Gaussian performs exceedingly well. Lattice strain also contributes to line broadening of a WAXS peak, which is not explicitly accounted for, and thus, for specimens with lattice strain, the lamellar thickness predicted by simulated annealing is expected to be a lower limit.

Thus, by a proper choice of variables ϕ_i and L_i, we seek to minimize the error function E defined as:

$$E(\phi_i, L_i) = \int_{2\theta=0}^{\pi} \left[\Psi(2\theta) - \sum_{i=1}^{n} \phi_i \psi_i(2\theta, L_i) \right]^2 \mathrm{d}2\theta \tag{5}$$

The minimization of the error function in equation 5 is a constrained minimization problem. To solve this problem, the technique of simulated annealing [48] can be used. Simulated annealing is an adaptation of the Metropolis-Hasting algorithm for global optimization, and is described briefly in the context of decomposing WAXS patterns.

The phase space for the system under consideration is the space defined by the various volume fractions and lamellar thicknesses as $S(\phi_i, L_i)$, subjected to the following constraints:

$$0 \leq \phi_i \leq 1,\ i = 1 \ldots n \tag{6}$$

$$0 \leq L_i \leq L_{max},\ i = 1 \ldots n \tag{7}$$

$$\sum_{i=1}^{n} \phi_i = 1 \tag{8}$$

Strictly speaking, the value of L_{max} is unbounded. However, from a practical standpoint, it is extremely rare to obtain lamellae thicker than about 1000 Å, and thus, the maximum lamellar thickness is fixed at a suitably large number, depending on the chemical composition of the polymer.

Once the phase space has been suitably constrained, the technique of simulated annealing begins by choosing a state S_j at random in the phase space defined above. The error function $E(S_j)$ is evaluated at this point. A trial new state S_{j+1} is generated as $S_{j+1} = S_j + \Delta S_j$. "Downhill" moves, or moves that decrease the value of E are always accepted, and "uphill" moves, or moves that increase the value of E are accepted with a probability P given by the Metropolis criterion. Thus, the probability with which the trial state is accepted is given by

$$P(T) = \min \left\{ 1, \exp \left[\frac{E_j - E_{j+1}}{\tilde{T}} \right] \right\} \tag{9}$$

where $\tilde{T}$ is called the temperature of the system. At the start of a simulated annealing run, the temperature is assigned a value of unity, and changed as $\tilde{T}_{new} = R\tilde{T}_{old}$ after every N_{trials} evaluations of E. Here, R is referred to as the "cooling rate". The choice of R and N_{trials} is not unique, and is dependant on the system under consideration. High values of R correspond to "quenching," and lead to the cooling algorithm getting stuck in local minima. Previously, $R = 0.9$ and $N_{trial} = 20$ have been used [47] with some success.

3. Application of the Simulated Annealing Method

In this section, a detailed step–by–step procedure for using the method of simulated annealing is detailed. It is a more generalized version of previously published work. [47] The following knowledge is required:

- Chemical composition of the polymer specimen – Generally known beforehand in most industrial and academic applications
- Different types of phases each chemical species is capable of forming – Available in the relevant literature
- Ideal WAXS pattern of an infinitely thick lamella of each phase – Can be obtained from the crystal structure, and computation of structure factors
- WAXS pattern of the amorphous phase of each chemical species – Can be taken as the WAXS pattern of purely amorphous material

For example, in analyzing the microstructure of blends of isotactic polyprolylene with polyethylene, the relevant crystalline phases of i-PP are the α, β, and γ phases, [49–52] while those of polyethylene are the orthorhombic and hexagonal phases. [10] The unit cell structure for each of these phases is available in the literature, and can be used to compute their WAXS pattern. What is really required for the simulated annealing algorithm is the ratio of peak heights of the ideal crystalline phase.

Once the ideal WAXS patterns of the individual phases are obtained, the cooling rate and number of trials to be attempted at a particular temperature are chosen, generally, by trial and error. A proper choice of R and N_{trials} will allow the simulated annealing algorithm to escape from local minima, while allowing the code to complete in a reasonable amount of time.

The decomposition of WAXS diffractograms by simulated annealing has been applied to i-PP [47]. The parameters specific to the simulated annealing algorithm were chosen as $R = 0.9$ and $N_{trials} = 20$. Experimental data was obtained from the literature. Specifically, the data published by Mezghani and Philips [34], and Broda [53–55], as these papers present extensive sets of WAXS data, along with independent measurements of phase content ϕ_i, and lamellar thickness L_i. As is customary, these papers report WAXS data on an arbitrary intensity scale. In situations where multiple sets of WAXS data are presented in the same figure, the curves are shifted by an unspecified amount towards higher intensities for the sake of readability. In order to overcome the effect of the unknown shift factor, all WAXS curves obtained from the literature (denoted as $\Psi'(2\theta)$) were shifted downward such that the lowest intensity value of each curve was zero. Thus, the curves used as input to the simulated annealing routine were obtained as:

$$\Psi(2\theta) = \Psi'(2\theta) - \min(\Psi'(2\theta)) \tag{10}$$

This transformation of the experimental data effectively reduces the contribution of the amorphous halo to the diffractogram and results in an underestimation of the amorphous content ϕ_A of the test specimen, while leaving the volume fractions of the crystalline phases *relative to each other* unchanged. The lamellar thicknesses L_i are also unaffected, except perhaps in the case of extremely low crystalline content material, as the transformation does not change the width of the crystalline peaks. A WAXS pattern that has been corrected for instrument broadening, but without the amorphous halo removed if used directly, without any transformation, will produce the correct values for the amorphous content. It is useful to analyze each WAXS pattern multiple times, each time with a different random seed.

Depending on the choice of R and N_{trials}, a small number of seeds may result in the reporting of a local minimum as the global minimum. Such false minima can often be distinguished by the discrepancy between the experimental data and the fit.

A comparison between experiment and fitting was made with WAXS data published by Broda [53–55]. These experiments conducted to obtain the WAXS data are described in great detail elsewhere [53]. Essentially, Mosten 52.945, a commercial isotactic polypropylene was extruded from the melt at two different temperatures, 210 °C and 250 °C, into air at 20 °C. Fibers were spun at seven different take-up velocities ranging from 100–1350 m/min. Non colored and colored fibers were produced. Colored fibers were produced by adding 0.5 wt. % of quinacridone pigment just before forming fibers. WAXS diffractograms of each of these specimens were obtained using an X-Ray diffractometer (HZG-4) in the 2θ range 5° to 35°. The lamellar thickness of these specimens was computed from SAXS data obtained using an MBraun SWAXS camera with a Kratky collimating system.

Tables 1 and 2 summarize the most probable lamellar thickness of the individual phases, and their volume fraction as obtained from simulated annealing for each of the experimental WAXS patterns, and this data on lamellar thickness can be compared with SAXS measurements. In SAXS, the lamellar thickness obtained is averaged over all phases, and in order to perfom a direct comparison, the overall most probable lamellar thickness predicted by simulated annealing was computed as:

$$\langle L \rangle = \frac{\sum\limits_{i=\alpha,\beta,\gamma} \phi_i L_i}{\sum\limits_{i=\alpha,\beta,\gamma} \phi_i} \tag{11}$$

Figure 1 shows a plot of $\langle L \rangle$ as a function of extrusion velocity for the different specimens. The experimentally obtained lamellar thicknesses were seen to lie in the range of 25 – 80 Å. With the exception of one outlier, the predictions from simulated annealing also fall in the same range. Given the small range and size of the error bars, it is difficult to determine whether $\langle L \rangle$ increases or decreases with take-up velocity. The lamellar thickness determined using simulated annealing seems to be slightly lower than the thickness determined by SAXS, indicating that the specimens might contain some residual lattice strain.

The fraction of amorphous content obtained from simulated annealing lies in the range 20–34% for all specimens. These values seem unrealistically low, and this is a consequence of the baseline correction that is employed in equation 10, and thus, a direct comparison between crystallinity indices is not possible. However, the content of the crystalline phases relative to each other is expected to match values obtained by other measurement techniques.

The fits obtained to the WAXS patterns published by Broda [53] are shown in figures 2 and 3. In a majority of the specimens, the fitted curves agree well with the experimental curve. In some cases, such as specimens 3–6 in figure 2 (b), there is some disparity between the fitted and experimental curves. Possible sources of error include minute errors in the crystal structure of the ideal phases, possible instrument error and parasitic scattering. However, in spite of this disparity, the lamellar thicknesses obtained seem realistic, and are in excellent agreement with the lamellar thicknesses obtained using SAXS.

Table 1. Most probable lamellar thickness of each phase predicted by simulated annealing

	Extrusion velocity (m/min)	Lamellar thickness L_i (Å) 210 °C α	β	γ	250 °C α	β	γ
Non colored	100	101±7	29±16	32±2	132±1	7±1	29±5
	200	96±10	23±17	51±9	79±2	12±2	33±3
	300	94±14	11±3	51±7	11±1	34±2	87±9
	400	81±10	21±10	48±10	11±4	61±4	35±2
	880	105±18	13±5	34±5	103±2	107±3	83±7
	1050	86±9	19±9	38±4	40±12	39±13	52±13
	1350	109±7	22±14	41±6	96±7	8±1	33±2
Colored with quinacridone	100	28±9	127±8	55±11	14±5	99±5	59±14
	200	64±14	105±7	35±13	90±10	98±3	26±3
	300	101±6	61±12	32±8	89±10	20±5	46±4
	400	106±13	17±8	63±11	94±11	9±2	57±12
	880	81±9	29±15	66±13	87±10	26±10	52±9
	1050	61±7	15±8	70±10	78±10	13±4	50±8
	1350	99±6	9±2	43±9	92±8	19±8	43±3

The overall lamellar thickness obtained in this study is in good agreement with the lamellar thickness obtained using SAXS. As stated earlier, the method employed in the present study does not explicitly consider the line broadening resulting from lattice strain, but lumps this effect along with line broadening resulting from finite lamellar size. This leads to an underestimation of the actual lamellar thickness. The separation of line broadening into contributions from lattice strain and finite lamellar thickness remains an open question. In addition to the overall lamellar thickness, simulated annealing predicts an average lamellar thickness for each individual phase. At the time of writing, validation of this finding is an open question.

While simulated annealined performs reasonably, it is important to point out some of the limitations and outstanding issues. The Scherrer constant K used in equation 4 is taken to be 0.93, following Scherrer's original work [41,42]. It is possible that for other polymers, a different value of K is the more proper one. A different value of K, used in conjunction with a different value of L that maintains the ratio K/L does not affect the full width at half maximum, w, of a peak. Thus, a different value of K is expected to correspondingly change the value of L, while leaving the phase content obtained from simulated annealing unchanged, as w depends on the ratio K/L.

If the baseline correction is employed, the crystallinity predicted by simulated annealing will be somewhat higher than typical crystallinity values seen in the literature. With a WAXS diffractogram corrected only for the fixed background intensity that is obtained from scattering by air, further baseline correction becomes unnecessary, and a more accurate

Table 2. Summary of the volume fraction of each phase predicted by simulated annealing for non-colored fibers, and fibers colored with quinacridone. Column denoted ϕ_A^E indicates experimental finding from Broda [53]. The discrepancy between columns ϕ_A and ϕ_A^E is a direct consequence of the baseline correction in equation 10

	Extrusion velocity (m/min)	Volume fraction ϕ_i (%) 210 °C					250 °C				
		ϕ_α	ϕ_β	ϕ_γ	ϕ_A	ϕ_A^E	α	β	γ	ϕ_A	ϕ_A^E
Non colored	100	28±3	12±2	37±3	21±2	48	18±2	25±2	31±9	24±5	59
	200	21±2	21±2	29±2	26±3	60	11±4	21±3	33±1	34±6	68
	300	15±2	21±3	35±2	28±1	66	31±3	6±2	29±5	31±2	82
	400	15±2	24±4	35±3	24±3	68	43±6	5±1	31±3	19±7	90
	880	15±5	17±2	46±3	20±6	67	10±1	29±7	11±6	48±11	90
	1050	17±2	20±3	37±3	24±2	65	20±3	16±3	30±4	32±3	88
	1350	20±2	29±2	24±4	26±4	47	18±1	21±1	33±1	26±2	59
Colored	100	33±7	30±3	19±3	18±5	45	41±4	24±3	14±2	22±4	49
	200	31±7	20±3	38±4	12±4	48	26±2	13±1	33±4	28±3	49
	300	25±3	14±4	36±3	25±2	49	26±3	19±4	26±4	29±4	50
	400	19±3	23±3	27±3	32±4	49	16±2	15±3	33±3	37±3	60
	880	22±3	17±4	28±4	33±1	49	13±3	24±5	26±4	36±5	66
	1050	18±3	27±3	20±2	34±4	48	11±2	20±3	38±2	31±2	68
	1350	26±3	16±3	29±1	29±3	48	15±2	20±4	34±5	31±5	58

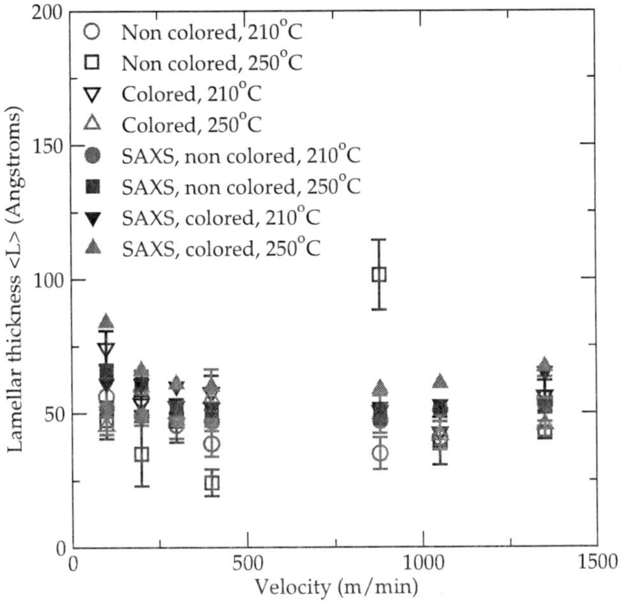

Figure 1. The overall most probable lamellar thickness predicted by simulated annealing as a function of take-up velocity (open symbols). Results obtained using independent SAXS measurements are also shown (filled symbols) [53].

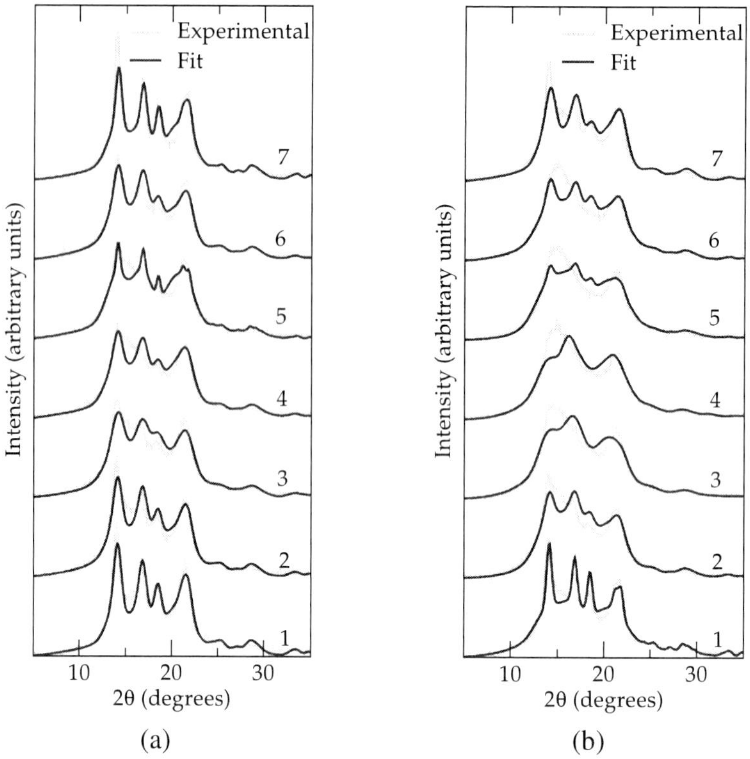

Figure 2. Fits to experimental data from Broda [53–55] for non-colored i-PP fibers (a) extruded at 210 °C (b) extruded at 250 °C at different velocities: 1 – 100 m/min, 2 – 200 m/min, 3 – 300 m/min, 4 – 400 m/min, 5 – 880 m/min, 6 – 1050 m/min, 7 – 1350 m/min.

value of crystallinity will be obtained. The accuracy of crystallinity predicted depends on the accuracy of the WAXS pattern of a purely amorphous specimen, and there is a certain amount of difficulty associated with obtaining a purely amorphous specimen of i-PP [56].

Simulated annealing aims to find a single point in the phase space $S(\phi_i, L_i)$ that minimizes the error E defined in equation 5. In principle, there can exist multiple, widely separated points in phase space that yield comparable values of the error E. In other words, given an experimental WAXS pattern, two widely different crystalline phase contents and lamellar thicknesses may be obtained that fit the experimental data. Such a situation would occur if the peaks of the ideal crystalline phases are all close to each other. For polymers such as i-PP, even though some peaks of each crystalline phase are close to the peaks of other crystalline phases, each crystalline phase has at least one distinct peak that is not present in any of the other crystalline phases. Indeed, both the α and γ phases have high-intensity peaks at 14.1° and 13.8°, and all three phases have moderate intensity peaks in the vicinity of 21.2°. Nevertheless, each phase has its own distinct peak (α at 18.55°, β at 16.05° and γ at 20.07°) of relatively high intensity. Thus, for example, when analyzing a specimen rich in γ-content, choosing a point in phase-space that is γ deprived yields a large value of error E, therby forcing the search algorithm to move towards γ-rich points. It is this feature of i-PP that allows simulated annealing to find solutions in phase space that are

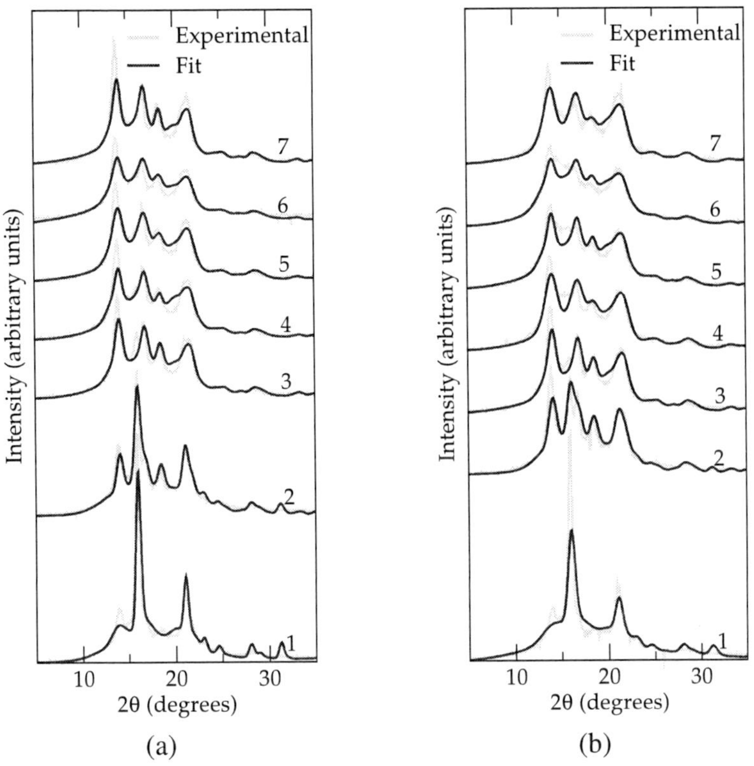

Figure 3. Fits to experimental WAXS data from Broda [53–55] for i-PP fibers colored with quinacridone pigment (a) extruded at 210 °C (b) extruded at 250 °C at different velocities: 1 – 100 m/min, 2 – 200 m/min, 3 – 300 m/min, 4 – 400 m/min, 5 – 880 m/min, 6 – 1050 m/min, 7 – 1350 m/min.

close to each other, and we expect the algorithm to yield better results for polymers that have high-intensity peaks that are widely separated on the 2θ axis. Other techniques, such as assigning a weighting factor to the high–intensity unique peaks can also be envisioned, and may become necessary, depending on the particular polymer specimen.

4. Appendix: Source Code

The source code included along with this chapter is written for analyzing isotactic polypropylene, and can be modified to suit other polymers. In the present form, the following files are required for proper functioning:

1. `ipp_simann.f90` : Fortran source, containing instructions on how to compile and use the code

2. `peaks-alpha; peaks-beta; peaks-gamma`: Intensity vs. 2-θ data for the ideal crystalline structure of the α, β, and γ phases of i-PP. Data on the unit cell dimensions of the ideal phases was obtained from the literature [7–9, 33, 34, 57–60]. Using this

information, structures were built using the commercial software Materials Studio, a product developed by Accelerys Inc. [61] The Reflex module of the software was then used to compute the structure factors F_{hkl}, and thereby the position (as a function of 2θ), and the relative intensity of the powder diffraction peaks of each of the crystalline phases.

3. `experiment.dat`: Experimental data

4. `random.seed`: File containing random seed number

4.1. `ipp_simann.f90`

Note: The first column is the line number, and can be useful in identifying long lines

```
1  !==================================================
2  program ipp_simann
3  !
4  !  AUTHOR:          Gopinath Subramanian
5  !  LAST MODIFIED:   May 16, 2010
6  !  CONTACT:         gop.sub@gmail.com
7  !
8  !
9  !
10 ! This code is an application of the simulated annealing algorithm to extract th
      phase content, and lamellar thickness
11 ! of the individual phases in isotactic polypropylene. The following paper
      describes the method in detail:
12 !
13 !   ''Simultaneous estimation of the phase content and lamellar thickness in
      isotactic
14 !     polypropylene by the simulated annealing of wide-angle X-ray scattering da
      ''
15 !
16 !     G. Subramanian and R. Ozisik, Journal of Applied Polymer Science, Vol. 117
      Iss. 4, 2368 -- 2394, 2010.
17 !
18 !
19 ! OVERVIEW
20 ! --------
21 !
22 ! * This code aims to minimize an objective function, which is the mean--square
      difference between experimental data,
23 !   and a trial WAXS pattern generated using knowledge of the crystal structure,
      and lamellar thickness of i-PP.
24 !
25 ! * Phase content is reported as volume fraction, and lamellar thickness is in
      Angstroms.
26 !
27 ! * Optimal solution and fit are written to files min-x and min-plot, respectively
      .
28 !
29 ! * The nomenclature used in the simulated annealing portion is intended to mate
      as closely as possible, that of
30 !   Corana et al., ACM Transactions on Mathematical Software, V. 15, I. 3, p. 28
      1989.
31 !
32 !       REQUIRED FILES:
33 !       ---------------
34 !
35 !         1. experiment.dat      - Two--column formatted experimental data. Column
      1: 2-theta; Column 2: Intensity
```

```
!
!                                      (a) File containing raw experimental data. If the baseline correction for scattering by air
!                                          is not done, the effect will be lumped along with the amorphous content.
!                                      (b) Normalization not required
!
!       2. (a) peaks-alpha --\
!          (b) peaks-beta     | - Peak positions and relative intensities of ideal alpha, beta, gamma phases of i-PP
!          (c) peaks-gamma --/
!
!
!       3. random.seed           - Seed for random number generator
!
!
! TIPS AND TRICKS
! ----------------
!
! (a) Computation time (and accuracy) increases with resolution of experimental data
! (b) It is best to analyze the same experimental data with multiple seeds, and compare results
! (c) Wavelength of the X-ray set to Cu - K-alpha radiation by default. This code has not been tested with any other radiation
! (d) The section marked "iPP-SIMANN PARAMETERS" contains values that control the speed and accuracy of the code
! (e) At multiple places in the code, commented write and pause statements have been placed. Uncomment them to examine
!     progress of the code in steps
!
!===================================================


        implicit none

! Simulated annealing variables. The main variables are:
!   f               : Objective function that is being minimized. In the paper, this is referred to as the Error function E
!   x               : x(1-3) -- volume fractions of alpha, beta, and gamma phases
!                     x(4-6) -- lamellar thickness of alpha, beta, and gamma phases
!   xtrial          : current value of the vector x(:) being examined
!   ub, lb          : Upper and lower bounds on x
!   xopt            : Value of x(:) that produces the minimum error, so far
!   vm              : Step length vector
        double precision, allocatable :: x(:), xtrial(:), ub(:), lb(:), xopt(:), vm(:), fstar(:), c(:)
        integer, allocatable          :: nacp(:)
        double precision              :: T, RT, f, ftrial, fopt, fdum, eps
        logical                       :: terminate, accept
        integer                       :: N, iseed, neps
        integer                       :: NS, NT, nacc, nobds, nfcn



! XRD variables.
!   th2             : 2-theta values of experimental data
!   xrd_expt        : Intensity values of experimental data
!   xrd_sim         : Intensity values, simulated
!   lambda          : X-ray wavelength, Cu K-alpha is default
        double precision, allocatable :: th2(:), xrd_expt(:), xrd_sim(:), xrd_sim_temp(:), xrd_amorphous(:)  ! Variables containing 2-theta, and intensity information
```

```fortran
      double precision, allocatable :: peaks_al(:,:), peaks_be(:,:), peaks_ga(:,:)
                                        ! Peaks data for alpha, beta and gamma phases
      double precision              :: lambda      ! X-ray wavelength
      integer                       :: npks_al, npks_be, npks_ga
      integer                       :: npoints

! Other general variables
      character(50)                 :: cdum
      integer                       :: i,j,k,l,readstat



      call setup_xrd()       ! Setup initial parameters - read experimental data, peaks data for ideal phases

      lambda  = 1.52         ! Wavelength of radiation used to collect XRD data

! Read random number seed, and initialize random number generator
      open(1,file='random.seed',status='old')
        read(1,*)iseed
      close(1)
      if ( iseed .gt. 0 ) iseed = -iseed
      iseed   = int( rand(iseed)*100 )

!==================================================
! iPP SIMANN PARAMETERS. Modify these to change speed/accuracy of the code
      eps     = 1D-15             !Termination criterion
      neps    = 5
      T       = 1.0               ! Start temperature
      RT      = 0.5               ! Temperature reduction factor
      NT      = max(100,5*N)      ! Number of evaluations per temperature step
      NS      = 20                ! Number of evaluations before changing vector step length
!==================================================


! Allocate memory, set bounds, and some more initialization
      N       = 6
      allocate(x(N),xtrial(N),ub(N),lb(N),xopt(N), vm(N), nacp(N), fstar(neps), c(N))

      ub(1)   = 1.0D0             ! phi-alpha
      lb(1)   = 0.0D0

      ub(2)   = 1.0D0             ! phi-beta
      lb(2)   = 0.0D0

      ub(3)   = 1.0D0             ! phi-gamma
      lb(3)   = 0.0D0

      ub(4)   = 1000.0D0          ! Lamellar size, alpha phase (Angstroms)
      lb(4)   = 0.0D0

      ub(5)   = 1000.0D0          ! Lamellar size, beta phase  (Angstroms)
      lb(5)   = 0.0D0

      ub(6)   = 1000.0D0          ! Lamellar size, gamma phase (Angstroms)
      lb(6)   = 0.0D0


      do i=1,N
        x(i) = lb(i) + rand(iseed) * (ub(i) - lb(i))
```

```
        enddo

        xopt     = x
        xtrial   = x
        fstar(:) = 1e20
        c(:)     = 2.0
        vm       = ub - lb
        f        = fcn(x)
        fopt     = f

        write(*,*)
        write(*,*)
        write(*,*)'------------------------------------'
        write(*,*)'Initial state:'
        write(*,*)
        write(*,*)'Volume fraction: '
        write(*,*)'  Alpha     = ',xopt(1)
        write(*,*)'  Beta      = ',xopt(2)
        write(*,*)'  Gamma     = ',xopt(3)
        write(*,*)'  Amorphous = ',1.0D0-(xopt(1) + xopt(2) + xopt(3))
        write(*,*)
        write(*,*)'Lamellar thickness (Angstroms): '
        write(*,*)'  Alpha     = ',xopt(4)
        write(*,*)'  Beta      = ',xopt(5)
        write(*,*)'  Gamma     = ',xopt(6)
        write(*,*)
        write(*,*)'Error: ',fopt
        write(*,*)'------------------------------------'

        terminate = .false.
        mainloop: do while( .not. terminate)

                nacc    = 0
                nobds   = 0
                nfcn    = 0

                do i=1,NT
                    nacp(:) = 0

                    do j=1,NS
                        do k=1,N

                            xtrial(k) = x(k) + (rand(iseed)*2.0 - 1.0) * vm(k)
                                                       ! Pick a trial x

                            if( xtrial(k) .gt. ub(k) .or. xtrial(k) .lt. lb(k) )
                                then          ! If trial is out of bounds,
                                 xtrial(k) = lb(k) + rand(iseed) * (ub(k) - lb(k))
                                                       ! pick a random value inside the
                                         bounds
                            endif

                            accept = .false.
                                ! Should I accept this trial?
                            ftrial = fcn(xtrial)
                            nfcn   = nfcn+1
                            if ( ftrial .lt. f ) then
                                accept=.true.
                                    ! Accept the move if function decreases
                            else
                                ! Otherwise, use Metropolis criteria
                                if( rand(iseed) .lt. dexp( (f-ftrial)/T ) ) then
                                    accept=.true.
                                endif
```

```
                endif

                if ( accept ) then

                    x(k)    = xtrial(k)
                    f       = ftrial
                    nacc    = nacc + 1
                    nacp(k) = nacp(k) + 1

                    if ( f .lt. fopt ) then    ! Ooo. How exciting. Found a new minimum.
                        fopt = f
                        xopt = x

                        open(712,file='min-x',status='replace')    ! Output minimum to file
                            write(712,*)'Volume fraction: '
                            write(712,*)'   Alpha     = ',x(1)
                            write(712,*)'   Beta      = ',x(2)
                            write(712,*)'   Gamma     = ',x(3)
                            write(712,*)'   Amorphous = ',1.0D0-(x(1) + x(2) + x(3))
                            write(712,*)
                            write(712,*)'Lamellar thickness (Angstroms): '
                            write(712,*)'   Alpha     = ',x(4)
                            write(712,*)'   Beta      = ',x(5)
                            write(712,*)'   Gamma     = ',x(6)
                        close(712)

                        open(712,file='min-plot',status='replace')
                            do l=1,npoints
                                write(712,*)th2(l),xrd_expt(l),xrd_sim(l)
                            enddo
                        close(712)

!                       write(*,*) 'NEW MINIMUM FOUND. '
!                       write(*,*)x
!                       write(*,*)f
!                       write(*,*)x-xtrial
!                       pause

                    endif
                else
                    ! If move is not accepted, reset xtrial
                    xtrial(k) = x(k)
                endif


            enddo
        enddo


        do l=1,N                         ! Adjust VM
            fdum=dfloat(nacp(l))/dfloat(NS)
            if ( fdum .gt. 0.6 ) then
                vm(l) = vm(l) * (1.0 + c(l)*(fdum -0.6)/0.4)
            else if ( fdum .lt. 0.4 ) then
                vm(l) = vm(l) / ( 1.0 + c(l) * (0.4 - fdum)/0.4 )
            endif
            fdum = ub(l) - lb(l)
            if ( vm(l) .gt. fdum ) vm(l) = fdum
        enddo

```

```
            enddo

            fstar(1:neps-1) = fstar(2:neps)               ! Should I terminate?
            fstar(neps)     = f
            terminate       = .true.
            do l=1,neps
               if ( abs( fstar(l) - f ) .gt. eps ) then
                  terminate = .false.
                  exit
               endif
            enddo


            T = T*RT      ! Reduce temperature



            write(*,*)'Temperature = ',T
            write(*,*)
            write(*,*)'Best fit parameters so far:'
            write(*,*)
            write(*,*)'Volume fraction: '
            write(*,*)'  Alpha     = ',xopt(1)
            write(*,*)'  Beta      = ',xopt(2)
            write(*,*)'  Gamma     = ',xopt(3)
            write(*,*)'  Amorphous = ',1.0D0-(xopt(1) + xopt(2) + xopt(3))
            write(*,*)
            write(*,*)'Lamellar thickness (Angstroms): '
            write(*,*)'  Alpha     = ',xopt(4)
            write(*,*)'  Beta      = ',xopt(5)
            write(*,*)'  Gamma     = ',xopt(6)
            write(*,*)
            write(*,*)'Error: ',fopt
            write(*,*)'----------------------------------------'
      enddo mainloop

      write(*,*)'=========================================================='
      write(*,*)'     BEST SOLUTION OBTAINED :'
      write(*,*)
      write(*,*)'Volume fraction: '
      write(*,*)'  Alpha     = ',xopt(1)
      write(*,*)'  Beta      = ',xopt(2)
      write(*,*)'  Gamma     = ',xopt(3)
      write(*,*)'  Amorphous = ',1.0D0-(xopt(1) + xopt(2) + xopt(3))
      write(*,*)
      write(*,*)'Lamellar thickness (Angstroms): '
      write(*,*)'  Alpha     = ',xopt(4)
      write(*,*)'  Beta      = ',xopt(5)
      write(*,*)'  Gamma     = ',xopt(6)
      write(*,*)
      write(*,*)'Error: ',fcn(xopt)
      write(*,*)'=========================================================='

   contains

!===================================================
   subroutine setup_xrd
!===================================================

      implicit none
      double precision :: fdum

      npoints = linesinfile('experiment.dat')
```

```
        allocate( th2(npoints), xrd_expt(npoints), xrd_sim(npoints), xrd_sim_temp(npoints), xrd_amorphous(npoints))

        open(1,file='experiment.dat',status='old')
          do i=1,npoints
            read(1,*)th2(i),xrd_expt(i)
          enddo
        close(1)

        call avint(th2,xrd_expt,th2(1),th2(npoints),fdum)        ! Normalize experimental data
        xrd_expt(:) = xrd_expt(:)/fdum

! Read peaks data for alpha, beta, gamma.
        npks_al = linesinfile('peaks-alpha')-1       ! Subtract 1 for header line
        npks_be = linesinfile('peaks-beta') -1
        npks_ga = linesinfile('peaks-gamma')-1

        allocate(peaks_al(npks_al,2),peaks_be(npks_be,2),peaks_ga(npks_ga,2))

        open(1,file='peaks-alpha',status='old')
          read(1,*)cdum
          do i=1,npks_al
            read(1,*)peaks_al(i,1),peaks_al(i,2)
          enddo
        close(1)

        open(1,file='peaks-beta',status='old')
          read(1,*)cdum
          do i=1,npks_be
            read(1,*)peaks_be(i,1),peaks_be(i,2)
          enddo
        close(1)

        open(1,file='peaks-gamma',status='old')
          read(1,*)cdum
          do i=1,npks_ga
            read(1,*)peaks_ga(i,1),peaks_ga(i,2)
          enddo
        close(1)


! Generate amorphous halo - Fit obtained from elastomeric i-PP data published in
! S. Mansel et al., Macromolecular Chemistry and Physics V. 200, p. 1292, 1999
        do i=1,npoints
          xrd_amorphous(i) = 0.120756*dexp(-0.69314718*((16.5122 - th2(i))/3.877585)**2)
        enddo



    end subroutine setup_xrd
!==================================================
    function fcn(x)

!     Function that computes the mean square error between experimental and trial XRD pattern
!==================================================

        implicit none
        double precision, intent(in) :: x(:)
        double precision             :: fcn
        integer                      :: nvars, i

        xrd_sim(:)=0.0
```

```

       if ( x(1)+x(2)+x(3) .gt. 1.0 ) then
          fcn = 42.0D5
       else
          call gen_xrd(peaks_al ,lambda ,x(4))
          xrd_sim(:)=xrd_sim(:) + x(1)*xrd_sim_temp(:)

          call gen_xrd(peaks_be ,lambda ,x(5))
          xrd_sim(:)=xrd_sim(:) + x(2)*xrd_sim_temp(:)

          call gen_xrd(peaks_ga ,lambda ,x(6))
          xrd_sim(:)=xrd_sim(:) + x(3)*xrd_sim_temp(:)

          xrd_sim(:)=xrd_sim(:) + (1.0-x(1)-x(2)-x(3))*xrd_amorphous(:)

          xrd_sim_temp(:)=xrd_sim(:) - xrd_expt(:)

          fcn=dot_product(xrd_sim_temp(:),xrd_sim_temp(:))

!     open(7,file='fort.7')
!     open(8,file='fort.8')
!     do i=1,size(xrd_sim(:))
!        write(7,*) th2(i),xrd_sim(i)
!        write(8,*) th2(i),xrd_expt(i)
!     enddo
!     close(7)
!     close(8)
!     write(*,*)'X-vector=',x
!     write(*,*)'fcn=',fcn
!     pause


        endif

        return
    end function fcn

!===============================================
    function rand(idum)
!
!   Random number generator
!   Using the numerical recipes code to generate double precision random numbers
!
!   Minimal random number generator of Park and Miller combined with a
!   Marsaglia shift sequence. Returns a uniform random deviate between 0.0
!   and 1.0 (exclusive of the endpoint values). This fully portable, scalar
!   generator has the traditional(not Fortran 90) calling sequence with a
!   random deviate as the returned function value: call with idum a negative
!   integer to initialize; thereafter, do not alter idum except to reinitialize.
!   The period of this generator is about 3.1x10^18.
!===============================================
        implicit none
        integer, parameter              :: K4B=selected_int_kind(9)
        integer(K4B), intent(inout)     :: idum
        double precision                :: rand


        integer(K4B), parameter :: IA=16807,IM=2147483647,IQ=127773,IR=2836
        double precision, save :: am
        integer(K4B), save :: ix=-1,iy=-1,k
        if (idum <= 0 .or. iy < 0) then    ! Initialize
          am  = nearest(1.0,-1.0)/IM
          iy  = ior(ieor(888889999,abs(idum)),1)
          ix  = ieor(777755555,abs(idum))
```

```
          idum = abs(idum)+1
        endif

        ix   = ieor(ix,ishft(ix,13))        ! Marsaglia shift sequence
        ix   = ieor(ix,ishft(ix,-17))
        ix   = ieor(ix,ishft(ix,5))
        k    = iy/IQ
        iy   = IA*(iy-k*IQ)-IR*k
        if (iy .lt. 0) iy=iy+IM
        rand = am*ior(iand(IM,ieor(ix,iy)),1)

    end function rand


!==================================================
    function linesinfile(cdum)
!==================================================

        implicit none
        character(LEN=*)     :: cdum
        integer              :: readstat, linesinfile
        character(2)         :: fdum

        linesinfile=0
        open(8675,file=cdum,status='old',IOSTAT=readstat)
           if (readstat .eq. 6) then
              linesinfile=-1
              return
           endif

           do while (readstat .ne. -1)
              read(8675,*,IOSTAT=readstat)fdum
              linesinfile=linesinfile+1
           enddo

           linesinfile=linesinfile-1
        close(8675)

    end function linesinfile

!==================================================
     subroutine avint ( xtab, ytab, a, b, result )
!
!    Rewrite of the AVINT subroutine to estimate the integral of unevenly spaced
!    data.
!
!    Table of X and Y values are in xtab and ytab.
!      * xtab must be monotonically increasing
!      * xtab and ytab must have the same size
!
!    Limits of integration are a and b
!==================================================

        implicit none

        integer              :: ntab, i, inlft, inrt, istart, istop
        double precision     :: a, b, ba, bb, bc, ca, cb, cc, fa, fb
        double precision     :: result, slope, syl, syl2, syl3, syu, syu2, syu3
        double precision     :: term1, term2, term3, total, x1, x12, x13, x2, x23, x3
        double precision     :: xtab(:), ytab(:)

        result = 0.0D+00
        ntab   = size(xtab)

```

```
      if ( a .eq. b ) then
         return
      endif

      if ( ntab .lt. 2 ) then
         write(*,*)
         write(*,*)'AVINT - Fatal error!'
         write(*,*)' NTAB is less than 3. NTAB = ',ntab
         stop
      endif

      do i = 2, ntab

         if ( xtab(i) .le. xtab(i-1) ) then
            write(*,*)
            write(*,*)'AVINT - Fatal error!'
            write(*,*)' XTAB is not monotonically increasing at i = ',i
            write(*,*)' XTAB(I-1) = ',xtab(i-1)
            write(*,*)' XTAB(I) =   ',xtab(i)
            stop
         endif

      enddo

      ! Special case for NTAB = 2.
      if ( ntab .eq. 2 ) then
         slope = ( ytab(2) - ytab(1) ) / ( xtab(2) - xtab(1) )
         fa    = ytab(1) + slope * ( a - xtab(1) )
         fb    = ytab(2) + slope * ( b - xtab(2) )
         result = 0.5D+00 * ( fa + fb ) * ( b - a )
         return
      endif

      if ( xtab(ntab-2) .lt. a .or. b .lt. xtab(3) ) then
         write (*,*) ' '
         write (*,*) 'AVINT - Fatal error!'
         write (*,*) ' There were less than 3 function values'
         write (*,*) ' between the limits of integration.'
         stop
      end if


      ! Find a and b in the xtab array
      i = 1
      do
         if ( a .le. xtab(i) ) then
            exit
         end if
         i = i + 1
      enddo

      inlft = i

      i = ntab
      do
         if ( xtab(i) .le. b ) then
            exit
         end if
         i = i - 1
      enddo
      inrt = i

      if ( inrt - inlft .lt. 2 ) then
         write ( *, '(a)' ) ' '
```

```
          write ( *, '(a)' ) 'AVINT - Fatal error!'
          write ( *, '(a)' ) '  There were less than 3 function values'
          write ( *, '(a)' ) '  between the limits of integration.'
          stop
       endif

       if ( inlft .eq. 1 ) then
          istart = 2
       else
          istart = inlft
       endif

       if ( inrt .eq. ntab ) then
          istop = ntab - 1
       else
          istop = inrt
       endif

       total = 0.0D+00

       syl  = a
       syl2 = syl  * syl
       syl3 = syl2 * syl

       do i = istart , istop

          x1 = xtab(i-1)
          x2 = xtab(i)
          x3 = xtab(i+1)

          x12 = x1 - x2
          x13 = x1 - x3
          x23 = x2 - x3

          term1 =   ( ytab(i-1) ) / ( x12 * x13 )
          term2 = - ( ytab(i)   ) / ( x12 * x23 )
          term3 =   ( ytab(i+1) ) / ( x13 * x23 )

          ba = term1 + term2 + term3
          bb = - ( x2 + x3 ) * term1 - ( x1 + x3 ) * term2 - ( x1 + x2 ) * term3
          bc = x2 * x3 * term1 + x1 * x3 * term2 + x1 * x2 * term3

          if ( i .eq. istart ) then
             ca = ba
             cb = bb
             cc = bc
          else
             ca = 0.5D+00 * ( ba + ca )
             cb = 0.5D+00 * ( bb + cb )
             cc = 0.5D+00 * ( bc + cc )
          endif

          syu  = x2
          syu2 = syu  * syu
          syu3 = syu2 * syu

          total = total + ca * ( syu3 - syl3 ) / 3.0D+00 &
                   + cb * ( syu2 - syl2 ) / 2.0D+00 &
                   + cc * ( syu  - syl )
          ca = ba
          cb = bb
          cc = bc

          syl  = syu
```

```
        syl2 = syu2
        syl3 = syu3

    enddo

    syu = b
    syu2 = syu * syu
    syu3 = syu2 * syu

    result = total + ca * ( syu3 - syl3 ) / 3.0D+00 &
             + cb * ( syu2 - syl2 ) / 2.0D+00 &
             + cc * ( syu  - syl  )

    return


  end subroutine avint


!==================================================
subroutine gen_xrd(peaks_in ,lambda, xsize)

! Generate XRD pattern of one phase. Inputs are peak locations , and relative heights
!==================================================

   implicit none

   double precision, intent(in)    :: peaks_in(:,:),lambda, xsize

   double precision                :: xsize_scherrer

   integer                         :: i,j,k, npeaks, ilarge
   double precision                :: fwhm, th2_rad , intres

   xsize_scherrer  = xsize * 3.1415926/180      ! Scherrer equation needs lamellar thickness in Angstrom-radians per degree
   npeaks          = size(peaks_in(:,1))
   xrd_sim_temp(:) = 0.0

   do i=1,npeaks
      do j=1,npoints

         th2_rad          = th2(j)*3.1415926/180.0
         fwhm             = 0.93 * lambda /( xsize_scherrer*cos(th2_rad/2))
         xrd_sim_temp(j)  = xrd_sim_temp(j) + peaks_in(i,2)*dexp(-0.693147180*(peaks_in(i,1) - th2(j))/(fwhm/2.0))**2)

      enddo
   enddo


   ! Normalize
   call avint(th2(:),xrd_sim_temp(:),th2(1),th2(npoints),intres)
   xrd_sim_temp(:) = xrd_sim_temp(:)/intres


!    open(999,file='xrd_sim.dat',status='replace')
!       do i=1,npoints
!          write(999,*)th2(i),xrd_sim_temp(i)
!       enddo
!    close(999)
!    pause

```

```
    return

end subroutine gen_xrd




end program ipp_simann
```

4.2. peaks-alpha

```
#2-th (deg)     Intensity
      14.1500000000          1.0000000000
      16.9000000000          0.6959000000
      18.5500000000          0.6516000000
      21.2500000000          0.2990000000
      21.8500000000          0.4306000000
      24.4500000000          0.0296000000
      25.2000000000          0.0326000000
      25.5000000000          0.0504000000
      27.1500000000          0.0489000000
      28.4500000000          0.0508000000
      28.5500000000          0.0475000000
      29.1000000000          0.0476000000
      29.2500000000          0.0209000000
      29.7500000000          0.0168000000
      29.8750000000          0.0149000000
      32.7000000000          0.0172000000
      32.8500000000          0.0171000000
      33.3000000000          0.0397000000
      33.4000000000          0.0154000000
      33.6750000000          0.0414000000
      34.9000000000          0.0611000000
      35.8500000000          0.0112000000
      36.9000000000          0.0253000000
      37.0000000000          0.0199000000
      37.5500000000          0.0100000000
      38.5750000000          0.0216000000
      41.1000000000          0.0137000000
      41.4500000000          0.0107000000
      42.4000000000          0.0172000000
      42.5000000000          0.0512000000
      42.6000000000          0.0245000000
      42.8000000000          0.0238000000
      42.9000000000          0.0140000000
      43.1500000000          0.0408000000
      43.2500000000          0.0288000000
      44.2000000000          0.0160000000
      44.3000000000          0.0211000000
      44.5000000000          0.0108000000
      44.6000000000          0.0151000000
```

4.3. peaks-beta

```
#2-th (deg)     Intensity
      16.0500000000          1.0000000000
      16.5000000000          0.0252000000
      21.1000000000          0.3485000000
      23.1000000000          0.0713000000
      24.6500000000          0.0484000000
      28.0000000000          0.0608000000
```

```
28.2500000000      0.0275000000
29.0000000000      0.0209000000
31.2500000000      0.0755000000
31.3500000000      0.0308000000
31.9500000000      0.0109000000
35.3500000000      0.0162000000
36.6000000000      0.0201000000
42.7500000000      0.0271000000
42.8500000000      0.0109000000
43.5500000000      0.0196000000
```

4.4. peaks-gamma

```
#2-th (deg)     Intensity
 8.3400000000      0.0102040820
13.8400000000      1.0000000000
15.0500000000      0.1428571400
16.7200000000      0.7959183700
17.2300000000      0.1224489800
18.3500000000      0.0000000000
20.0700000000      0.7551020400
21.2200000000      0.3877551000
21.8800000000      0.6734693900
23.3500000000      0.0000000000
24.3500000000      0.0510204080
25.2000000000      0.1122449000
26.9500000000      0.0102040820
27.5500000000      0.0051020408
27.6600000000      0.0051020408
27.8800000000      0.0102040820
28.8300000000      0.1224489800
29.0100000000      0.0306122450
29.6300000000      0.0000000000
29.6900000000      0.0000000000
30.3700000000      0.0102040820
33.3500000000      0.0102040820
34.6200000000      0.0135714290
34.6900000000      0.0135714290
34.7500000000      0.0135714290
```

4.5. experiment.dat

```
10.1328000000      0.0186026000
10.5221000000      0.0216175000
10.9538000000      0.0199642000
11.3860000000      0.0214163000
11.7750000000      0.0228785000
12.2072000000      0.0243306000
12.5968000000      0.0288983000
12.9866000000      0.0350186000
13.2050000000      0.0489429000
13.3370000000      0.0628873000
13.4261000000      0.0783945000
13.5151000000      0.0939018000
13.6039000000      0.1078560000
13.6930000000      0.1233630000
13.7821000000      0.1388710000
13.9144000000      0.1543680000
14.1303000000      0.1543170000
14.1717000000      0.1434380000
14.2127000000      0.1310060000
14.2540000000      0.1201270000
14.2956000000      0.1108010000
```

14.3793000000	0.0952533000
14.4630000000	0.0797058000
14.5470000000	0.0657111000
14.8037000000	0.0516761000
15.1480000000	0.0438320000
15.5356000000	0.0375305000
15.9681000000	0.0405354000
16.2716000000	0.0466759000
16.4470000000	0.0621630000
16.4926000000	0.0761275000
16.5817000000	0.0916347000
16.6287000000	0.1133630000
16.6708000000	0.1071420000
16.7601000000	0.1242020000
16.8025000000	0.1195340000
16.8865000000	0.1055390000
16.9270000000	0.0900015000
17.0107000000	0.0744540000
17.0944000000	0.0589066000
17.3080000000	0.0448817000
17.6530000000	0.0416957000
18.0415000000	0.0400525000
18.3889000000	0.0508411000
18.6057000000	0.0554490000
18.9062000000	0.0445094000
19.2949000000	0.0444189000
19.6418000000	0.0521021000
19.9885000000	0.0582325000
20.1178000000	0.0566496000
20.4617000000	0.0472527000
20.7652000000	0.0533932000
20.9404000000	0.0673276000
21.1156000000	0.0812619000
21.3734000000	0.0734379000
21.7200000000	0.0795683000
21.8072000000	0.0842064000
21.9786000000	0.0764025000
22.1058000000	0.0623977000
22.2762000000	0.0483829000
22.4465000000	0.0343680000
22.5748000000	0.0265742000
22.7905000000	0.0249712000
23.1787000000	0.0217751000
23.6104000000	0.0201218000
24.0423000000	0.0200212000
24.2154000000	0.0215337000
24.4743000000	0.0199206000
24.8628000000	0.0182773000
25.1227000000	0.0228751000
25.2958000000	0.0243876000
25.4680000000	0.0212419000
25.7264000000	0.0165234000
25.8554000000	0.0133877000
26.1581000000	0.0148700000
26.4172000000	0.0148096000
26.6332000000	0.0147593000
26.8492000000	0.0147090000
27.2817000000	0.0177139000
27.5403000000	0.0145481000
27.8864000000	0.0175730000
28.3186000000	0.0190252000
28.7082000000	0.0235928000
29.1396000000	0.0203867000
29.5278000000	0.0171907000

```
29.9160000000              0.0139947000
```

4.6. `random.seed`

```
81652
```

References

[1] J. R. Isasi, J. A. Haigh, J. T. Graham, L. Mandelkern, and R. G. Alamo. *Polymer*, **41**(25):8813–8823, December 2000.

[2] G R Strobl and M Schneider. *Journal of Polymer Science Part B: Polymer Physics*, **18**:1343, 1980.

[3] L. Mandelkern. *Polymer Journal*, **17**:337, 1985.

[4] G. Farrow and I. M. Ward. *Polymer*, **1**:330, 1960.

[5] Yasuhiro Takahashi and Hiroyuki Tadokoro. *Macromolecules*, **6**(5):672–675, September 1973.

[6] P. Lightfoot, M. A. Mehta, and P. G. Bruce. *Science*, **262**(5135):883–885, November 1993.

[7] B. Lotz and J. C. Wittmann. *Polymer*, **37**:4979, 1996.

[8] D. Dorset, M. McCourt, S. Kopp, M. Schumacher, T. Okihara, and B. Lotz. *Polymer*, **39**:6331–6337, 1998.

[9] Stefano V. Meille, S. Bruckner, and W. Porzio. *Macromolecules*, **23**:4114–4121, 1990.

[10] L Fontana, D Q Vinh, M Santoro, S Scandolo, F A Gorelli, R Bini, and M Hanfland. *Physical Review B*, **75**:174112, 2007.

[11] Koji Nishida, Takashi Konishi, Toshiji Kanaya, and Keisuke Kaji. *Polymer*, **45**:1433–1437.

[12] J L Way and J R Atkinson. *Journal of Materials Science*, **6**(2):102–109, 1971.

[13] T. J. Bessell, D. Hull, and J. B. Shortall. *Journal of Materials Science*, **10**(7):1127–1136, July 1975.

[14] J. M. Schultz. *Polymer Engineering & Science*, **24**(10):770–785, 1984.

[15] J T Yeh and J Runt. *Journal of Materials Science*, **24**:2637–2642, 1989.

[16] H. Sue, J. Earls, and R. Hefner. *Journal of Materials Science*, **32**(15):4031–4037, August 1997.

[17] Sang D. Park, Mitsugu Todo, Kazuo Arakawa, and Masaaki Koganemaru. *Polymer*, **47**(4):1357–1363, February 2006.

[18] S. Grzybowski, P. Zubielik, and E. Kuffel. *IEEE Transactions on Power Delivery*, **4**(3):1507–1512, 1989.

[19] T. Tanaka. *IEEE Transactions on Electrical Insulation*, **27**(3):424–431, 1992.

[20] Y. Zhou, N. H. Wang, P. Yan, X. D. Liang, and Z. C. Guan. *Journal of Electrostatics*, **57**(3-4):381–388, March 2003.

[21] T. Tanaka, G. C. Montanari, and R. Mulhaupt. *IEEE Transactions on Dielectrics and Electrical Insulation*, **11**(5):763–784, 2004.

[22] T. Nakagawa, T. Nakiri, R. Hosoya, and Y. Tajitsu. *IEEE Transactions on Industry Applications*, **40**(4):1020–1024, 2004.

[23] Zhiyong Xia, Hung Jue Sue, Zhigang Wang, Carlos A Avila-Orta, and Benjamin S. Hsiao. *Journal of Macromolecular Science, Part B: Physics*, **40**:625–638, 2001.

[24] Z. G. Wang, B. S. Hsiao, B. X. Fu, L. Liu, F. Yeh, B. B. Sauer, H. Chang, and J. M. Schultz. *Polymer*, **41**(5):1791–1797, 2000.

[25] Carla Marega, Antonio Marigo, Gianmatteo Cingano, Roberto Zannetti, and Guglielmo Paganetto. *Polymer*, **37**(25):5549–5557, 1996.

[26] Rajesh H. Somani, Benjamin S. Hsiao, Aurora Nogales, Srivatsan Srinivas, Andy H. Tsou, Igors Sics, Francisco J. Balta-Calleja, and Tiberio A. Ezquerra. *Macromolecules*, **33**(25):9385–9394, 2000.

[27] Kyo Jin Ihn, Masaki Tsuji, Akiyoshi Kawaguchi, and Kenichi Katayama. *Bulletin of the Institute for Chemical Research, Kyoto University*, **68**:30–40, 1990.

[28] B H Sohn and S H Yun. *Polymer*, **43**:2507–2512), 2002.

[29] D Trifonova, J Varga, and G J Vancso. *Polymer Bulletin*, **41**:341–348, 1998.

[30] Lin Li, Chi Ming Chan, King Lun Yeung, Jian Xiong Li, Kai Mo Ng, and Yuguo Lei. *Macromolecules*, **34**:316–325, 2001.

[31] Günter Reiter, Gilles Castelein, and Jens Uwe Sommer. *Physical Review Letters*, **86**:5918–5921, 2000.

[32] Hongyi Zhou and G. L. Wilkes. *Polymer*, **38**(23):5735–5747, 1997.

[33] K. Mezghani. *Polymer*, **38**:5725–5733, 1997.

[34] Khaled Mezghani and Paul J. Phillips. *Polymer*, **39**:3735–3744, 1998.

[35] W Ruland. *Acta Crystallographica*, **14**:1180, 1961.

[36] W Ruland. *Polymer*, **5**:89, 1964.

[37] C. G. Vonk. *Journal of Applied Crystallography*, **6**:81, 1973.

[38] S. Kavesh and J. M. Schultz. *Journal of Polymer Science Part A-2*, **9**:85, 1971.

[39] José R. Isasi, Leo Mandelkern, María J. Galante, and Rufina G. Alamo. *Journal of Polymer Science Part B: Polymer Physics*, **37**(4):323–334, 1999.

[40] A. T. Jones, J. M. Aizlewood, and D. R. Becket. *Makromolekular Chemie*, **75**:134, 1964.

[41] P Scherrer. *Göttinger Nachrichten Gesellschaft*, **2**:98, 1918.

[42] A. L. Patterson. *Physical Review*, **56**:978–982, 1939.

[43] B D Cullity and S R Stock. *Elements of X-Ray Diffraction*. Pearson Education, Limited, 2001.

[44] H. P. Klug and Alexander L. E. *X-Ray Diffraction Procedures*. Wiley, 1974.

[45] A Sanchez-Bajo and F. L. Cumbrera. *Journal of Applied Crystallography*, **30**:427–430, 1997.

[46] H. Wang and J. Zhou. *Journal of Applied Crystallography*, **38**:830–832, 2005.

[47] Gopinath Subramanian and Rahmi Ozisik. Simultaneous estimation of the phase content and lamellar thickness in isotactic polypropylene by the simulated annealing of wide-angle x-ray scattering data. *Journal of Applied Polymer Science*, **117**(4):2386–2394, 2010.

[48] S. Kirkpatrick, C. D. Gelatt, and M. P. Vecchi. *Science*, **220**:671–680, 1983.

[49] R Krache, R Benavente, J M Lopez-Majada, J M Perena, M L Cerrada, and E. Perezet. *Macromolecules*, **40**(19):6871, 2007.

[50] E Nedkov and T. Dobreva. *European Polymer Journal*, **40**.

[51] K Busse and J Kressler. *Macromolecules*, **33**:8775, 2000.

[52] E J Addink and J Bientema. *Polymer*, **2**:185, 1961.

[53] Jan Broda. *Journal of Applied Polymer Science*, **89**(12):3364–3370, 2003.

[54] J. Broda. *Polymer*, **44**(5):1619–1629, March 2003.

[55] Jan Broda. *Crystal Growth & Design*, **4**(6):1277–1282, November 2004.

[56] R. J. Davies, N. E. Zafeiropoulos, K. Schneider, S. V. Roth, M. Burghammer, C. Riekel, J. C. Kotek, and M. Stamm. *Colloid & Polymer Science*, **282**(8):854–866, June 2004.

[57] H. Awaya. *Polymer*, **26**:591, 1988.

[58] Stefano Meille and Sergio Bruckner. *Nature*, **340**(6233):455–457, 1989.

[59] R. A. Campbell and P. J. Philips. *Polymer*, **34**:4809, 1993.

[60] J. Suhm. *Journal of Materials Chemistry*, **8**:553, 1998.

[61] http://accelrys.com/products/materials studio/.

In: X-Ray Scattering
Editor: Christopher M. Bauwens
ISBN: 978-1-61324-326-8

Chapter 7

DEPTH PROFILE ANALYSIS OF SURFACE LAYER STRUCTURE USING X-RAY DIFFRACTION AND X-RAY REFLECTIVITY AT SMALL GLANCING ANGLES OF INCIDENCE

Yoshikazu Fujii*
Kobe University, Kobe, Japan

ABSTRACT

In this chapter, two recent analysis method of depth profile of surface layer structure using x-ray scattering at small glancing angles of incidence are presented. In the first, the analysis using x-ray diffraction is presented, and next, the analysis using x-ray reflectivity is presented. In the first, the analysis method for the depth profile of poly-crystalline surface using x-ray diffraction at small glancing angles of incidence is discussed. The peak profiles of the diffracted x-ray intensity are investigated as function of incidence angles at small glancing angle incidence of x-rays on a poly-crystalline surface. The intensity of x-ray propagation in surface layer materials characterized by a complex refractive index that changes continuously with depth is derived, and with use of the result, an analyzing method for evaluating the structure of the poly-crystalline layers and the distribution of crystal grain size in the surface layers is studied. The dependence of the diffracted x-ray intensities on the glancing angle lead to the depth profile of the poly-crystalline layers of oxidized iron with accuracy of the order of nanometers. The dependence of the peak width of diffracted x-rays on the glancing angle lead to the depth profile of the crystal grain size of iron in the surface layer. The dependence of the diffraction angle of diffracted x-rays on the glancing angle lead to the depth profile of the strain distribution in the surface layer. The derived analyzing method can be applied to the residual stress distribution analysis of the surface layer materials of which densities change continuously in depth as multi thin films, compound plating layers. Next, recent analysis of depth profile of surface layer using x-ray reflectivity is discussed. In the previous study, the x-ray reflectivity has been calculated based on the Parratt formalism, accounting for the effect of roughness by the theory of Nevot-Croce. However, the

* E-mail address: fujiiyos@kobe-u.ac.jp

calculated result showed a strange phenomenon in that the amplitude of the oscillation due to interference effects increases in the case of a specific roughness of the surface. The strange result had its origin in a used equation due to a serious mistake in which the Fresnel transmission coefficient in the reflectivity equation is increased at a rough interface, and the increase in the transmission coefficient completely overpowers any decrease in the value of the reflection coefficient because of a lack of consideration of diffuse scattering. The mistake in Nevot and Croce's treatment originates in the fact that the modified Fresnel coefficients were calculated based on the theory which contains the x-ray energy conservation rule at surface and interface. In this chapter, a new accurate formalism that corrects this mistake is presented. The new accurate formalism derives an accurate analysis of the x-ray reflectivity from a multilayer surface, taking into account the effect of roughness-induced diffuse scattering. The calculated reflectivity by this accurate reflectivity equation should enable the structure of buried interfaces to be analyzed more accurately.

1. Introduction

X-rays scattered from a material surface at a glancing angle provide useful information on the structure of the surface layer.[1-13] When x-rays are applied to a material surface at a small glancing angle of incidence, the intensity of x-rays scattered on the surface is the sum of the x-rays that scattered by atoms only on the surface, and the contribution of the atoms at each depth to the x-rays intensity varies with the glancing angle, so the depth profile of the surface layers can be found by analyzing the incidence angle dependence of the scattered x-rays.

The peak profiles of the diffracted x-ray intensity are investigated as function of incidence angles at small glancing angle incidence of x-rays on a poly-crystalline surface. The intensity of x-ray propagation in surface layer materials characterized by a complex refractive index that changes continuously with depth is derived, and with use of the result, an analyzing method for evaluating the structure of the poly-crystalline layers and the distribution of crystal grain size in the surface layers is studied. The dependence of the diffracted x-ray intensities on the glancing angle lead to the depth profile of the poly-crystalline layers of oxidized iron with accuracy of the order of nanometers. The dependence of the peak width of diffracted x-rays on the glancing angle lead to the depth profile of the crystal grain size of iron in the surface layer. The dependence of the diffraction angle of diffracted x-rays on the glancing angle lead to the depth profile of the strain distribution in the surface layer.

The depth profile of the surface layers can be found by analyzing the incidence angle dependence of the scattered x-rays[1-6] and x-ray reflectometry has become a powerful tool for investigations on rough surface and interface structures.[7,8] When a x-ray beam is incident at a glancing angle on a rough surface, coherent scattering gives rise to a specularly reflected beam and a single transmitted refracted beam, whereas incoherent scattering causes diffusely scattered radiation. In the calculation of x-ray reflectivity, most attention is paid to coherent scattering. In addition, several theories exist to describe the influence of roughness on x-ray scattering.[9-10] In many previous studies, the x-ray reflectivity was calculated based on the Parratt formalism[1], coupled with the use of the theory of Nevot and Croce to include roughness.[2] However, the calculated results of the x-ray reflectivity done in this way often showed strange results where the amplitude of the oscillation due to the interference effects would increase for a rougher surface. This strange result disagrees with reality and suggests that there is a problem in how the effect of roughness is incorporated into the Parratt formalism, and hence that the effect of roughness is not accurately represented in the theoretical formula of the x-ray reflectivity. Because the x-ray scattering vector in a specular reflectivity measurement is normal to the surface, it provides the density profile solely in the direction perpendicular to surface. Specular reflectivity measurements can yield the magnitude of the average roughness perpendicular to surface and interfaces, but cannot give information about the lateral extent of the roughness. In previous studies, the effect of roughness on the calculation of the x-ray reflectivity only took into account the effect of the density changes of the medium in a direction normal to the surface and interface. On the other hand, diffuse scattering can provide information about the lateral extent of the roughness. In contrast to previous calculations of the x-ray reflectivity, in the present analysis, the effect of a decrease in the intensity of penetrated x-rays due to diffuse scattering at a rough surface and rough interface is considered. In this chapter, a new accurate formalism that corrects this mistake is presented. The new accurate formalism derives an accurate analysis of the x-ray reflectivity from a multilayer surface, taking into account the effect of roughness-induced diffuse scattering. The calculated reflectivity by this accurate reflectivity equation should enable the structure of buried interfaces to be analyzed more accurately.

2. Depth Profile Analysis of Surface Layer Using X-Ray Diffraction at Small Glancing Angle of Incidence

An analysing method for the depth profile of poly-crystalline surface using x-ray diffraction at small glancing angles of incidence is discussed. The peak profiles of the diffracted x-ray intensity are investigated as function of incidence angles at small glancing angle incidence of x-rays on a poly-crystalline surface. The intensity of x-ray propagation in surface layer materials characterized by a complex refractive index that changes continuously with depth is derived, and with use of the result, the structure of the poly-crystalline layers and the distribution of crystal grain size in the surface layers are evaluated.

2.1. Experimental of X-Ray Diffraction at Small Glancing Angle of Incidence

The intensities of the diffracted x-rays on a polycrystalline iron surface at the various small glancing angles of incidence are measured. The polycrystalline surface of pure iron (ferrite, 99.9% purity) was mechanically polished and annealed at 400°C for 5 minutes. The pre-heating forms a thin amorphous oxide layer on the iron surface. Small glancing angle x-ray scattering on polycrystalline iron surface was measured by using synchrotron radiation at Hyogo Prefecture beam-line BL24 of SPring-8. X-rays (10keV) from a standard in-vacuum undulator are monochromatized by a double-crystal monochromator in the preceding optics hutch and then are introduced into the experimental hutch. Figure 1 shows a schematic view of the experimental arrangement. The x-ray beam irradiates upon a specimen surface at a glancing angle α through slits 1 and 2, as shown in Fig. 1. The cross-section of the x-ray beam behind the slits is 0.1mm wide and 0.8mm high. Scattered x-rays from the surface at exit angle β from 0° to 0.4° are received in θ direction by slits 3 and then are received by a scintillation counter. The acceptance angle of the scattered x-ray beams limited by slits 3 is 1 mrad in the θ direction and 7 mrad in the β direction as shown in Fig. 1. The irradiated surface region by the x-rays is from 0.8mm×3mm to 0.8mm×14mm, which is changed by a glancing angle α.

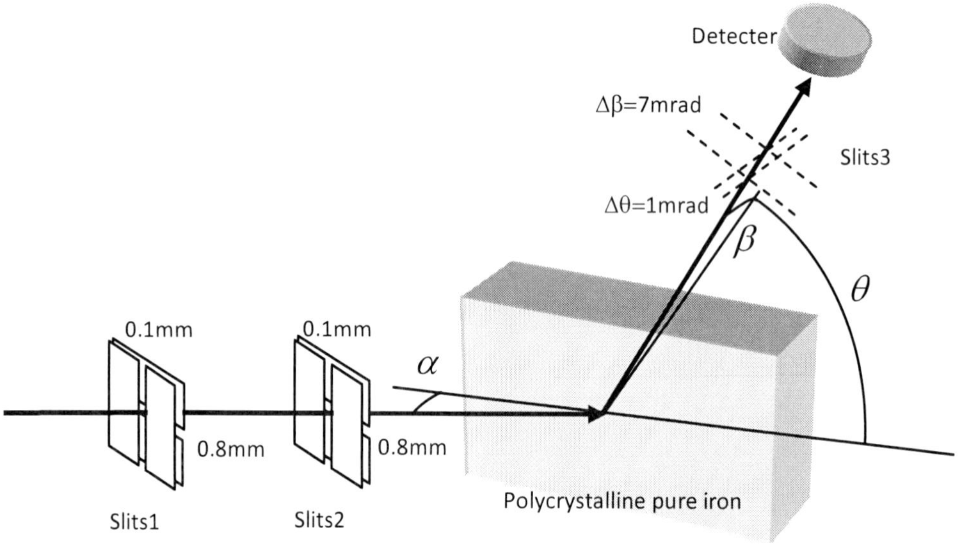

Figure 1. The schematic view of the experimental arrangement.

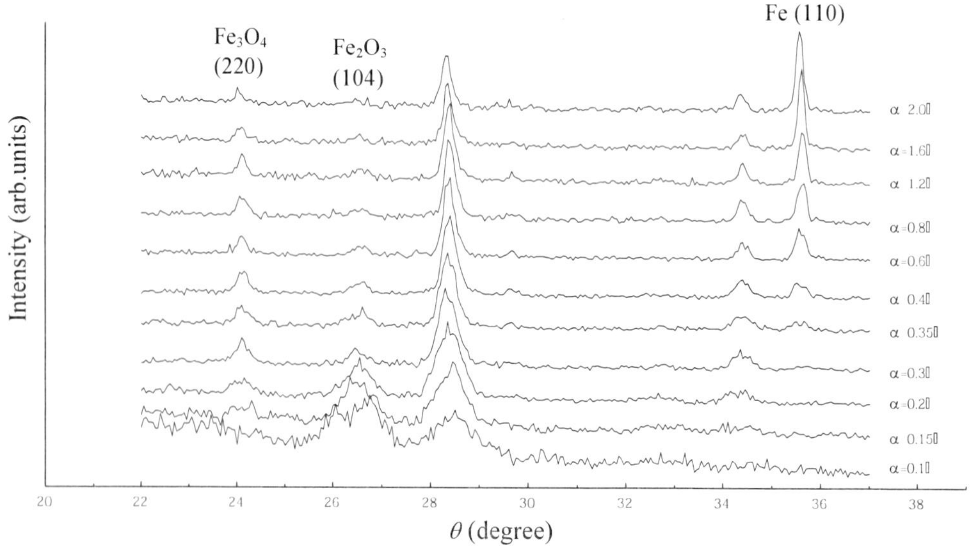

Figure 2. Diffraction patterns at several angles of incidence.

Angular distributions of the scattered x-ray intensities were measured at several glancing angles of incidence. Figure 2 shows x-ray diffraction patterns at several angles of incidence. Next, peak intensities of diffracted x-rays are derived by gauss fitting of the experimental results of the peak profile with correction for a broadening due to acceptance angle. Figure 3 shows a change in the peak intensities corresponding to Fe_2O_3, Fe_3O_4, and Fe as a function of incidence angle: the peak intensity of Fe_2O_3 is high at 0.2° of incidence and low at larger incidence angles; the peak intensity of Fe_3O_4 is high at 0.3° and decreases at larger incidence angles; the peak intensity of Fe increases at a larger incidence angle. Analyzing these intensity profiles leads to the contribution of the crystals of each depth under the surface.

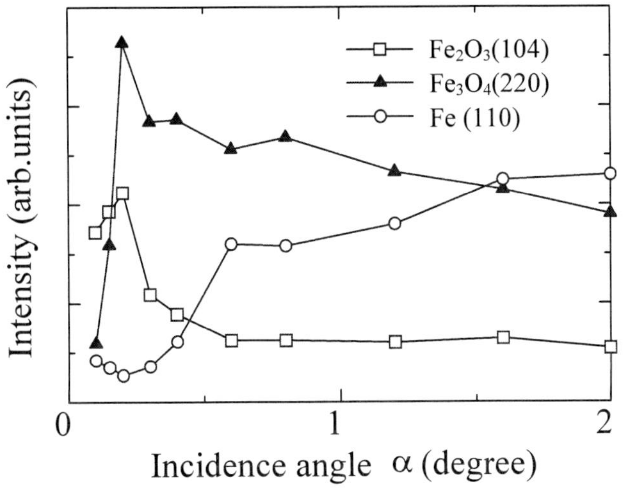

Figure 3. The peak intensities of diffracted x-rays as a function of the angle of incidence.

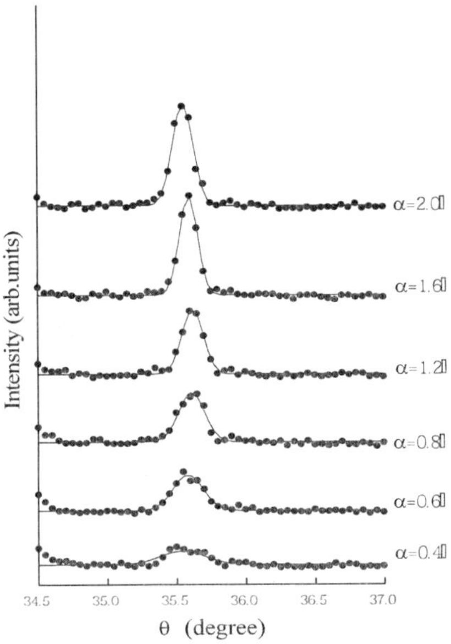

Figure 4. Peak profiles of diffracted x-rays from Fe(110) at several angles of incidence.

Figure 4 shows x-ray diffraction peak profiles from Fe(110) at several angles of incidence. Small shift of peak position by changing the incident angle can be seen in Fig. 4. This is the shift by deviation between θ and diffraction angle θ_d as shown in the equation; $\cos\theta_d = \cos\theta\cos(\alpha'+\beta)$, where α' is the refracted angle passing below the surface as approximately shown in the equation; $\alpha' \cong \sqrt{\alpha^2 - \alpha_c^2}$, and the direction of the refracted x-ray is $(\cos\alpha', 0, -\sin\alpha')$, and the direction of the detected x-ray is $(\cos\theta\cos\beta, \sin\theta, \cos\theta\sin\beta)$, where the critical angle α_c to the total external reflection is about 0.4.

The peak width δθ of diffracted x-rays as a function of incident angle is shown in Figure 5. The experimental results of the peak width δθ of diffracted x-rays are broaden due to acceptance angle; $[(\alpha'+\beta)/\tan\theta]\Delta\beta + \Delta\theta$, where β=3.5mrad, Δβ=7mrad and Δθ=1mrad. The peak width $\delta\theta_d$ of diffracted x-rays shows the data of which a broadening due to acceptance angle was corrected as the following, $\delta\theta_d = \sqrt{\delta\theta^2 - \{[(\alpha'+\beta)/\tan\theta]\Delta\beta + \Delta\theta\}^2}$. The peak width is corresponding to the size of Fe crystal grain in the surface layer. Wide peak at smaller incidence angle show that Fe crystal grain is smaller nearby the surface, because the contribution from the surface top layer is large at a smaller glancing angle. Analyzing these peak profiles leads to the depth profile of the crystal grain size under the surface.

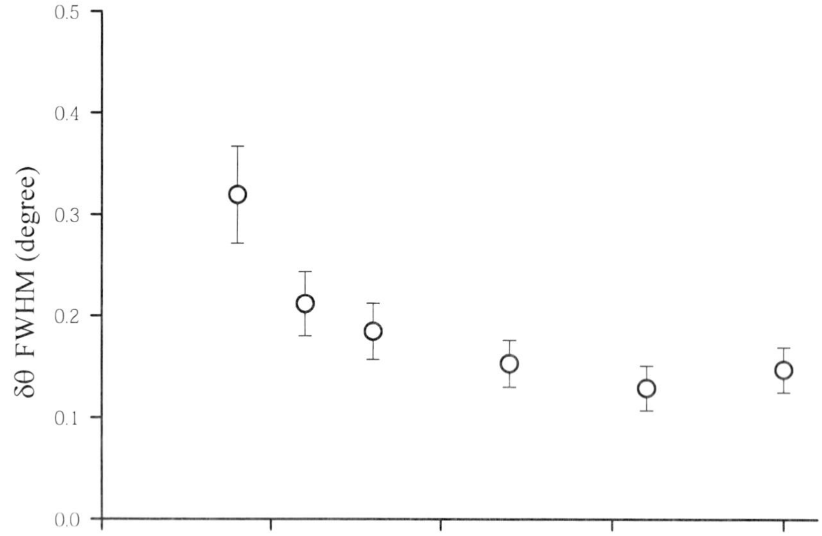

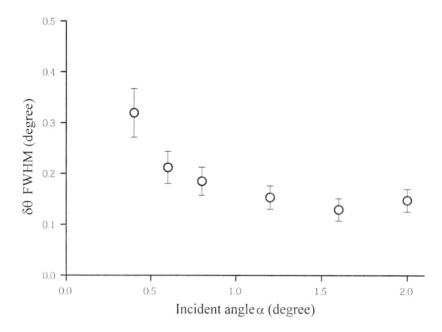

Figure 5. FWHM of the diffracted x-ray peaks from Fe(110) at several glancing angles.

2.2. Analysis of the Intensity of X-Rays Propagating in the Material

The intensity of x-rays, *i.e.*, the electric and magnetic field propagating in the material, can be obtained using Maxwell's equations.[14] The effects of x-rays on the material are characterized by complex refractive index n, which changes continuously with depth. The material for which the density changes continuously with depth is divided into N layers.

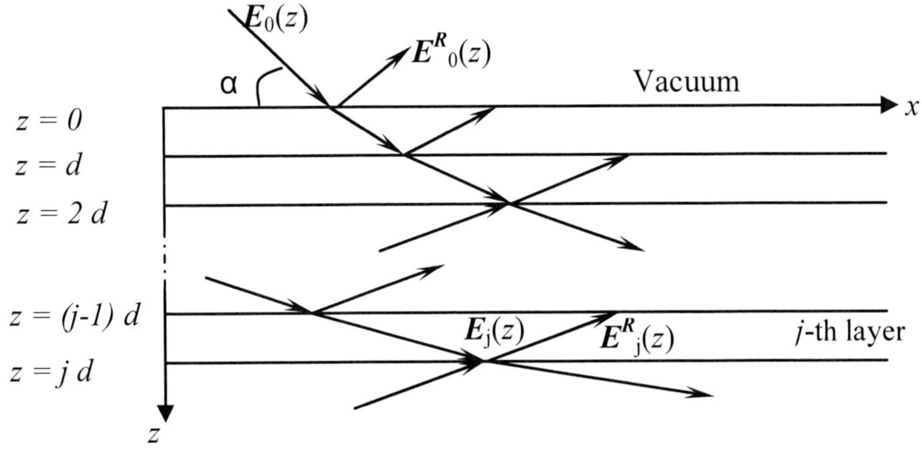

Figure 6. Multilayer representation for calculation of refractive x-ray intensity

Figure 6 is a multilayer representation of the calculation of refractive x-ray intensity. The vacuum is denoted as $j = 0$ and $n_0 = 1$. The thickness of one layer is d and only the thickness of the bottom layer is assumed to be infinite.

The reflectance of multiplayer system, consisting of N layers can be calculated using the recursive formalism given by Parratt.[1] Let n_j be the refractive index of the j-th layer, defined as

$$n_j = 1 - \delta_j - i\beta_j , \qquad (1)$$

where δ_j and β_j are the real and imaginary parts of the refractive index. These optical constants related to the atomic scattering factor and electron density of the j-th layer material. For x-rays of wavelength λ, the optical constants of the j-th layer material consisting of N_{ij} atoms per unit volume can be expressed as

$$\delta_j = \frac{\lambda^2 r_e}{2\pi} \sum_i f_{1i} N_{ij} , \quad \beta_j = \frac{\lambda^2 r_e}{2\pi} \sum_i f_{2i} N_{ij} , \qquad (2)$$

where r_e is the classical electron radius and f_{1i} and f_{2i} are the real and imaginary parts of the atomic scattering factor of the i-th element atom, respectively.

The electric field of x-ray radiation at glancing angle of incidence α is expressed as

$$\boldsymbol{E}_0(z) = \boldsymbol{A}_0 \exp[i(\boldsymbol{k}_0 \cdot \boldsymbol{r} - \omega t)] . \qquad (3)$$

Incident radiation is usually decomposed into two geometries to simplify the analysis, one with incident electric field $\boldsymbol{E}$ parallel to the plane of incidence (p-polarization) and one with $\boldsymbol{E}$ perpendicular to that plane (s-polarization). An arbitrary incident wave can be represented in terms of these two polarizations. E_{0x} and E_{0z} are p-polarization, and E_{0y} is s-polarization, and those components of the amplitude's electric vector are expressed as

$$A_{0x} = -A_{0p} \sin\alpha \ , \ A_{0y} = A_{0s} \ , \ A_{0z} = A_{0p} \cos\alpha \ . \tag{4}$$

The components of wave vector of incidence x-rays are

$$k_{0x} = \frac{2\pi}{\lambda}\cos\alpha \ , \ k_{0y} = 0 \ , \ k_{0z} = \frac{2\pi}{\lambda}\sin\alpha \ . \tag{5}$$

The electric field of x-rays propagating in the j-th layer material is shown as

$$\boldsymbol{E}_j(z) = \boldsymbol{A}_j \exp[\, i(\boldsymbol{k}_j \cdot \boldsymbol{r} - k_{jz}(j-1)d - \omega t)] \ . \tag{6}$$

The amplitude $\boldsymbol{A}_j$ of j–th layer is derived from continuity equations for the interface between the j-1 and j layer as shown by

$$\boldsymbol{A}_1 = \boldsymbol{\Phi}_1 \boldsymbol{A}_0 \ , \ \boldsymbol{A}_j = \boldsymbol{\Phi}_j \boldsymbol{A}_{j-1} \exp(ik_{j-1,z}d), \tag{7}$$

the Fresnel coefficient tensor for refraction on the interface between j-1 and j layer is given by

$$\begin{aligned}
\boldsymbol{\Phi}_{j,xx} &= \frac{2k_{j,z}\boldsymbol{k}_{j-1}\cdot\boldsymbol{k}_{j-1}}{k_{j-1,z}\boldsymbol{k}_j\cdot\boldsymbol{k}_j + k_{j,z}\boldsymbol{k}_{j-1}\cdot\boldsymbol{k}_{j-1}} \\
\boldsymbol{\Phi}_{j,yy} &= \frac{2k_{j-1,z}}{k_{j-1,z} + k_{j,z}} \\
\boldsymbol{\Phi}_{j,zz} &= \frac{2k_{j-1,z}\boldsymbol{k}_{j-1}\cdot\boldsymbol{k}_{j-1}}{k_{j-1,z}\boldsymbol{k}_j\cdot\boldsymbol{k}_j + k_{j,z}\boldsymbol{k}_{j-1}\cdot\boldsymbol{k}_{j-1}} \\
\boldsymbol{\Phi}_{j,xy} &= \boldsymbol{\Phi}_{j,yx} = \boldsymbol{\Phi}_{j,yz} = \boldsymbol{\Phi}_{j,zy} = \boldsymbol{\Phi}_{j,zx} = \boldsymbol{\Phi}_{j,xz} = 0
\end{aligned} \ . \tag{8}$$

The wave vector $\boldsymbol{k}_j$ of the j-th layer is related to the refractive index n_j of the j-th layer as shown in the following

$$\frac{\boldsymbol{k}_j\cdot\boldsymbol{k}_j}{n_j^2} = \frac{\omega^2}{c^2} = const \tag{9}$$

and the x,y-direction components of wave vector are constant, then the z-direction component of wave vector of the j-th layer is shown as

$$k_{j,z} = \sqrt{n_j^2\boldsymbol{k}_0\cdot\boldsymbol{k}_0 - k_{0,x}^2} \ , \tag{10}$$

and the Fresnel coefficient for refraction on the interface between j-1 and j layer is given by

$$\begin{aligned}\boldsymbol{\Phi}_{j,xx} &= \frac{2k_{j,z}n_{j-1}^{2}}{k_{j-1,z}n_{j}^{2}+k_{j,z}n_{j-1}^{2}}\\ \boldsymbol{\Phi}_{j,yy} &= \frac{2k_{j-1,z}}{k_{j-1,z}+k_{j,z}}\\ \boldsymbol{\Phi}_{j,zz} &= \frac{2k_{j-1,z}n_{j-1}^{2}}{k_{j-1,z}n_{j}^{2}+k_{j,z}n_{j-1}^{2}}\end{aligned} \quad . \tag{11}$$

By the condition of incident x-ray, the z-direction component of wave vector of the *j*-th layer is shown as

$$k_{j,z} = \frac{2\pi}{\lambda}\sqrt{n_j^{\ 2} - \cos^2\alpha} = \frac{2\pi}{\lambda}\left(a_j + ib_j\right) . \tag{12}$$

Using these equations, the electric field of x-ray propagating in *j*-th layer is expressed as

$$\boldsymbol{E}_j(z) = (\prod_{J=1}^{j}\boldsymbol{\Phi}_J)\boldsymbol{A}_0 \exp[i(k_{0,x}x + d\sum_{J=1}^{j-1}k_{J,z} + k_{j,z}(z-(j-1)d) - \omega t)] \ , \tag{13}$$

and the intensity of the refractive x-ray in j-th layer at depth z is shown as

$$I(z) = \left|(\prod_{J=1}^{j}\boldsymbol{\Phi}_J)\boldsymbol{A}_0^{*}(\prod_{J=1}^{j}\boldsymbol{\Phi}_J)\boldsymbol{A}_0\right| \cdot \exp\{ -d\frac{4\pi}{\lambda}\sum_{J=1}^{j-1}b_J - \frac{4\pi}{\lambda}b_j(z-(j-1)d) \} \cdot \tag{14}$$

2.3. Depth Profiling of Poly-Crystalline Layers Structure

We assume that the diffraction peak intensities corresponding to Fe_2O_3, Fe_3O_4 and Fe are in proportion to the sum of the refractive x-ray intensity in each layer of the polycrystalline specimen, as shown in the following equation

$$\mathrm{I}_M = \int_0^{\infty} \rho_M(z) I(z)\,dz \ , \tag{15}$$

where $\rho_M(z)$ is the density of the polycrystalline M at the depth z. The surface and interface of the polycrystalline specimen have a roughness characterized with the root mean square deviation σ of the surface or interface with respect to a flat surface or interface. The structural diagram of polycrystalline oxide layers on the iron surface is shown in Fig. 7. A schematic diagram of the depth profile of the density of the polycrystalline specimen is shown in Fig. 8. The density of the polycrystalline M at depth z_1 to z_1+D with interface roughness σ_1 andσ_2 and thickness of D is expressed as

$$\rho_M(z) = \int_{-\infty}^{z} f(x-z_1,\sigma_1)dz \cdot \left\{1 - \int_{-\infty}^{z} f(x-z_1-D,\sigma_2)dz\right\} , \tag{16}$$

$$f(x,\sigma)=\frac{1}{\sqrt{2\pi}\sigma}\exp(-\frac{x^2}{2\sigma^2})\ . \qquad (17)$$

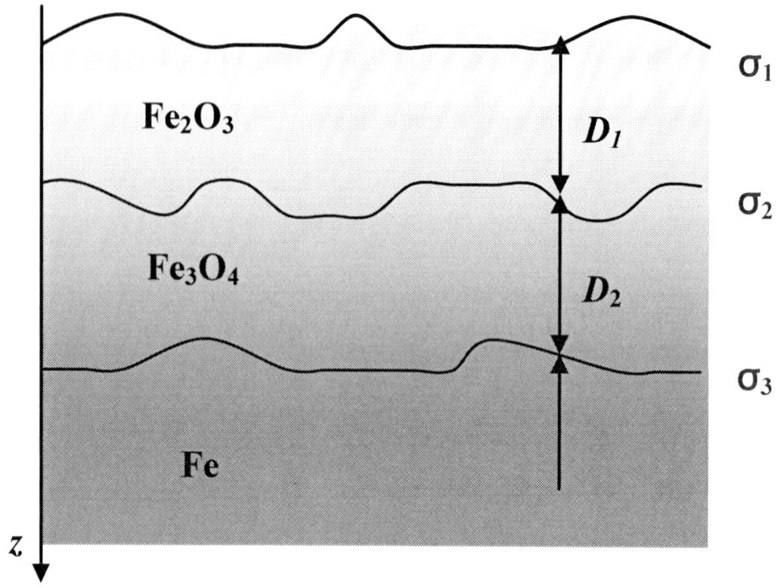

Figure 7. Structural diagram of polycrystalline oxide layers on the iron surface.

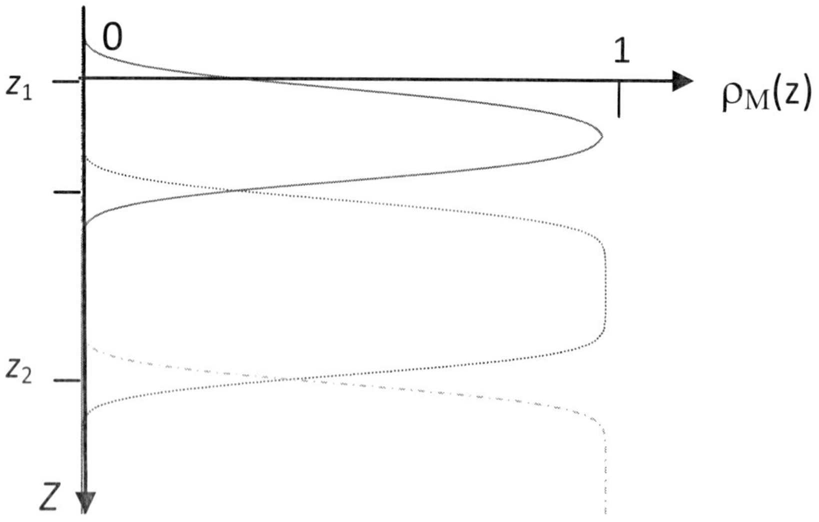

Figure 8. Schematic diagram of the depth profile of the density of the polycrystalline specimen.

We calculated the diffraction intensities corresponding to Fe_2O_3, Fe_3O_4 and Fe for this model with a rough surface and rough interfaces between each layer. In the calculation, the thickness d of one layer of multilayer is 0.01nm. Parameters of the fit are the thickness D_1 of Fe_2O_3 , D_2 of Fe_3O_4 , and surface roughness σ_1, the interface roughnesses σ_2 and σ_3.

Figure 9 shows the calculated intensities of x-rays from top layer and from 2nd layer at changing the thickness D1 of top layer, when surface and interface roughnesses σ are 5nm. X-ray intensities from 2nd layer at the angle of incidence smaller than the critical angle decrease rapidly when the thickness D1 of top layer is larger. This character is shown as same in the experimental results of incident angle dependence of the diffracted x-ray intensities for Fe_3O_4 (220). Then it was found that the Fe_3O_4 layer is 2nd layer and is below the Fe_2O_3 layer.

The calculated results of the incident angle dependence of the intensity for Fe_2O_3, Fe_3O_4 and Fe are compared with each experimental result. The contribution from the thin top layer is large at a smaller glancing angle, and contribution from the substrate Fe becomes large with increasing incident angle.

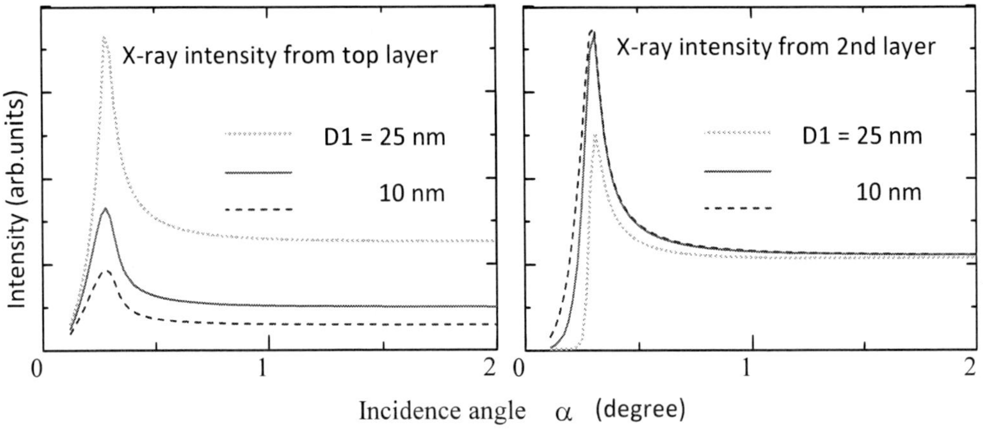

Figure 9. Calculated intensities from top layer and 2nd layer at several thicknesses D1 of top layer.

The analytical results that show the best fit with the experimental results are shown in the Figure 10. The fitting parameters are shown in Table 1.

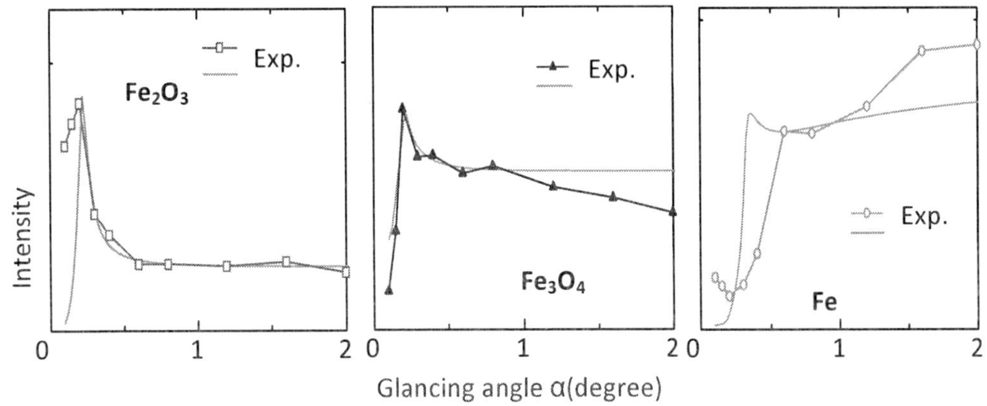

Figure 10. Comparison between the calculated and experimental results of the diffracted x-ray intensities for Fe_2O_3 (104), Fe_3O_4 (220), and Fe (110).

Table 1. Parameters of the fit on depth profile of polycrystalline oxide layers on the iron surface

	Thickness D (nm)	Roughness σ(nm)
		$\sigma_1 = 0.3$
Fe_2O_3	4.0	
		$\sigma_2 = 2.0$
Fe_3O_4	8.5	
		$\sigma_3 = 1.0$

The fitting accuracy with an error ratio 10% is ~ 1nm on the thickness of each layer. It was found that the thickness of each layer effects the ratio of the intensities at the small and large glancing angles The fitting accuracy of surface roughness σ_1 and the interface roughness σ_2 is ~ 0.2nm. It was found that the roughness affects the ratio of the intensities at the critical angle and the large glancing angle. Thus the thickness and interface roughness of the surface oxidized iron layers (Fe_2O_3 and Fe_3O_4) were determined by analyzing the dependency of the diffracted x-rays intensities on the incidence. The depth profile of the polycrystalline layers of oxidized iron was evaluated to accuracy of the order of nanometers. Because the refractive indices of Fe_2O_3 and Fe_3O_4 are close to each other, this structure cannot be measured in good precision by the usual x-ray reflectivity measurement. The incidence angle dependence of the diffracted x-rays on polycrystalline specimen gives good information for investigating the depth profile of the surface structure of such material.

2.4. Depth Profiling of the Crystal Grain Size nearby the Surface

The peak width of diffracted x-rays of Fe(110) is corresponding to the size of Fe crystal grain in the surface layer. Wide peak at smaller incidence angle shows that Fe crystal grain is smaller nearby the surface. Analyzing these peak profiles leads to the depth profile of the crystal grain size under the surface. The analysis was done on the supposition that the forms of the crystal grains and the direction of the crystal grains are distributed uniformly even in a surface, because x-rays are irradiating the over million crystal grains, and because there is no report that the direction of the crystal grain of the iron is not distributed uniformly in a surface.

When x-rays are applied to a crystal of lattice constant a_o and grain size L_o , intensity distribution of diffracted x-rays in θ_d direction from the crystal is shown by Laue function as

$$\frac{\sin^2\left(KL_o/2\right)}{\sin^2\left(Ka_o/2\right)}, \tag{18}$$

where

$$K = 2k_o \sin\left(\theta_d/2\right). \tag{19}$$

When x-rays are applied to polycrystalline in a surface layer, the diffraction intensity is the sum of the diffracted x-rays from the polycrystalline in each layer. Then the peak profile of the diffracted x-rays is shown in the following equation

$$I_M(K) = \int_0^\infty S_M(z) \frac{\sin^2\left(KL(z)/2\right)}{\sin^2\left(Ka_o/2\right)} I(z)dz \quad , \tag{20}$$

where $S_M(z)$ is the density of the polycrystalline M at the depth z. The grain size at the depth z was supposed to be L(z). $I_M(K)$ has diffraction peaks as Laue function like. I(z) is function of the incidence angle α . Therefore when L(z) is not constant in depth, $I_M(K)$ becomes function of α . The incident angle dependence of the peak width of diffracted x-rays gives good information to investigate the depth profile of the crystal grain size in the surface layer. The investigated iron specimen surface in this study has polycrystalline oxide layers. The thickness of the surface oxidized iron layers (Fe_2O_3 and Fe_3O_4) were evaluated 4.0nm and 8.5nm, respectively. In the case that some materials exist in the surface layer, the refractive index n(z) change with the density in the depth z, and the intensity I(z) of the refractive x-rays in depth z is calculated with using the refractive index n(z).

In Fig. 5, wide peak at smaller incidence angle shows that Fe crystal grain is smaller nearby the surface, because the contribution from the surface top layer is large at a smaller glancing angle. Then the size of the crystal grain in the surface top layer is made L_s, the size of the crystal grain in deep layer is made L_D, and the characteristics depth of the surface condition is made D. With using these characteristic parameters we express the crystal grain size L(z) of each depth under the surface as shown in the following equation,

$$L(z) = L_D + (L_s - L_D)\exp\left(-\frac{z}{D}\right) \cdot \tag{21}$$

The peak profile of the diffracted x-rays, which simulates the observation, is calculated by eq. (20) with the using the grain size model of eq. (21), and the calculated results of the incident angle dependence of the peak width are compared with the experimental result.

Figure 11 shows comparison between the calculated and experimental results of the peak width of diffracted x-rays for Fe(110), where the experimental results show the peak width $\delta\theta_d$ of which a broadening due to acceptance angle was corrected. Parameters that reproduce the experimental result were L_D=85nm, L_S=17nm and D=150nm, respectively. The evaluated result shows that the crystal grain under surface is very small. This result is consistent that the grain size in mechanically polished surface is usually smaller than those in bulk, but this analyzed results seems too small. It is said industrially that a crystal grain in the neighborhood on the surface breaks and becomes small multiple grains as so called crystallites, when surface is made smoothly by mechanically polishing. The result in this study agrees with what is said industrially, and those contents were proved. But, it is big still in comparison with the scale of the surface roughness. These results show that, though a crystal grain around the surface is broken, a crystal grain itself is shaved, and it is being made smooth by mechanically polishing the surface. The result of evaluated grain size L_D in deep layer was smaller than those of bulk. It shows that the evaluated results from x-ray diffraction shows the crystallites size in the crystal grain under mechanically polished iron surface.

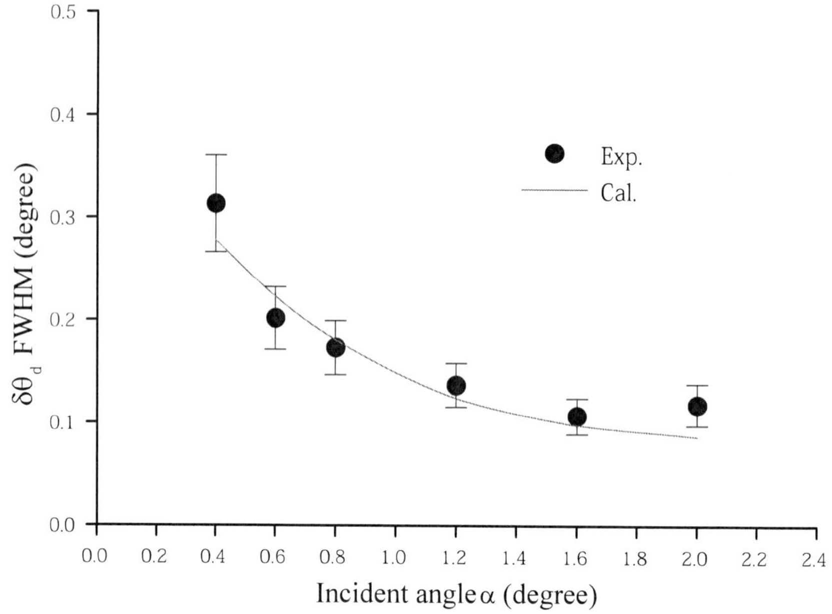

Figure 11. Comparison between the calculated and experimental results of the peak width of diffracted x-ray for Fe(110).

3. Depth Profile Analysis of Strain Distribution in Surface Layer Using X-Ray Diffraction at Small Glancing Angle of Incidence

In this section, diffracted x-rays at small glancing angle of incidence on surface materials are investigated as function of incidence angles for the studies of residual stresses of surface layers. And analyzing method for evaluating the depth profiles of the strain distribution in the surface layer using x-ray diffraction at small glancing angles of incidence is discussed.

Compound plating is used for the purpose of the strength improvement on the surface of the material, and accurately knowing the residual stress in the compound plating becomes important. Residual stresses in poly-crystalline materials were measured using the glancing incidence x-ray diffraction. In the previous studies, the in-depth distribution of residual stresses was estimated with the use of the term of penetration depth.[15,16] These estimations cannot be applied to the residual stress distribution analysis of the surface layer materials as multi thin films and compound plating layers with rough surface and rough interfaces because densities of the surface layer materials change continuously in depth. We therefore derived the x-ray intensity propagating during the surface layer materials of which density changes continuously in depth, and calculated the dependence of the diffracted x-ray intensity on the glancing angle.[5,6] With the use of the analyzing method, the depth profiles of the strain distribution in the surface layer were studied.[17] The derived analyzing method can be applied to the strain distribution analysis of the surface layer materials of which densities change continuously in depth as multi thin films, compound plating layers, and the surface with roughness and layers structure.

3.1. Experimental for the Depth Profile Analysis of Strain Distribution in Surface Layer by X-Ray Diffraction at Small Glancing Angle of Incidence

Intensities of the diffracted x-rays on the chromium coating steel were measured at several glancing angles α_0 of incidence with Synchrotron radiation at SPring-8 BL24XU. Thickness of chromium layer is about 2 μm. Figure 12 shows a schematic view of the experimental arrangement. Figure 13 shows the intensity distributions of diffracted x-rays by poly-crystalline chromium in the surface layer.

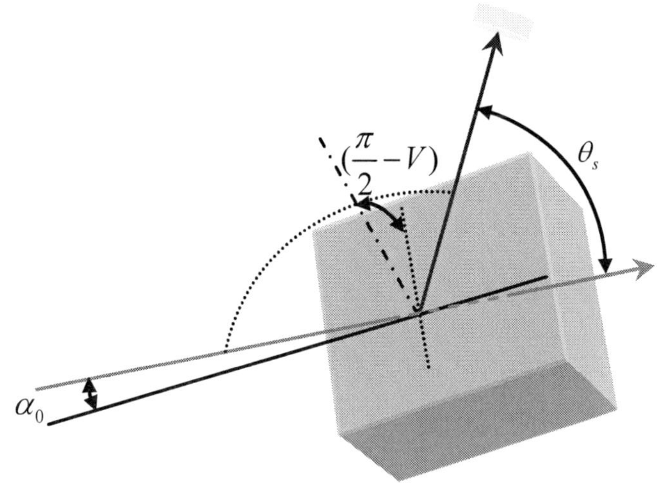

Figure 12. The schematic view of the experimental arrangement.

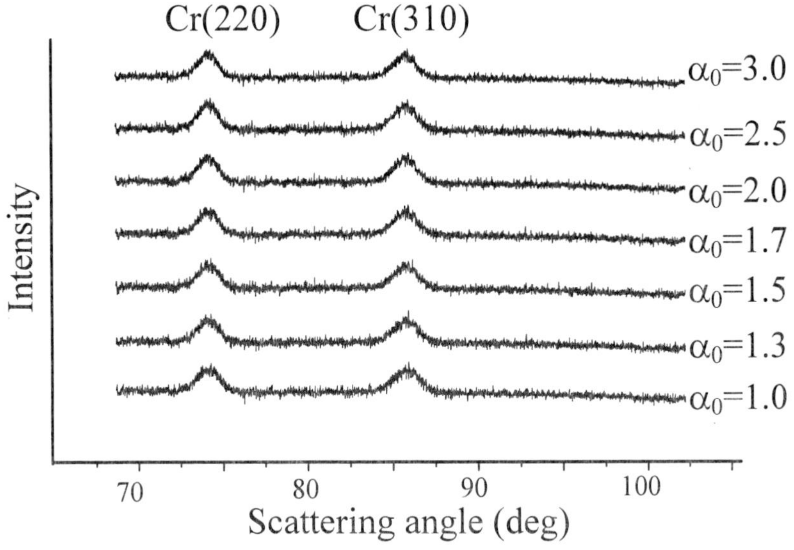

Figure 13. Intensity distributions of scattered x-rays at several glancing angles of incidence.

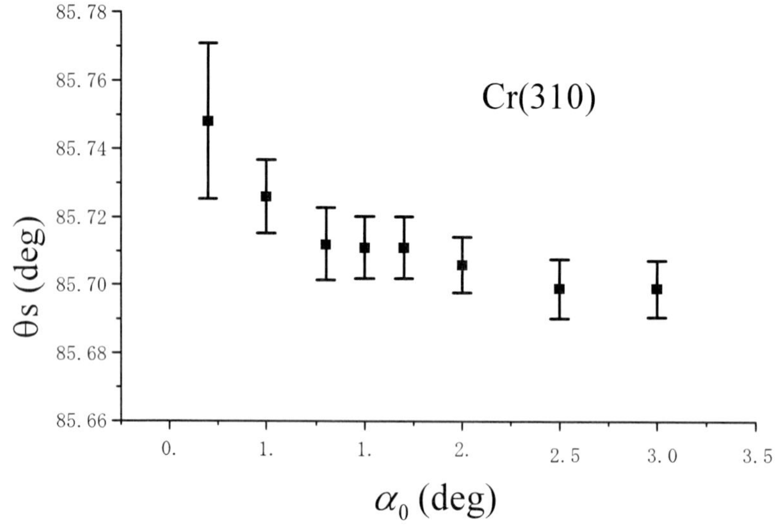

Figure 14. Observed scattering angle of diffracted x-rays from Cr(310).

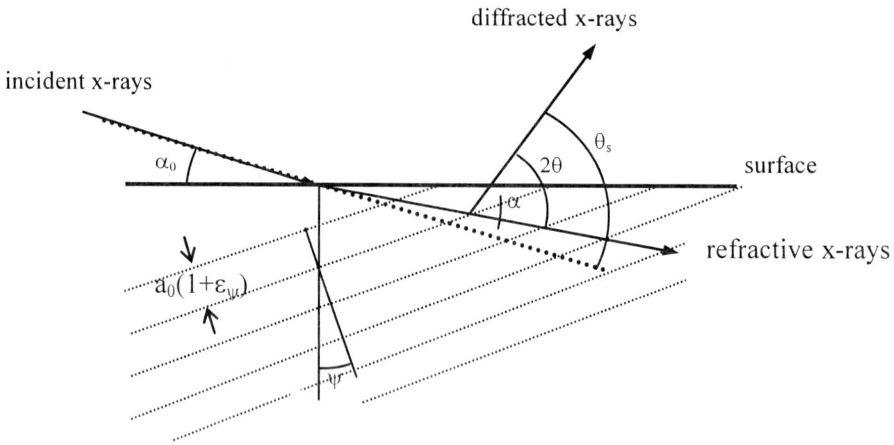

Figure 15. The schematic of glancing incidence x-ray diffraction at incident angle α_0.

The scattering angles of x-rays diffracted by poly-crystalline chromium in the surface layer are shown in Fig. 14. The increase in the scattered angle θ_s at smaller incident angle α_0 includes the effect of refraction as shown in Fig. 15. Analyzing the incidence angle dependence of the diffraction angle, the depth profiles of the strain distribution in the surface layer can be derived.

3.2. Depth Profile Analysis of Strain Distribution in Surface Layer

In the condition of small glancing angle of incidence, most of the x-rays are scattered by only shallow regions near the surface. In the region, the stress σ_z in the direction z normal to the surface can be neglected, *i.e.* $\sigma_z = 0$. Here the isotropic stress, *i.e.* $\sigma_x = \sigma_y = \sigma$ is assumed, the strain $\varepsilon_z = -2\nu/E \cdot \sigma$, where ν is the Poisson's ratio and E is the Young's modulus.

On the stress estimation by using x-ray diffraction at small glancing angle incidence, $\sin^2\psi$ diagram method is useful, but it becomes difficult to apply the method due to its nonlinearity when a sample has stress gradient in the surface, *i.e.* the co-axial stress σ becomes $\sigma(z)$; the in-depth distribution of residual stresses. This difficulty is due to the changing strain with depth dependence to the surface, *i.e.* the strain $\varepsilon_z = \varepsilon_z(z)$. In the previous studies, the residual stresses $<\sigma>$ was estimated with the use of the term of x-ray penetration depth T as the following equation, [15,16]

$$<\sigma>=\frac{1}{T}\int_0^h \sigma(z)\exp(-z/T)dz \ . \tag{22}$$

The strain ε_ψ in the direction normal to the diffraction plane is,

$$\varepsilon_\psi(z) = -\frac{1+\nu}{2\nu}\varepsilon_z(z)\sin^2\psi + \varepsilon_z(z) \ . \tag{23}$$

Scattered angle θ_s of the diffracted x-ray is

$$\theta_s = 2(\theta_0 - \varepsilon_\psi(z)\tan\theta_0) \ , \tag{24}$$

where θ_0 is the Bragg angle at stress free and the effect of refraction was neglected. Using above relation, the residual stresses $<\sigma>$ in the previous studies was estimated as the following equation,

$$\theta_s = 2\theta_0 + 2\tan\theta_0\left(\frac{2\nu}{E} - \frac{1+\upsilon}{E}\sin^2\psi\right)\frac{1}{T}\int_0^h \sigma(z)\exp(-z/T)dz \ , \tag{25}$$

where the penetration depth T was approximately shown as $\sin(\alpha_0)/\mu$, where μ is a linear absorption coefficient. These estimations cannot be applied to the surface layer materials of which densities change in depth z as the surface with roughness and layers structure because T is not constant. We therefore characterized the refractive index of the surface layer materials with complex refractive index, which changes continuously in depth, and derived the x-ray intensity propagating during the surface layer materials. The intensity of x-rays, *i.e.*, the electric and magnetic field propagating in the material, can be obtained using Maxwell's equations.[14] The effects of x-rays on the material are characterized by complex refractive index n, which changes continuously with depth. The material for which the density changes continuously with depth is divided into thin N layers. The reflectance of multiplayer system, consisting of N layers can be calculated using the recursive formalism given by Parratt.[1] Derivation of the electric field of x-ray propagating in *j*-th layer is explained in previous section.

The electric field of x-ray radiation at glancing angle of incidence α_0 is expressed as

$$\boldsymbol{E}_0(z) = \boldsymbol{A}_0 \exp[i(\boldsymbol{k}_0 \cdot \boldsymbol{r} - \omega t)] . \tag{26}$$

The amplitude $\boldsymbol{A}_j$ of j–th layer is derived from continuity equations for the interface between the j-1 and j layer as shown by

$$\boldsymbol{A}_1 = \boldsymbol{\Phi}_1 \boldsymbol{A}_0 \ , \ \boldsymbol{A}_j = \boldsymbol{\Phi}_j \boldsymbol{A}_{j-1} \exp(ik_{j-1,z} d), \tag{27}$$

By the condition of incident x-ray, the z-direction component of wave vector of the j-th layer is shown as

$$k_{j,z} = \frac{2\pi}{\lambda}\sqrt{n_j^2 - \cos^2\alpha_0} = \frac{2\pi}{\lambda}(a_j + ib_j) \ . \tag{28}$$

The Fresnel coefficient tensor for refraction on the interface between j-1 and j layer is given by

$$\begin{aligned}
\boldsymbol{\Phi}_{j,xx} &= \frac{2n_{j\text{-}1}^2\sqrt{n_j^2 - \cos^2\alpha_0}}{n_j^2\sqrt{n_{j\text{-}1}^2 - \cos^2\alpha_0} + n_{j\text{-}1}^2\sqrt{n_j^2 - \cos^2\alpha_0}} = \frac{2n_{j\text{-}1}^2(a_j + ib_j)}{n_j^2(a_{j\text{-}1} + ib_{j\text{-}1}) + n_{j\text{-}1}^2(a_j + ib_j)} \\
\boldsymbol{\Phi}_{j,yy} &= \frac{2\sqrt{n_{j\text{-}1}^2 - \cos^2\alpha_0}}{\sqrt{n_{j\text{-}1}^2 - \cos^2\alpha_0} + \sqrt{n_j^2 - \cos^2\alpha_0}} = \frac{2(a_{j\text{-}1} + ib_{j\text{-}1})}{(a_{j\text{-}1} + ib_{j\text{-}1}) + (a_j + ib_j)} \\
\boldsymbol{\Phi}_{j,zz} &= \frac{2n_{j\text{-}1}^2\sqrt{n_{j\text{-}1}^2 - \cos^2\alpha_0}}{n_j^2\sqrt{n_{j\text{-}1}^2 - \cos^2\alpha_0} + n_{j\text{-}1}^2\sqrt{n_j^2 - \cos^2\alpha_0}} = \frac{2n_{j\text{-}1}^2(a_{j\text{-}1} + ib_{j\text{-}1})}{n_j^2(a_{j\text{-}1} + ib_{j\text{-}1}) + n_{j\text{-}1}^2(a_j + ib_j)}
\end{aligned} \tag{29}$$

Using these equations, the electric field of x-ray propagating in j-th layer is expressed as

$$\boldsymbol{E}_j(z) = (\prod_{J=1}^{j} \boldsymbol{\Phi}_J)\boldsymbol{A}_0 \exp[i(k_{0,x}x + d\sum_{J=1}^{j-1} k_{J,z} + k_{j,z}(z - (j-1)d) - \omega t)] \ , \tag{30}$$

and the intensity of the refractive x-ray in j-th layer at depth z is shown as

$$I(z) = \left|(\prod_{J=1}^{j} \boldsymbol{\Phi}_J)\boldsymbol{A}_0 \| (\prod_{J=1}^{j} \boldsymbol{\Phi}_J)\boldsymbol{A}_0\right| \cdot \exp\{ -d\frac{4\pi}{\lambda}\sum_{J=1}^{j-1} b_J - \frac{4\pi}{\lambda} b_j(z - (j-1)d) \} . \tag{31}$$

On the other hand, scattered angle θ_S of the diffracted x-ray from the crystal plane at the depth z is

$$\theta_s = 2\theta_0 - 2\varepsilon_{\psi}(z)\tan\theta_0 + \alpha_0 - \alpha(z) \ , \tag{32}$$

where θ_0 is the Bragg angle at stress free and $\alpha(z)$ is the refracting angle at the depth z.

The diffraction peak intensities are in proportion to the sum of the refractive x-ray intensity in each layer of those polycrystalline. Then the peak angle of the diffracted x-rays is shown in the following equation,

$$\theta_S(\alpha_0) = 2\theta_0 + \alpha_0 + \frac{\int_0^h [2\tan\theta_0(\frac{1+\nu}{2\nu}\sin^2\psi - 1)\varepsilon_Z(z) - \alpha(z)]I(z)dz}{\int_0^h I(z)dz}, \quad (33)$$

where I(z), $\alpha(z)$ and ψ are function of the incident angle α_0, therefore θ_s is function of α_0.

The incident angle dependence of the scattered angle $\theta_s(\alpha_0)$ of the diffracted x-rays can give good information to investigate the depth profile of the strain distribution in the surface layer materials of which densities change continuously in depth as the surface with roughness and layers structure. The refractive index n(z) change with the density $\rho(z)$ in the depth z, and the intensity I(z) of the refractive x-ray in depth z is calculated with using the refractive index n(z). In the case that some materials M exist in the surface layer, the peak angle of the diffracted x-rays from the material is

$$\theta_S(\alpha_0) = 2\theta_0 + \alpha_0 + \frac{\int_0^h [2\tan\theta_0(\frac{1+\nu}{2\nu}\sin^2\psi - 1)\varepsilon_Z(z) - \alpha(z)]\rho_M(z)I(z)dz}{\int_0^h \rho_M(z)I(z)dz}, \quad (34)$$

where $\rho_M(z)$ is the density of the polycrystalline material M in the depth z.

In the calculation of scattered angle θ_s from the chromium coating, we considered the surface roughness. The surface roughness is characterized with the root mean square deviation σ of the surface with respect to a flat surface. The density $\rho_M(z)$ of the polycrystalline chromium coating in the depth z with the surface roughness of σ is expressed as

$$\rho_M(z) = \int_{-\infty}^{z} \frac{1}{\sqrt{2\pi}\sigma}\exp(-\frac{x^2}{2\sigma^2})dx \quad (35)$$

Then the effect of the surface roughness is reflected in the refractive index n(z) and the intensity I(z) of the refractive x-ray. The estimation with using the term of penetration depth T in the previous studies cannot be applied to such material with the surface roughness since T depend on the incident angle. The calculation for the depth profiling of the strain is performed with using eq. (34).

At first, the incidence angle dependence of scattered angle is calculated with the constant strain ε_Z as shown in Fig. 16. By comparison with calculated and observed scattering angle, the strain seems to decrease at smaller angle of incidence.

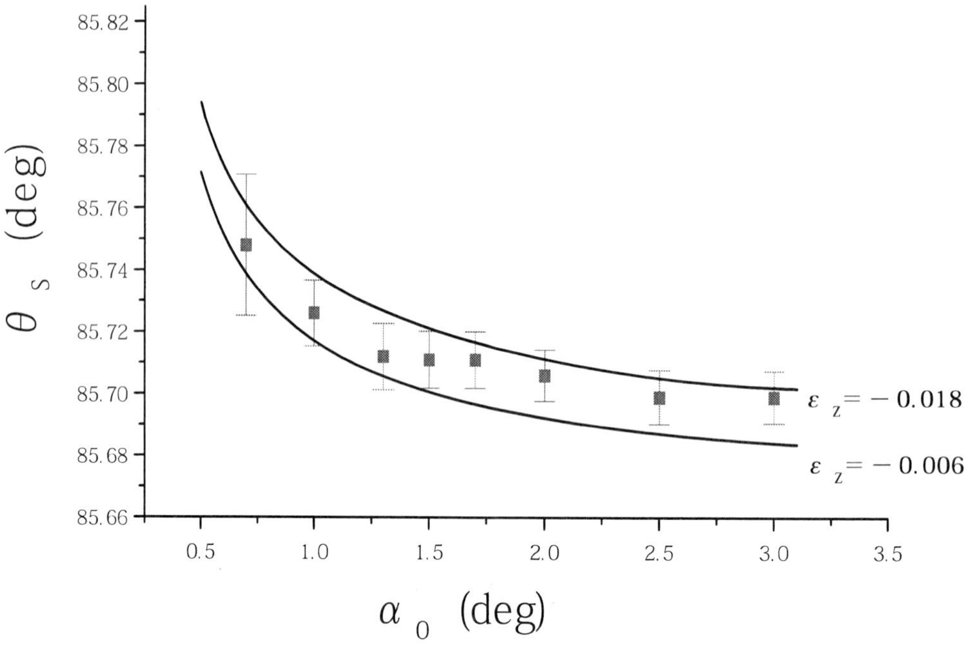

Figure 16. Incidence angle dependence of calculated and observed scattering angle. The solid lines are calculated result with the constant strain ε_z

From the calculations for several models of the strain distribution, the strain distribution that reproduces the experimental results of the incidence angle dependence was derived. The strain distribution in the surface layer of the chromium coating is found as the following

$$\varepsilon_z = -0.019\left\{1 - \exp\left(-\frac{z}{r}\right)\right\}, \tag{36}$$

where r is 80nm, and in the calculation the surface roughness σ of 100nm in eq.(35) is considered. This roughness corresponds to the roughness of substrate steel surface. The same level depth into which the strain changes as the surface roughness consents. Comparison between calculated and observed scattering angle of Cr(310) is shown in Fig. 17. The solid line is calculated result. The calculated result reproduces the experiment result, and the depth profile of the strain distribution in the surface poly-crystalline layers was evaluated with a high accuracy.

In concluding, the strain distribution in the surface layer of the chromium coating steel is investigated with the use of x-ray diffraction at small glancing angle of incidence. Analyzing the incidence angle dependence of the scattered angles of diffracted x-rays, the depth profile of the strain in the surface layer is derived. The depth profile of the strain is accurately obtained with surface roughness. The derived analyzing method can be applied to the residual stress distribution analysis of the surface layer materials of which densities change continuously in depth as the surface with roughness and layers structure.

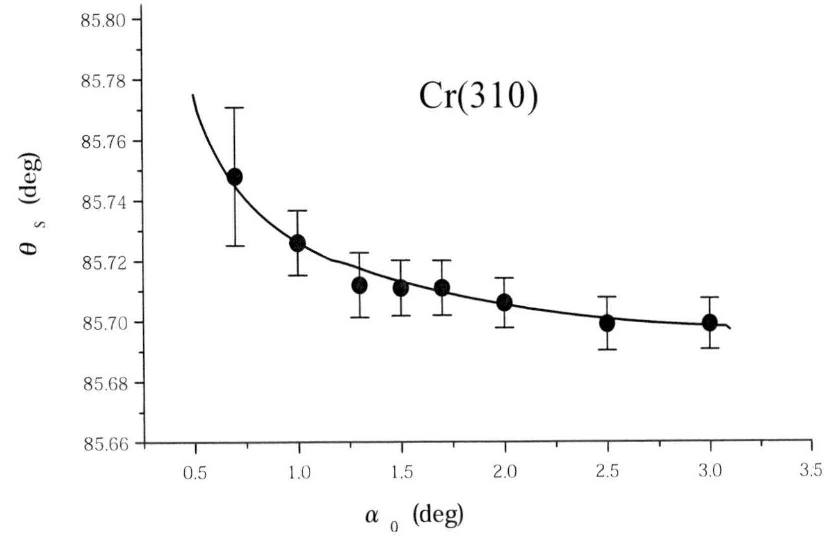

Figure 17. Comparison between calculated and observed scattering angle of Cr(310). The solid line is calculated result with the strain distribution in the surface layer.

4. Depth Profile Analysis of Surface Layer Using X-Ray Reflectivity at Small Glancing Angle of Incidence

The depth profile of the surface layers can be found by analyzing the incidence angle dependence of the scattered x-rays[1-6] and x-ray reflectometry has become a powerful tool for investigations on rough surface and interface structures.[7,8] In the calculation of x-ray reflectivity, most attention is paid to coherent scattering. In addition, several theories exist to describe the influence of roughness on x-ray scattering.[9-13] In many previous studies, the x-ray reflectivity was calculated based on the Parratt formalism[1], coupled with the use of the theory of Nevot and Croce to include roughness.[2] However, the calculated results of the x-ray reflectivity done in this way often showed strange results where the amplitude of the oscillation due to the interference effects would increase for a rougher surface. This strange result disagrees with reality and suggests that there is a problem in how the effect of roughness is incorporated into the Parratt formalism, and hence that the effect of roughness is not accurately represented in the theoretical formula of the x-ray reflectivity. Because the x-ray scattering vector in a specular reflectivity measurement is normal to the surface, it provides the density profile solely in the direction perpendicular to surface. Specular reflectivity measurements can yield the magnitude of the average roughness perpendicular to surface and interfaces, but cannot give information about the lateral extent of the roughness. In previous studies, the effect of roughness on the calculation of the x-ray reflectivity only took into account the effect of the density changes of the medium in a direction normal to the surface and interface. On the other hand, diffuse scattering can provide information about the lateral extent of the roughness. In contrast to previous calculations of the x-ray reflectivity, in the present analysis we consider the effect of a decrease in the intensity of penetrated x-rays due to diffuse scattering at a rough surface and rough interface.

4.1. X-Ray Reflectivity from a Multilayer with a Flat Surface and Flat Interfaces

The intensity of x-rays propagating in the surface layers of a material, *i.e.*, the electric and magnetic fields, can be obtained from Maxwell's equations.[14] The effects of the material on the x-ray intensity are characterized by a complex refractive index n, which varies with depth. We divide a material in which the density changes continuously with depth into N layers with an index j. The complex refractive index of the j-th layer is n_j. The vacuum is denoted as $j = 0$ and $n_0 = 1$. The thickness of the j-th layer is h_j, the thickness of the bottom layer being assumed to be infinite.

The reflectance of an N-layer multilayer system can be calculated using the recursive formalism given by Parratt.[1] In the following, we show in detail the process of obtaining Parratt's expression and, further, show that this expression requires conservation of energy at the interface. We go on to show that the dispersion of the energy by interface roughness cannot be correctly accounted for Parratt's expression.

Following that approach, let n_j be the refractive index of the j-th layer, defined as

$$n_j = 1 - \delta_j - i\beta_j\,, \tag{37}$$

where δ_j and β_j are the real and imaginary parts of the refractive index. These optical constants are related to the atomic scattering factor and electron density of the j-th layer material. For x-rays of wavelength λ, the optical constants of the j-th layer material consisting of N_{ij} atoms per unit volume can be expressed as

$$\delta_j = \frac{\lambda^2 r_e}{2\pi}\sum_i f_{1i} N_{ij}\,,\quad \beta_j = \frac{\lambda^2 r_e}{2\pi}\sum_i f_{2i} N_{ij}\,, \tag{38}$$

where r_e is the classical electron radius and f_{1i} and f_{2i} are the real and imaginary parts of the atomic scattering factor of the i-th element atom, respectively.

We take the vertical direction to the surface as the z axis, with the positive direction pointing towards the bulk. The scattering plane is made the x-z plane. The wave vector $\boldsymbol{k}_j$ of the j-th layer is related to the refractive index n_j of the j-th layer by

$$\frac{\boldsymbol{k}_j \cdot \boldsymbol{k}_j}{n_j^2} = \frac{\omega^2}{c^2} = const\,, \tag{39}$$

and, as this necessitates that the x,y-direction components of the wave vector are constant, then the z-direction component of the wave vector of the j-th layer is

$$k_{j,z} = \sqrt{n_j^{\,2}\boldsymbol{k}_0 \cdot \boldsymbol{k}_0 - k_{0,x}^{\,2}}\ . \tag{40}$$

In the 0-th layer, *i.e.*, in vacuum,

$$n_0 = 1 \ , \ \boldsymbol{k}_0 \cdot \boldsymbol{k}_0 = k^2 \ , \ k = \frac{2\pi}{\lambda} = \frac{\omega}{c} \ . \tag{41}$$

In the j-th layer, the components of the wave vector are

$$k_{j,x} = k\cos\theta \ , \ k_{j,y} = 0 \ , \ k_{j,z} = k\sqrt{n_j^{\,2} - \cos^2\theta} \ . \tag{42}$$

The electric field of x-ray radiation at a glancing angle of incidence θ is expressed as

$$\boldsymbol{E}_0(z) = \boldsymbol{A}_0 \exp[\,i(\boldsymbol{k}_0 \cdot \boldsymbol{r} - \omega t)] \ . \tag{43}$$

The incident radiation is usually decomposed into two geometries to simplify the analysis, one with the incident electric field E parallel to the plane of incidence (p-polarization) and one with E perpendicular to that plane (s-polarization). An arbitrary incident wave can be represented in terms of these two polarizations. Thus, E_{0x} and E_{0z} correspond to p-polarization, and E_{0y} to s-polarization; those components of the amplitude's electric vector are expressed as

$$A_{0x} = -A_{0p}\sin\theta \ , \ A_{0y} = A_{0s} \ , \ A_{0z} = A_{0p}\cos\theta \ . \tag{44}$$

The components of the wave vector of the incident x-rays are

$$k_{0x} = k\cos\theta \ , \ k_{0y} = 0 \ , \ k_{0z} = k\sin\theta \ , \tag{45}$$

The electric field of reflected x-ray radiation of exit angle □ is expressed as

$$\boldsymbol{E}'_0(z) = \boldsymbol{A}'_0 \exp[\,i(\boldsymbol{k}'_0 \cdot \boldsymbol{r} - \omega t)] \ . \tag{46}$$

where

$$k'_{0x} = k_{0x} \ , \ k'_{0y} = 0 \ , \ k'_{0z} = -k_{0z} \ . \tag{47}$$

Because an x-ray is a transverse wave, the amplitude and the wave vector are orthogonal as follows,

$$\boldsymbol{A}_j \cdot \boldsymbol{k}_j = 0 \quad , \quad \boldsymbol{A}'_j \cdot \boldsymbol{k}'_j = 0 \ . \tag{48}$$

We consider the relation of the electric field $\boldsymbol{E}_0$ of x-rays incident at a flat surface from vacuum, the electric field $\boldsymbol{E}_1$ of x-rays propagating in the first layer material, the electric field $\boldsymbol{E}'_0$ of x-rays reflected from the surface exit to vacuum, and the electric field $\boldsymbol{E}'_1$ of x-rays propagating toward to the surface in the first layer material.

The electric fields $\boldsymbol{E}_1$, $\boldsymbol{E}'_1$ in the first layer material below the surface are expressed as

$$\boldsymbol{E}_1(z) = \boldsymbol{A}_1 \exp[i(\boldsymbol{k}_1 \cdot \boldsymbol{r} - \omega t)] \;,\; \boldsymbol{E}'_1(z) = \boldsymbol{A}'_1 \exp[i(\boldsymbol{k}'_1 \cdot \boldsymbol{r} - \omega t)]. \tag{49}$$

where

$$k'_{1x} = k_{1x} \;,\; k'_{1y} = 0 \;,\; k'_{1z} = -k_{1z} \;. \tag{50}$$

$$k_{1,x} = k\cos\theta \;,\; k_{1y} = 0,\; k_{1,z} = k\sqrt{n_1^{\,2} - \cos^2\theta} \;. \tag{51}$$

The relation of the amplitudes A_0 , A'_0 , A_1 , and A'_1 can be found from the continuity equations of the electric fields for the interface between the 0-th and 1-th layers as follows

$$A_{0,x} + A'_{0,x} = A_{1,x} + A'_{1,x} \;,\; A_{0,y} + A'_{0,y} = A_{1,y} + A'_{1,y} \;, \tag{52}$$

$$k_{0,x}A_{0,x} + k'_{0,x}\,A'_{0,x} = k_{1,x}A_{1,x} + k'_{1,x}\,A'_{1,x} \;, \tag{53}$$

$$k_{0,y}A_{0,y} + k'_{0,y}\,A'_{0,y} = k_{1,y}A_{1,y} + k'_{1,y}\,A'_{1,y} \;, \tag{54}$$

Another relation of the amplitudes A_0 , A'_0 , A_1 , and A'_1 can be found from the continuity equations of the magnetic fields for the interface between the 0-th and 1-th layers are shown below

$$k_{0,z}A_{0,y} - k_{0,y}A_{0,z} + k'_{0,z}\,A'_{0,y} - k'_{0,y}\,A'_{0,z} = k_{1,z}A_{1,y} - k_{1,y}A_{1,z} + k'_{1,z}\,A'_{1,y} - k'_{1,y}\,A'_{1,z} \;, \tag{55}$$

$$k_{0,z}A_{0,x} - k_{0,x}A_{0,z} + k'_{0,z}\,A'_{0,x} - k'_{0,x}\,A'_{0,z} = k_{1,z}A_{1,x} - k_{1,x}A_{1,z} + k'_{1,z}\,A'_{1,x} - k'_{1,x}\,A'_{1,z} \;, \tag{56}$$

From the above equations, the**se** amplitudes are related by the Fresnel coefficient tensor □ for refraction and the Fresnel coefficient tensor Ψ for reflection as follows

$$\begin{pmatrix} \boldsymbol{A}'_0 \\ \boldsymbol{A}_1 \end{pmatrix} = \begin{pmatrix} \boldsymbol{\Psi}_{0,1} & \boldsymbol{\Phi}_{1,0} \\ \boldsymbol{\Phi}_{0,1} & \boldsymbol{\Psi}_{1,0} \end{pmatrix} \begin{pmatrix} \boldsymbol{A}_0 \\ \boldsymbol{A}'_1 \end{pmatrix} . \tag{57}$$

Here, the Fresnel coefficient tensor Φ for refraction at the interface between the 0-th and 1-th layers is given by

$$\begin{aligned}
\Phi_{0,1,xx} &= \frac{2k_{1,z}\boldsymbol{k}_0 \square \boldsymbol{k}_0}{k_{0,z}\boldsymbol{k}_1 \square \boldsymbol{k}_1 + k_{1,z}\boldsymbol{k}_0 \square \boldsymbol{k}_0} & \Phi_{1,0,xx} &= \frac{2k_{0,z}\boldsymbol{k}_1 \square \boldsymbol{k}_1}{k_{0,z}\boldsymbol{k}_1 \square \boldsymbol{k}_1 + k_{1,z}\boldsymbol{k}_0 \square \boldsymbol{k}_0} \\
\Phi_{0,1,yy} &= \frac{2k_{0,z}}{k_{0,z} + k_{1,z}} & \Phi_{1,0,yy} &= \frac{2k_{1,z}}{k_{1,z} + k_{0,z}} \\
\Phi_{0,1,zz} &= \frac{2k_{0,z}\boldsymbol{k}_0 \square \boldsymbol{k}_0}{k_{0,z}\boldsymbol{k}_1 \square \boldsymbol{k}_1 + k_{1,z}\boldsymbol{k}_0 \square \boldsymbol{k}_0} \;, & \Phi_{1,0,zz} &= \frac{2k_{1,z}\boldsymbol{k}_1 \square \boldsymbol{k}_1}{k_{0,z}\boldsymbol{k}_1 \square \boldsymbol{k}_1 + k_{1,z}\boldsymbol{k}_0 \square \boldsymbol{k}_0} \;. \\
\Phi_{0,1,xy} &= \Phi_{0,1,yx} = \Phi_{0,1,yz} = 0 & \Phi_{1,0,xy} &= \Phi_{1,0,yx} = \Phi_{1,0,yz} = 0 \\
\Phi_{0,1,zy} &= \Phi_{0,1,zx} = \Phi_{0,1,xz} = 0 & \Phi_{1,0,zy} &= \Phi_{1,0,zx} = \Phi_{1,0,xz} = 0
\end{aligned} \tag{58}$$

The Fresnel coefficient tensor Ψ for reflection from the interface between the 0-th and 1-th layers is given by

$$
\begin{aligned}
\Psi_{0,1,xx} &= \frac{k_{1,z}\mathbf{k}_0\cdot\mathbf{k}_0 - k_{0,z}\mathbf{k}_1\cdot\mathbf{k}_1}{k_{0,z}\mathbf{k}_1\cdot\mathbf{k}_1 + k_{1,z}\mathbf{k}_0\cdot\mathbf{k}_0} & \Psi_{1,0,xx} &= \frac{k_{0,z}\mathbf{k}_1\cdot\mathbf{k}_1 - k_{1,z}\mathbf{k}_0\cdot\mathbf{k}_0}{k_{0,z}\mathbf{k}_1\cdot\mathbf{k}_1 + k_{1,z}\mathbf{k}_0\cdot\mathbf{k}_0} \\
\Psi_{0,1,yy} &= \frac{k_{0,z} - k_{1,z}}{k_{0,z} + k_{1,z}} & \Psi_{1,0,yy} &= \frac{k_{1,z} - k_{0,z}}{k_{0,z} + k_{1,z}} \\
\Psi_{0,1,zz} &= -\frac{k_{1,z}\mathbf{k}_0\cdot\mathbf{k}_0 - k_{0,z}\mathbf{k}_1\cdot\mathbf{k}_1}{k_{0,z}\mathbf{k}_1\cdot\mathbf{k}_1 + k_{1,z}\mathbf{k}_0\cdot\mathbf{k}_0}, & \Psi_{1,0,zz} &= \frac{k_{1,z}\mathbf{k}_0\cdot\mathbf{k}_0 - k_{0,z}\mathbf{k}_1\cdot\mathbf{k}_1}{k_{0,z}\mathbf{k}_1\cdot\mathbf{k}_1 + k_{1,z}\mathbf{k}_0\cdot\mathbf{k}_0}\,. \\
\Psi_{0,1,xy} &= \Psi_{0,1,yx} = \Psi_{0,1,yz} = 0 & \Psi_{1,0,xy} &= \Psi_{1,0,yx} = \Psi_{1,0,yz} = 0 \\
\Psi_{0,1,zy} &= \Psi_{0,1,zx} = \Psi_{0,1,xz} = 0 & \Psi_{1,0,zy} &= \Psi_{1,0,zx} = \Psi_{1,0,xz} = 0
\end{aligned}
\tag{59}
$$

Here, we consider the reflection from a flat surface of a single layer. The reflection coefficient is defined as the ratio $R_{0,1}$ of the reflected electric field to the incident electric field at the surface of the material. The reflection coefficient $R_{0,1}$ from a single-layer flat surface is equal to the Fresnel coefficient Ψ_{01} for reflection, as the following shows

$$\mathbf{A'}_0 = \mathbf{R}_{0,1}\mathbf{A}_0 = \mathbf{\Psi}_{0,1}\mathbf{A}_0\,. \tag{60}$$

In general, when x-rays that are linearly polarized at an angle χ impinge on the surface at an angle of incidence θ, the components of the amplitude's electric vector are expressed as

$$\mathbf{A}_0 = \begin{pmatrix} A_{0x} \\ A_{0y} \\ A_{0z} \end{pmatrix} = \begin{pmatrix} -A_{0p}\sin\theta \\ A_{0s} \\ A_{0p}\cos\theta \end{pmatrix}, \quad \begin{pmatrix} A_{0p} \\ A_{0s} \end{pmatrix} = \begin{pmatrix} A_0\sin\chi \\ A_0\cos\chi \end{pmatrix}. \tag{61}$$

the amplitudes of reflected x-ray radiation are expressed as

$$\mathbf{A'}_0 = \begin{pmatrix} A'_{0x} \\ A'_{0y} \\ A'_{0z} \end{pmatrix} = \begin{pmatrix} \Psi_{0,1,xx} & 0 & 0 \\ 0 & \Psi_{0,1,yy} & 0 \\ 0 & 0 & \Psi_{0,1,zz} \end{pmatrix} \begin{pmatrix} A_{0x} \\ A_{0y} \\ A_{0z} \end{pmatrix}. \tag{62}$$

$$\mathbf{A'}_0 = \begin{pmatrix} A'_{0x} \\ A'_{0y} \\ A'_{0z} \end{pmatrix} = A_0 \begin{pmatrix} -\Psi_{0,1,xx}\sin\chi\sin\theta \\ \Psi_{0,1,yy}\cos\chi \\ \Psi_{0,1,zz}\sin\chi\cos\theta \end{pmatrix}. \tag{63}$$

The x-ray reflectivity R is ,

$$R = \left|\mathbf{R}_{0,1}\right| = \frac{\left|\mathbf{A'}_0\cdot\mathbf{A'}_0\right|}{\left|\mathbf{A}_0\cdot\mathbf{A}_0\right|}, \tag{64}$$

then

$$R = \Psi_{0,1,xx}\Psi^*_{0,1,xx}\sin^2\chi\sin^2\theta + \Psi_{0,1,yy}\Psi^*_{0,1,yy}\cos^2\chi + \Psi_{0,1,zz}\Psi^*_{0,1,zz}\sin^2\chi\cos^2\theta\ , \tag{65}$$

where,

$$\begin{aligned}
\Psi_{0,1,xx} &= \frac{k_{1,z}n_0^{\,2} - k_{0,z}n_1^{\,2}}{k_{0,z}n_1^{\,2} + k_{1,z}n_0^{\,2}} \\
\Psi_{0,1,yy} &= \frac{k_{0,z} - k_{1,z}}{k_{0,z} + k_{1,z}} \\
\Psi_{0,1,zz} &= -\frac{k_{1,z}n_0^{\,2} - k_{0,z}n_1^{\,2}}{k_{0,z}n_1^{\,2} + k_{1,z}n_0^{\,2}}
\end{aligned}\quad , \tag{66}$$

$$\begin{aligned}
\Psi_{0,1,xx}\Psi^*_{0,1,xx} &= \frac{-k_{0,z}n_1^{\,2} + k_{1,z}}{k_{0,z}n_1^{\,2} + k_{1,z}}\,\frac{-k_{0,z}n^*{}_1^{\,2} + k^*{}_{1,z}}{k_{0,z}n^*{}_1^{\,2} + k^*{}_{1,z}} \\
\Psi_{0,1,yy}\Psi^*_{0,1,yy} &= \frac{k_{0,z} - k_{1,z}}{k_{0,z} + k_{1,z}}\,\frac{k_{0,z} - k^*{}_{1,z}}{k_{0,z} + k^*{}_{1,z}} \\
\Psi_{0,1,zz}\Psi^*_{0,1,zz} &= \frac{k_{0,z}n_1^{\,2} - k_{1,z}}{k_{0,z}n_1^{\,2} + k_{1,z}}\,\frac{k_{0,z}n^*{}_1^{\,2} - k^*{}_{1,z}}{k_{0,z}n^*{}_1^{\,2} + k^*{}_{1,z}} = \Psi_{0,1,xx}\Psi^*_{0,1,xx}
\end{aligned}\quad , \tag{67}$$

then,

$$R = \Psi_{0,1,yy}\Psi^*_{0,1,yy}\cos^2\chi + \Psi_{0,1,zz}\Psi^*_{0,1,zz}\sin^2\chi\ , \tag{68}$$

Taking an average for χ ,

$$R = \left\langle \Psi_{0,1,yy}\Psi^*_{0,1,yy}\cos^2\chi + \Psi_{0,1,zz}\Psi^*_{0,1,zz}\sin^2\chi \right\rangle_\chi\ , \tag{69}$$

then

$$R = (\Psi_{0,1,yy}\Psi^*_{0,1,yy} + \Psi_{0,1,zz}\Psi^*_{0,1,zz})/2\ , \tag{70}$$

For the reflectivity in the case of s-polarized x-rays incident,

$$R = \Psi_{0,1,yy}\Psi^*_{0,1,yy}\ . \tag{71}$$

Next, we consider the reflection from a flat surface of **a** multilayer with flat interfaces. We consider the electric field $\boldsymbol{E}_{j\text{-}1}$ of x-rays propagating in the j-1-th layer material, the electric field $\boldsymbol{E}_j$ of x-rays propagating in the j-th layer material, and the electric field $\boldsymbol{E'}_{j\text{-}1}$ of x-rays reflected from the j-th layer material at z=$z_{j\text{-}1,j}$ of the interface between the j-1-th layer and j-th layers.

The electric fields $\boldsymbol{E}_{j-1}$, $\boldsymbol{E'}_{j-1}$ on the interface between j-1-th layer and j-th layer and the electric fields $\boldsymbol{E}_j$, $\boldsymbol{E'}_j$ below the interface between j-1-th layer and j-th layer are expressed as

$$\boldsymbol{E}_{j-1}(z_{j-1,j}) = \boldsymbol{A}_{j-1}\exp[i(k_{j-1,x}x + k_{j-1,y}y + k_{j-1,z}h_{j-1} - \omega t)] ,$$
$$\boldsymbol{E}'_{j-1}(z_{j-1,j}) = \boldsymbol{A}'_{j-1}\exp[i(k_{j-1,x}x + k_{j-1,y}y - k_{j-1,z}h_{j-1} - \omega t)] ,$$
$$\boldsymbol{E}_j(z_{j-1,j}) = \boldsymbol{A}_j\exp[i(k_{j,x}x + k_{j,y}y - \omega t)] , \ \boldsymbol{E}'_j(z_{j-1,j}) = \boldsymbol{A}'_j\exp[i(k_{j,x}x + k_{j,y}y - \omega t)] \cdot \quad (72)$$

The electric fields of x-rays at the interface between the j-1-th layer and j-th layer can be formally expressed as follows

$$\boldsymbol{E}_j(z_{j-1,j}) = \boldsymbol{\Phi}_{j-1,j}\boldsymbol{E}_{j-1}(z_{j-1,j}) + \boldsymbol{\Psi}_{j,j-1}\boldsymbol{E}'_j(z_{j-1,j}), \quad (73)$$
$$\boldsymbol{E}'_{j-1}(z_{j-1,j}) = \boldsymbol{\Psi}_{j-1,j}\boldsymbol{E}_{j-1}(z_{j-1,j}) + \boldsymbol{\Phi}_{j,j-1}\boldsymbol{E}'_j(z_{j-1,j}) , \quad (74)$$

where $\Psi_{j-1,j}$ is the Fresnel coefficient tensor for reflection from the interface between the j-1 and j layers, and $\Phi_{j-1,j}$ is the Fresnel coefficient tensor for refraction at the interface between the j-1 and j layers. In addition, the electric field within the j-th layer varies with depth h_j as follows

$$\boldsymbol{E}_j(z_{j,j+1}) = \boldsymbol{E}_j(z_{j-1,j})\exp(ik_{j,z}h_j), \quad (75)$$
$$\boldsymbol{E}'_j(z_{j,j+1}) = \boldsymbol{E}'_j(z_{j-1,j})\exp(-ik_{j,z}h_j) . \quad (76)$$

The amplitudes A_j and A'_j at the j-th layer are derived from the above equations for the interface between the j-1 and j layers as follows

$$\boldsymbol{A}'_{j-1}\exp(-ik_{j-1,z}h_{j-1}) = \boldsymbol{\Psi}_{j-1,j}\boldsymbol{A}_{j-1}\exp(ik_{j-1,z}h_{j-1}) + \boldsymbol{\Phi}_{j,j-1}\boldsymbol{A}'_j , \quad (77)$$
$$\boldsymbol{A}_j = \boldsymbol{\Phi}_{j-1,j}\boldsymbol{A}_{j-1}\exp(ik_{j-1,z}h_{j-1}) + \boldsymbol{\Psi}_{j,j-1}\boldsymbol{A}'_j . \quad (78)$$

This relation is expressed by the following matrix

$$\begin{pmatrix} \boldsymbol{A}'_{j-1}\exp(-ik_{j-1,z}h_{j-1}) \\ \boldsymbol{A}_j \end{pmatrix} = \begin{pmatrix} \boldsymbol{\Psi}_{j-1,j} & \boldsymbol{\Phi}_{j,j-1} \\ \boldsymbol{\Phi}_{j-1,j} & \boldsymbol{\Psi}_{j,j-1} \end{pmatrix}\begin{pmatrix} \boldsymbol{A}_{j-1}\exp(ik_{j-1,z}h_{j-1}) \\ \boldsymbol{A}'_j \end{pmatrix}, \quad (79)$$

Here, the Fresnel coefficient tensor Φ for refraction at the interface between the j-1-th and j-th layers is given by

$$
\begin{aligned}
&\Phi_{j-1,j,xx} = \frac{2k_{j,z}\boldsymbol{k}_{j-1}\cdot\boldsymbol{k}_{j-1}}{k_{j-1,z}\boldsymbol{k}_j\cdot\boldsymbol{k}_j + k_{j,z}\boldsymbol{k}_{j-1}\cdot\boldsymbol{k}_{j-1}} \qquad \Phi_{j,j-1,xx} = \frac{2k_{j-1,z}\boldsymbol{k}_j\cdot\boldsymbol{k}_j}{k_{j-1,z}\boldsymbol{k}_j\cdot\boldsymbol{k}_j + k_{j,z}\boldsymbol{k}_{j-1}\cdot\boldsymbol{k}_{j-1}}\\
&\Phi_{j-1,j,yy} = \frac{2k_{j-1,z}}{k_{j-1,z}+k_{j,z}} \qquad \Phi_{j,j-1,yy} = \frac{2k_{j,z}}{k_{j,z}+k_{j-1,z}}\\
&\Phi_{j-1,j,zz} = \frac{2k_{j-1,z}\boldsymbol{k}_{j-1}\cdot\boldsymbol{k}_{j-1}}{k_{j-1,z}\boldsymbol{k}_j\cdot\boldsymbol{k}_j + k_{j,z}\boldsymbol{k}_{j-1}\cdot\boldsymbol{k}_{j-1}}, \quad \Phi_{j,j-1,zz} = \frac{2k_{j,z}\boldsymbol{k}_j\cdot\boldsymbol{k}_j}{k_{j-1,z}\boldsymbol{k}_j\cdot\boldsymbol{k}_j + k_{j,z}\boldsymbol{k}_{j-1}\cdot\boldsymbol{k}_{j-1}}. \qquad (80)\\
&\Phi_{j-1,j,xy} = \Phi_{j-1,j,yx} = \Phi_{j-1,j,yz} = 0 \qquad \Phi_{j,j-1,xy} = \Phi_{j,j-1,yx} = \Phi_{j,j-1,yz} = 0\\
&\Phi_{j-1,j,zy} = \Phi_{j-1,j,zx} = \Phi_{j-1,j,xz} = 0 \qquad \Phi_{j,j-1,zy} = \Phi_{j,j-1,zx} = \Phi_{j,j-1,xz} = 0
\end{aligned}
$$

The Fresnel coefficient tensor Ψ for reflection from the interface between the j-1 and j layers is given by

$$
\begin{aligned}
&\Psi_{j-1,j,xx} = \frac{k_{j,z}\boldsymbol{k}_{j-1}\cdot\boldsymbol{k}_{j-1} - k_{j-1,z}\boldsymbol{k}_j\cdot\boldsymbol{k}_j}{k_{j-1,z}\boldsymbol{k}_j\cdot\boldsymbol{k}_j + k_{j,z}\boldsymbol{k}_{j-1}\cdot\boldsymbol{k}_{j-1}} \qquad \Psi_{j,j-1,xx} = \frac{k_{j-1,z}\boldsymbol{k}_j\cdot\boldsymbol{k}_j - k_{j,z}\boldsymbol{k}_{j-1}\cdot\boldsymbol{k}_{j-1}}{k_{j-1,z}\boldsymbol{k}_j\cdot\boldsymbol{k}_j + k_{j,z}\boldsymbol{k}_{j-1}\cdot\boldsymbol{k}_{j-1}}\\
&\Psi_{j-1,j,yy} = \frac{k_{j-1,z}-k_{j,z}}{k_{j-1,z}+k_{j,z}} \qquad \Psi_{j,j-1,yy} = \frac{k_{j,z}-k_{j-1,z}}{k_{j-1,z}+k_{j,z}}\\
&\Psi_{j-1,j,zz} = -\frac{k_{j,z}\boldsymbol{k}_{j-1}\cdot\boldsymbol{k}_{j-1} - k_{j-1,z}\boldsymbol{k}_j\cdot\boldsymbol{k}_j}{k_{j-1,z}\boldsymbol{k}_j\cdot\boldsymbol{k}_j + k_{j,z}\boldsymbol{k}_{j-1}\cdot\boldsymbol{k}_{j-1}}, \quad \Psi_{j,j-1,zz} = \frac{k_{j,z}\boldsymbol{k}_{j-1}\cdot\boldsymbol{k}_{j-1} - k_{j-1,z}\boldsymbol{k}_j\cdot\boldsymbol{k}_j}{k_{j-1,z}\boldsymbol{k}_j\cdot\boldsymbol{k}_j + k_{j,z}\boldsymbol{k}_{j-1}\cdot\boldsymbol{k}_{j-1}}. \qquad (81)\\
&\Psi_{j-1,j,xy} = \Psi_{j-1,j,yx} = \Psi_{j-1,j,yz} = 0 \qquad \Psi_{j,j-1,xy} = \Psi_{j,j-1,yx} = \Psi_{j,j-1,yz} = 0\\
&\Psi_{j-1,j,zy} = \Psi_{j-1,j,zx} = \Psi_{j-1,j,xz} = 0 \qquad \Psi_{j,j-1,zy} = \Psi_{j,j-1,zx} = \Psi_{j,j-1,xz} = 0
\end{aligned}
$$

The amplitudes $\boldsymbol{A}_{j-1}$ and $\boldsymbol{A'}_{j-1}$ of the electric fields $\boldsymbol{E}_{j-1}$, $\boldsymbol{E'}_{j-1}$ at the j-th layer and the amplitudes $\boldsymbol{A}_j$ and $\boldsymbol{A'}_j$ of the electric fields $\boldsymbol{E}_j$, $\boldsymbol{E'}_j$ at the j+1-th layer are related by the following equations;

$$
\begin{pmatrix} \boldsymbol{\Phi}_{j-1,j} & 0 \\ 0 & \boldsymbol{\Phi}_{j-1,j} \end{pmatrix}\begin{pmatrix} \boldsymbol{A}_{j-1} \\ \boldsymbol{A'}_{j-1} \end{pmatrix} = \begin{pmatrix} \exp(-ik_{j-1,z}h_{j-1}) & 0 \\ 0 & \exp(ik_{j-1,z}h_{j-1}) \end{pmatrix}\begin{pmatrix} 1 & -\boldsymbol{\Psi}_{j,j-1} \\ \boldsymbol{\Psi}_{j-1,j} & \left(\boldsymbol{\Phi}_{j-1,j}\boldsymbol{\Phi}_{j,j-1} - \boldsymbol{\Psi}_{j-1,j}\boldsymbol{\Psi}_{j,j-1}\right) \end{pmatrix}\begin{pmatrix} \boldsymbol{A}_j \\ \boldsymbol{A'}_j \end{pmatrix}, \qquad (82)
$$

For s-polarization, the Fresnel coefficients are,

$$
\Phi_{j-1,j,yy} = \frac{2k_{j-1,z}}{k_{j-1,z}+k_{j,z}}, \quad \Phi_{j,j-1,yy} = \frac{2k_{j,z}}{k_{j-1,z}+k_{j,z}}. \qquad (83)
$$

$$
\Psi_{j-1,j,yy} = \frac{k_{j-1,z}-k_{j,z}}{k_{j-1,z}+k_{j,z}}, \quad \Psi_{j,j-1,yy} = \frac{k_{j,z}-k_{j-1,z}}{k_{j-1,z}+k_{j,z}}. \qquad (84)
$$

Then, the relations between the amplitudes A_{j-1}, A'_{j-1}, A_j, and A'_j at the interface of the j-1-th and j-th layers are expressed as follows,

$$\begin{pmatrix} A_{j-1} \\ A'_{j-1} \end{pmatrix} = \begin{pmatrix} \exp(-ik_{j-1,z}h_{j-1}) & 0 \\ 0 & \exp(ik_{j-1,z}h_{j-1}) \end{pmatrix} \begin{pmatrix} \dfrac{k_{j-1,z}+k_{j,z}}{2k_{j-1,z}} & \dfrac{k_{j-1,z}-k_{j,z}}{2k_{j-1,z}} \\ \dfrac{k_{j-1,z}-k_{j,z}}{2k_{j-1,z}} & \dfrac{k_{j-1,z}+k_{j,z}}{2k_{j-1,z}} \end{pmatrix} \begin{pmatrix} A_j \\ A'_j \end{pmatrix} . \tag{85}$$

The reflection coefficient is defined as the ratio $\boldsymbol{R}_{0,1}$ of the reflected electric field to the incident electric field at the surface of the material and is given by,

$$\boldsymbol{A'}_0 = \boldsymbol{R}_{0,1}\boldsymbol{A}_0 \cdot \tag{86}$$

The reflection coefficient $\boldsymbol{R}_{j\text{-}1,j}$ of the electric field $\boldsymbol{E'}_{j\text{-}1}$ to the electric field $\boldsymbol{E}_{j\text{-}1}$ at the interface of j-1-th layer and j-th layer is ,

$$\boldsymbol{A'}_{j\text{-}1} = \boldsymbol{R}_{j-1,j}\boldsymbol{A}_{j\text{-}1} , \tag{87}$$

and the ratio $R_{j\text{-}1,j}$ is related to the ratio $R_{j,j+1}$ as follows,

$$\boldsymbol{R}_{j-1,j} = \frac{\boldsymbol{\Psi}_{j-1,j} + (\boldsymbol{\Phi}_{j-1,j}\boldsymbol{\Phi}_{j,j-1} - \boldsymbol{\Psi}_{j-1,j}\boldsymbol{\Psi}_{j,j-1})\boldsymbol{R}_{j,j+1}}{1-\boldsymbol{\Psi}_{j,j-1}\boldsymbol{R}_{j,j+1}} \exp(2ik_{j-1,z}h_{j-1}) \cdot \tag{88}$$

Here, from the relation between the Fresnel coefficient for reflection and the Fresnel coefficient for refraction,

$$\boldsymbol{\Phi}_{j-1,j}\boldsymbol{\Phi}_{j,j-1} - \boldsymbol{\Psi}_{j-1,j}\boldsymbol{\Psi}_{j,j-1} = 1 , \tag{89}$$

$$\boldsymbol{\Psi}_{j-1,j} = -\boldsymbol{\Psi}_{j,j-1} \cdot \tag{90}$$

we can formulate the following relationship

$$\boldsymbol{R}_{j-1,j} = \frac{\boldsymbol{\Psi}_{j-1,j} + \boldsymbol{R}_{j,j+1}}{1+\boldsymbol{\Psi}_{j-1,j}\boldsymbol{R}_{j,j+1}} \exp(2ik_{j-1,z}h_{j-1}) \cdot \tag{91}$$

It is reasonable to assume that no wave will be reflected back from the substrate, so that,

$$\boldsymbol{R}_{N,N+1} = 0 \cdot \tag{92}$$

Then, the x-ray reflectivity is simply,

$$R = \left|R_{0,1}\right|^2 . \tag{93}$$

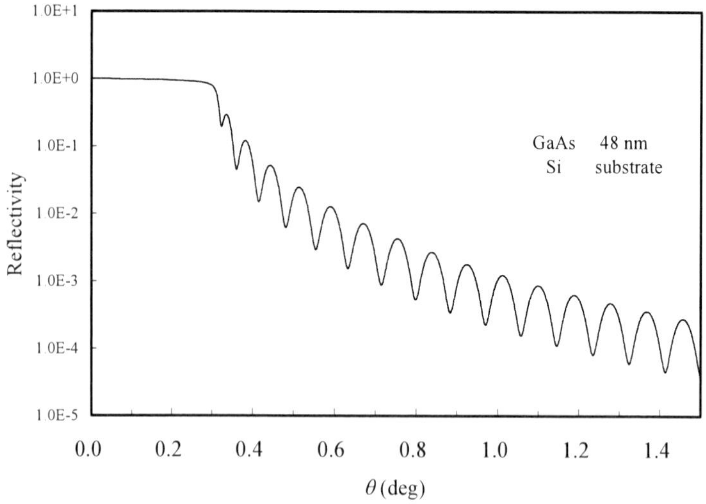

Figure 18. Calculated reflectivity from a silicon wafer covered with a thin (48 nm) GaAs film, not including any roughness effects.

Figure 18 shows the result of a calculation based on these expressions of the reflectivity of x-rays (wave length 0.154 nm) from a silicon wafer covered with a GaAs film with a thickness of 48 nm. The decrease in reflectivity for angles larger than the total reflection critical angle shows oscillations. These oscillations are caused by interference between x-rays that reflect from the surface and those that reflect from the interface.

4.2. Previous Calculations of X-Ray Reflectivity When Roughness Exists at the Surface and Interface

When the surface and interface have roughness, the Fresnel coefficient for reflection is reduced by the roughness.[7-10,19-21] The effect of the roughness was previously put into the calculation based on the theory of Nevot and Croce.[2] The effect of such roughness was taken into account only through the effect of the changes in density of the medium in a vertical direction to the surface and interface. With the use of relevant roughness parameters like the root-mean-square (rms) roughness $\sigma_{j-1,j}$ of the j-th layer, the reduced Fresnel reflection coefficient Ψ' for s-polarization is transformed as shown below,

$$\Psi'_{j,j-1} = \Psi_{j,j-1} \exp(-2k_{j,z} k_{j-1,z} \sigma^2_{j,j-1}) \ , \quad (94)$$

and the x-ray reflectivity is calculated using the following equation,

$$R_{j-1,j} = \frac{R_{j,j+1} + \Psi'_{j-1,j}}{1 + R_{j,j+1} \Psi'_{j-1,j}} \exp(-2ik_{j-1,z} h_{j-1}) \ . \quad (95)$$

Figure 19 shows the reflectivity from the same type of a GaAs-covered silicon wafer as in Figure 18, but now dashed line shows the calculated result in the case that the surface has an rms roughness of 4 nm, and dotted line shows the equivalent result when the surface and interface both have an rms roughness of 4 nm. In the latter case, the reflectivity curve (dots) decreases more quickly than that in Figure 18. However, the ratio of the oscillation amplitude to the value of the reflectivity does not decrease. It seems unnatural that the effect of interference does not also decrease at a rough surface and interface, because the amount of coherent x-rays should reduce due to diffuse scattering at a rough surface and interface.

In the reflectivity curve (dashed line) for a surface roughness of 4 nm and with a flat interface, the ratio of the oscillation amplitude to the size of the reflectivity near an angle of incidence of 0.36° is much larger than the reflectivity of the flat surface in Figure 18. It seems very strange that the interference effects would increase so much at a rough surface.

We now consider the above strange result of the x-ray reflectivity which was calculated based on the Parratt formalism[1] with the use of the Nevot and Croce approach to account for roughness.[2] In that calculation, the x-ray reflectivity is derived using the relation of the reflection coefficient $R_{j-1,j}$ and $R_{j,j+1}$ as follows,

$$R_{j-1,j} = \frac{R_{j,j+1} + \Psi'_{j-1,j}}{1 + R_{j,j+1}\Psi'_{j-1,j}} \exp(-2ik_{j-1,z}h_{j-1}) \text{ ,} \tag{96}$$

where the reduced Fresnel reflection coefficient Ψ' that takes into account the effect of the roughness is as shown below,

$$\Psi'_{j,j-1} = \Psi_{j,j-1} \exp(-2k_{j,z}k_{j-1,z}\sigma^2_{j,j-1}) \text{ .} \tag{97}$$

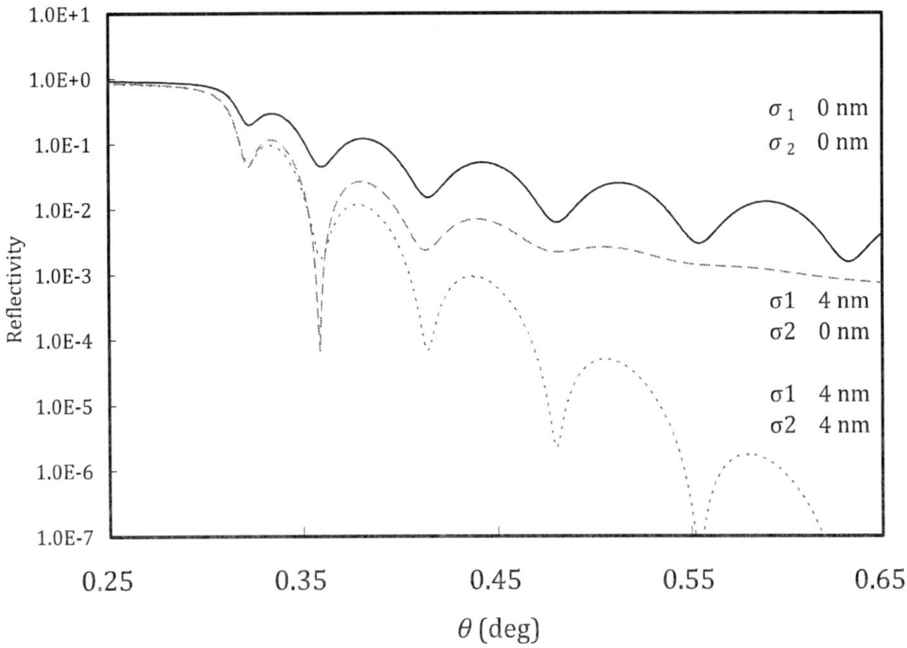

Figure 19. Calculated reflectivities from a GaAs layer with a thickness of 48 nm on a Si substrate. The solid curve is for a flat surface and a flat interface. The dashed curve is for a surface roughness σ1 of 4 nm and a flat interface, while the dotted curve is for a surface roughness σ1 of 4 nm and interface roughness σ2 of 4 nm.

However, the relationship between the reflection coefficients $R_{j\text{-}1,j}$ and $R_{j,j+1}$ was originally derived as the following equation,

$$\boldsymbol{R}_{j-1,j} = \frac{\boldsymbol{\Psi}'_{j-1,j} + (\boldsymbol{\Phi}'_{j-1,j}\boldsymbol{\Phi}'_{j,j-1} - \boldsymbol{\Psi}'_{j-1,j}\boldsymbol{\Psi}'_{j,j-1})\boldsymbol{R}_{j,j+1}}{1 - \boldsymbol{\Psi}'_{j,j-1}\boldsymbol{R}_{j,j+1}} \exp(2ik_{j-1,z}h_{j-1}) \ , \tag{98}$$

Here, the following conditional relations between the Fresnel coefficient for reflection and refraction are relevant to the above equation,

$$\boldsymbol{\Phi}'_{j-1,j}\boldsymbol{\Phi}'_{j,j-1} - \boldsymbol{\Psi}'_{j-1,j}\boldsymbol{\Psi}'_{j,j-1} = 1 \ , \tag{99}$$

and,

$$\boldsymbol{\Psi}'_{j-1,j} = -\boldsymbol{\Psi}'_{j,j-1} \ . \tag{100}$$

then,

$$\boldsymbol{\Phi}'_{j-1,j}\boldsymbol{\Phi}'_{j,j-1} + \boldsymbol{\Psi}'^{2}_{j,j-1} = 1 \ , \tag{101}$$

i.e.,

$$\boldsymbol{\Phi}'_{j-1,j}\boldsymbol{\Phi}'_{j,j-1} = 1 - \boldsymbol{\Psi}'^{2}_{j,j-1} \ . \tag{102}$$

The Fresnel coefficients for refraction at the rough interface are derived using the Fresnel reflection coefficient Ψ as follows,

$$\boldsymbol{\Phi}'_{j-1,j}\boldsymbol{\Phi}'_{j,j-1} - \boldsymbol{\Phi}_{j-1,j}\boldsymbol{\Phi}_{j,j-1} = \boldsymbol{\Psi}^{2}_{j,j-1}\left(1 - \exp(-2k_{j,z}k_{j-1,z}\sigma^{2}_{j,j-1})\right) > 0 \ , \tag{103}$$

$$\boldsymbol{\Phi}'_{j-1,j}\boldsymbol{\Phi}'_{j,j-1} = \boldsymbol{\Phi}_{j-1,j}\boldsymbol{\Phi}_{j,j-1} + \left(1 - \boldsymbol{\Phi}_{j-1,j}\boldsymbol{\Phi}_{j,j-1}\right)\left(1 - \exp(-2k_{j,z}k_{j-1,z}\sigma^{2}_{j,j-1})\right) \ . \tag{104}$$

Therefore, the Fresnel coefficients for refraction at the rough interface are necessarily larger than the Fresnel coefficient for refraction at the flat interface. The resulting increase in the transmission coefficient completely overpowers any decrease in the value of the reflection coefficient. These coefficients for refraction obviously contain a mistake because the penetration of x-rays should decrease at a rough interface because of diffuse scattering. We propose that the unnatural results in the previous calculation of the x-ray reflectivity originate from the fact that diffuse scattering was not considered. In fact equation (99) contains the x-ray energy conservation rule at the interface as the following identity equation for the Fresnel coefficient,

$$\Phi_{j-1,j}\Phi_{j,j-1} - \Psi_{j-1,j}\Psi_{j,j-1} = \Phi_{j-1,j}\Phi_{j,j-1} + \Psi_{j-1,j}^{\ 2} = 1 \ . \tag{105}$$

Here, we consider the energy flow of the x-ray. In electromagnetic radiation, ***E***, ***H***, the energy flow is equal to the Poynting vector

$$p = \frac{1}{4}\left(E^* \times H + E \times H^*\right), \tag{106}$$

where

$$H = \sqrt{\frac{\varepsilon}{\mu}}\frac{k}{k} \times E , \tag{107}$$

and □and □ are the dielectric and magnetic permeability. The Poynting vector is therefore

$$p = \frac{1}{4}\sqrt{\frac{\varepsilon}{\mu}}\left(E^* \times \left(\frac{k}{k} \times E\right) + E \times \left(\frac{k}{k} \times E\right)^*\right) = \frac{1}{4}\sqrt{\frac{\varepsilon}{\mu}}\left(\frac{k}{k}E^* \cdot E + \frac{k^*}{k}E \cdot E^*\right) = \frac{1}{2\varpi\mu}\frac{k + k^*}{2}|E|^2 . \tag{108}$$

Then, the Poynting vector that crosses the interface is

$$\int pdS = \int \frac{1}{2\mu\omega}\frac{k + k^*}{2}|E|^2 dS = \frac{1}{2\mu\omega}\int \frac{k + k^*}{2}|E|^2 dS = \frac{1}{2\mu\omega}\frac{k_z + k_z^*}{2}|A|^2 . \tag{109}$$

The amplitudes A_{j-1} and A'_{j-1} of the electric fields E_{j-1}, E'_{j-1} at the j-th layer and amplitudes A_j and A'_j of the electric fields E_j, E'_j at the j+1-th layer are related by the following equations;

$$\begin{pmatrix} \Phi_{j-1,j} & 0 \\ 0 & \Phi_{j-1,j} \end{pmatrix}\begin{pmatrix} A_{j-1} \\ A'_{j-1} \end{pmatrix} = \begin{pmatrix} \exp(-ik_{j-1,z}h_{j-1}) & 0 \\ 0 & \exp(ik_{j-1,z}h_{j-1}) \end{pmatrix}\begin{pmatrix} 1 & -\Psi_{j,j-1} \\ \Psi_{j-1,j} & (\Phi_{j-1,j}\Phi_{j,j-1} - \Psi_{j-1,j}\Psi_{j,j-1}) \end{pmatrix}\begin{pmatrix} A_j \\ A'_j \end{pmatrix} , \tag{110}$$

When

$$\Phi_{j-1,j}\Phi_{j,j-1} - \Psi_{j-1,j}\Psi_{j,j-1} = 1 , \Psi_{j-1,j} = -\Psi_{j,j-1} , \tag{111}$$

we can describe the above equation as,

$$\begin{pmatrix} \Phi_{j-1,j} & 0 \\ 0 & \Phi_{j-1,j} \end{pmatrix}\begin{pmatrix} A_{j-1} & A'^*_{j-1} \\ A'_{j-1} & A^*_{j-1} \end{pmatrix} = \begin{pmatrix} \exp(-ik_{j-1,z}h_{j-1}) & 0 \\ 0 & \exp(ik_{j-1,z}h_{j-1}) \end{pmatrix}\begin{pmatrix} 1 & \Psi_{j-1,j} \\ \Psi_{j-1,j} & 1 \end{pmatrix}\begin{pmatrix} A_j & A'^*_j \\ A'_j & A^*_j \end{pmatrix} , \tag{112}$$

From the determinant of the refraction matrix,

$$\begin{vmatrix} \Phi_{j-1,j} & 0 \\ 0 & \Phi_{j-1,j} \end{vmatrix}\left(|A_{j-1}|^2 - |A'_{j-1}|^2\right) = \begin{vmatrix} \exp(-ik_{j-1,z}h_{j-1}) & 0 \\ 0 & \exp(ik_{j-1,z}h_{j-1}) \end{vmatrix}\begin{vmatrix} 1 & \Psi_{j-1,j} \\ \Psi_{j-1,j} & 1 \end{vmatrix}\left(|A_j|^2 - |A'_j|^2\right) , \tag{113}$$

then

$$\boldsymbol{\Phi}^2_{j-1,j}\left(\left|\boldsymbol{A}_{j-1}\right|^2-\left|\boldsymbol{A}'_{j-1}\right|^2\right)=\left(1-\boldsymbol{\Psi}^2_{j-1,j}\right)\left(\left|\boldsymbol{A}_j\right|^2-\left|\boldsymbol{A}'_j\right|^2\right), \tag{114}$$

$$\boldsymbol{\Phi}_{j-1,j}\left(\left|\boldsymbol{A}_{j-1}\right|^2-\left|\boldsymbol{A}'_{j-1}\right|^2\right)=\boldsymbol{\Phi}_{j,j-1}\left(\left|\boldsymbol{A}_j\right|^2-\left|\boldsymbol{A}'_j\right|^2\right), \tag{115}$$

$$\frac{2k_{j\text{-}1,z}}{k_{j\text{-}1,z}+k_{j,z}}\left(\left|\boldsymbol{A}_{j-1}\right|^2-\left|\boldsymbol{A}'_{j-1}\right|^2\right)=\frac{2k_{j,z}}{k_{j\text{-}1,z}+k_{j,z}}\left(\left|\boldsymbol{A}_j\right|^2-\left|\boldsymbol{A}'_j\right|^2\right), \tag{116}$$

$$k_{j\text{-}1,z}\left|\boldsymbol{A}_{j-1}\right|^2-k_{j\text{-}1,z}\left|\boldsymbol{A}'_{j-1}\right|^2=k_{j,z}\left|\boldsymbol{A}_j\right|^2-k_{j,z}\left|\boldsymbol{A}'_j\right|^2, \tag{117}$$

i.e., the x-ray energy flow is conserved at the interface. When the Fresnel coefficients at the rough interface obeys the following equations,

$$\boldsymbol{\Phi}'_{j-1,j}\boldsymbol{\Phi}'_{j,j-1}-\boldsymbol{\Psi}'_{j-1,j}\boldsymbol{\Psi}'_{j,j-1}=1\;,\;\boldsymbol{\Psi}'_{j-1,j}=-\boldsymbol{\Psi}'_{j,j-1}, \tag{118}$$

these coefficients fulfil x-ray energy flow conservation at the interface, and so diffuse scattering was not considered at the rough interface.

This conservation expression should not apply any longer when the Fresnel reflection coefficient is replaced by the reduced coefficient Ψ' when there is roughening at the interface. Therefore, calculating the reflectivity using this reduced Fresnel reflection coefficient Ψ' in equation (97) will incorrectly increase the Fresnel transmission coefficient Φ', *i.e.*, $\Phi<\Phi'$.

The penetration of x-rays should decrease at a rough interface because of diffuse scattering. Therefore, the identity equation for the Fresnel coefficients become,

$$\Phi'_{j-1,j}\Phi'_{j,j-1}-\Psi'_{j-1,j}\Psi'_{j,j-1}=\Phi'_{j-1,j}\Phi'_{j,j-1}+\Psi'^{\,2}_{j-1,j}=1-D^2<1\,. \tag{119}$$

where D^2 is a decrease due to diffuse scattering. Then, in the calculation of x-ray reflectivity when there is roughening at the surface or the interface, the Fresnel transmission coefficient Φ' should be used for the reduced coefficient. Several theories exist to describe the influence of roughness on x-ray scattering.[7-10] When the surface and interface are both rough, the Fresnel coefficient for refraction has been derived in several theories.[9,10,19,20]

4.3. The Refractive Fresnel Coefficient of a Rough Interface in Previous Calculations

Initially, we consider the reduced Fresnel coefficient, which is known as the Croce-Nevot factor. When the z position of the interface of 0-th layer and 1-th layer $z_{0,1}$ fluctuates vertically as a function of the lateral position because of the interface roughness, the relations between the amplitudes A_0, A'_0, A_1, and A'_1 are derived by the use of the Fresnel coefficient tensor Φ for refraction and the Fresnel coefficient tensor Ψ for reflection as follows

$$A_1\exp(ik_{1,z}z_{0,1})=\boldsymbol{\Phi}_{0,1}A_0\exp(ik_{0,z}z_{0,1})+\boldsymbol{\Psi}_{1,0}A'_1\exp(-ik_{1,z}z_{0,1}) \tag{120}$$

$$A'_0 \exp(-ik_{0,z}z_{0,1}) = \boldsymbol{\Psi}_{0,1}A_0 \exp(ik_{0,z}z_{0,1}) + \boldsymbol{\Phi}_{1,0}A'_1 \exp(-ik_{1,z}z_{0,1}) \tag{121}$$

then,

$$\begin{pmatrix} \boldsymbol{\Phi}_{0,1}\exp(ik_{0,z}z_{0,1}) & 0 \\ \boldsymbol{\Psi}_{0,1}\exp(ik_{0,z}z_{0,1}) & -\exp(-ik_{0,z}z_{0,1}) \end{pmatrix}\begin{pmatrix} A_0 \\ A'_0 \end{pmatrix} = \begin{pmatrix} \exp(ik_{1,z}z_{0,1}) & -\boldsymbol{\Psi}_{1,0}\exp(-ik_{1,z}z_{0,1}) \\ 0 & -\boldsymbol{\Phi}_{1,0}\exp(-ik_{1,z}z_{0,1}) \end{pmatrix}\begin{pmatrix} A_1 \\ A'_1 \end{pmatrix} \tag{122}$$

$$\begin{pmatrix} A_0 \\ A'_0 \end{pmatrix} = \frac{1}{\boldsymbol{\Phi}_{0,1}}\begin{pmatrix} \exp(-ik_{0,z}z_{0,1}) & 0 \\ \boldsymbol{\Psi}_{0,1}\exp(ik_{0,z}z_{0,1}) & -\boldsymbol{\Phi}_{0,1}\exp(ik_{0,z}z_{0,1}) \end{pmatrix}\begin{pmatrix} \exp(ik_{1,z}z_{0,1}) & -\boldsymbol{\Psi}_{1,0}\exp(-ik_{1,z}z_{0,1}) \\ 0 & -\boldsymbol{\Phi}_{1,0}\exp(-ik_{1,z}z_{0,1}) \end{pmatrix}\begin{pmatrix} A_1 \\ A'_1 \end{pmatrix} \tag{123}$$

$$\begin{pmatrix} A_0 \\ A'_0 \end{pmatrix} = \frac{1}{\boldsymbol{\Phi}_{0,1}}\begin{pmatrix} \exp(-i(k_{0,z}-k_{1,z})z_{0,1}) & -\boldsymbol{\Psi}_{1,0}\exp(-i(k_{0,z}+k_{1,z})z_{0,1}) \\ \boldsymbol{\Psi}_{0,1}\exp(i(k_{0,z}+k_{1,z})z_{0,1}) & (\boldsymbol{\Phi}_{0,1}\boldsymbol{\Phi}_{1,0}-\boldsymbol{\Psi}_{0,1}\boldsymbol{\Psi}_{1,0})\exp(i(k_{0,z}-k_{1,z})z_{0,1}) \end{pmatrix}\begin{pmatrix} A_1 \\ A'_1 \end{pmatrix} \tag{124}$$

where

$$\Psi_{0,1,yy} = \frac{k_{0,z}-k_{1,z}}{k_{0,z}+k_{1,z}}, \quad \Psi_{1,0yy} = \frac{k_{1,z}-k_{0,z}}{k_{0,z}+k_{1,z}} \tag{125}$$

$$\Phi_{0,1,yy} = \frac{2k_{0,z}}{k_{0,z}+k_{1,z}}, \quad \Phi_{1,0,yy} = \frac{2k_{1,z}}{k_{1,z}+k_{0,z}} \tag{126}$$

then

$$\begin{pmatrix} A_0 \\ A'_0 \end{pmatrix} = \begin{pmatrix} \frac{k_{0,z}+k_{1,z}}{2k_{0,z}}\exp(-i(k_{0,z}-k_{1,z})z_{0,1}) & \frac{k_{0,z}-k_{1,z}}{2k_{0,z}}\exp(-i(k_{0,z}+k_{1,z})z_{0,1}) \\ \frac{k_{0,z}-k_{1,z}}{2k_{0,z}}\exp(i(k_{0,z}+k_{1,z})z_{0,1}) & \frac{k_{0,z}+k_{1,z}}{2k_{0,z}}\exp(i(k_{0,z}-k_{1,z})z_{0,1}) \end{pmatrix}\begin{pmatrix} A_1 \\ A'_1 \end{pmatrix} \tag{127}$$

We take the average value of the matrix over the whole area coherently illuminated by the incident x-ray beam. This leads to

$$\begin{pmatrix} A_0 \\ A'_0 \end{pmatrix} = \begin{pmatrix} \frac{k_{0,z}+k_{1,z}}{2k_{0,z}}\langle\exp(-i(k_{0,z}-k_{1,z})z_{0,1})\rangle & \frac{k_{0,z}-k_{1,z}}{2k_{0,z}}\langle\exp(-i(k_{0,z}+k_{1,z})z_{0,1})\rangle \\ \frac{k_{0,z}-k_{1,z}}{2k_{0,z}}\langle\exp(i(k_{0,z}+k_{1,z})z_{0,1})\rangle & \frac{k_{0,z}+k_{1,z}}{2k_{0,z}}\langle\exp(i(k_{0,z}-k_{1,z})z_{0,1})\rangle \end{pmatrix}\begin{pmatrix} A_1 \\ A'_1 \end{pmatrix} \tag{128}$$

For Gaussian statistics of standard deviation value □ ,

$$g(z) = \frac{1}{\sqrt{2\pi}\sigma}\exp(-\frac{z^2}{2\sigma^2}) \tag{129}$$

$$\langle f(z)\rangle = \int_{-\infty}^{\infty} g(z)f(z)dz = \int_{-\infty}^{\infty}\frac{1}{\sqrt{2\pi}\sigma}\exp(-\frac{z^2}{2\sigma^2})f(z)dz \tag{130}$$

$$\left\langle \exp(ikz_{0,1}) \right\rangle = \int_{-\infty}^{\infty} g(z_{0,1}) \exp(ikz_{0,1}) dz_{0,1} = \int_{-\infty}^{\infty} \frac{1}{\sqrt{2\pi}\sigma_{0,1}} \exp(-\frac{z_{0,1}^{\ 2}}{2\sigma_{0,1}^{\ 2}}) \exp(ikz_{0,1}) dz_{0,1} = \exp\left(-\frac{1}{2} k^2 \sigma_{0,1}^{\ 2} \right) \tag{131}$$

$$\begin{pmatrix} A_0 \\ A'_0 \end{pmatrix} = \begin{pmatrix} \frac{k_{0,z}+k_{1,z}}{2k_{0,z}} \exp(-\frac{1}{2}(k_{0,z}-k_{1,z})^2 \sigma_{0,1}^{\ 2}) & \frac{k_{0,z}-k_{1,z}}{2k_{0,z}} \exp(-\frac{1}{2}(k_{0,z}+k_{1,z})^2 \sigma_{0,1}^{\ 2}) \\ \frac{k_{0,z}-k_{1,z}}{2k_{0,z}} \exp(-\frac{1}{2}(k_{0,z}+k_{1,z})^2 \sigma_{0,1}^{\ 2}) & \frac{k_{0,z}+k_{1,z}}{2k_{0,z}} \exp(-\frac{1}{2}(k_{0,z}-k_{1,z})^2 \sigma_{0,1}^{\ 2}) \end{pmatrix} \begin{pmatrix} A_1 \\ A'_1 \end{pmatrix} \tag{132}$$

Therefore

$$\begin{pmatrix} A'_0 \\ A_1 \end{pmatrix} = \begin{pmatrix} \frac{k_{0,z}-k_{1,z}}{k_{0,z}+k_{1,z}} \exp(-2k_{0,z}k_{1,z}\sigma_{0,1}^{\ 2}) & \frac{2k_{1,z}}{k_{0,z}+k_{1,z}} \exp(\frac{1}{2}(k_{0,z}-k_{1,z})^2 \sigma_{0,1}^{\ 2}) \left(\frac{(k_{0,z}+k_{1,z})^2}{4k_{0,z}k_{1,z}} \exp(-(k_{0,z}-k_{1,z})^2 \sigma_{0,1}^{\ 2}) - \frac{(k_{0,z}-k_{1,z})^2}{4k_{0,z}k_{1,z}} \exp(-(k_{0,z}+k_{1,z})^2 \sigma_{0,1}^{\ 2}) \right) \\ \frac{2k_{0,z}}{k_{0,z}+k_{1,z}} \exp(\frac{1}{2}(k_{0,z}-k_{1,z})^2 \sigma_{0,1}^{\ 2}) & \frac{k_{1,z}-k_{0,z}}{k_{0,z}+k_{1,z}} \exp(-2k_{0,z}k_{1,z}\sigma_{0,1}^{\ 2}) \end{pmatrix} \begin{pmatrix} A_0 \\ A'_1 \end{pmatrix} \tag{133}$$

$$\begin{pmatrix} A'_0 \\ A_1 \end{pmatrix} = \begin{pmatrix} \boldsymbol{\Psi}'_{0,1} & \boldsymbol{\Phi}'_{1,0} \\ \boldsymbol{\Phi}'_{0,1} & \boldsymbol{\Psi}'_{1,0} \end{pmatrix} \begin{pmatrix} A_0 \\ A'_1 \end{pmatrix} \tag{134}$$

Then the Fresnel reflection coefficients Ψ' are reduced as follows

$$\Psi'_{0,1} = \Psi_{0,1} \exp(-2k_{0,z}k_{1,z}\sigma_{0,1}^2) \quad , \quad \Psi'_{1,0} = -\Psi_{0,1} \ . \tag{135}$$

However, the Fresnel refraction coefficients Φ' increase as follows

$$\boldsymbol{\Phi}'_{0,1} = \boldsymbol{\Phi}_{0,1} \exp(\frac{1}{2}(k_{0,z}-k_{1,z})^2 \sigma_{0,1}^{\ 2}) \tag{136}$$

$$\boldsymbol{\Phi}'_{1,0} = \boldsymbol{\Phi}_{1,0} \exp(\frac{1}{2}(k_{0,z}-k_{1,z})^2 \sigma_{0,1}^{\ 2}) \left(\frac{(k_{0,z}+k_{1,z})^2}{4k_{0,z}k_{1,z}} \exp(-(k_{0,z}-k_{1,z})^2 \sigma_{0,1}^{\ 2}) - \frac{(k_{0,z}-k_{1,z})^2}{4k_{0,z}k_{1,z}} \exp(-(k_{0,z}+k_{1,z})^2 \sigma_{0,1}^{\ 2}) \right) \tag{137}$$

The modified Fresnel refraction coefficients $\Phi'_{0,1}$ corresponds to equation (10.29) in p.200 of Holy[7], equation (8.24) in p.242 of Daillant[8] and equation (1.117) in p.29 of Sakurai[21]. However, no one obtained the expression corresponding to $\Phi'_{1,0}$. It is peculiar that $\Phi'_{1,0}$ and $\Phi'_{0,1}$ are asymmetrical. It comes to cause a different result if 1-th layer and 0-th layers are replaced and calculated. Therefore this derived Φ' should not be used to calculate the reflectivity from rough surfaces and interfaces.

The derived Fresnel refraction coefficients Φ' increase. This increase in the transmission coefficient completely overpowers any decrease in the value of the reflection coefficient as the following,

$$\boldsymbol{\Phi}'_{0,1} \boldsymbol{\Phi}'_{1,0} - \boldsymbol{\Psi}'_{0,1} \boldsymbol{\Psi}'_{1,0} = 1 \tag{138}$$

$$\boldsymbol{\Phi}'_{j-1,j} \boldsymbol{\Phi}'_{j,j-1} - \boldsymbol{\Phi}_{j-1,j} \boldsymbol{\Phi}_{j,j-1} = \boldsymbol{\Psi}^2_{j,j-1} \left(1 - \exp(-2k_{j,z}k_{j-1,z}\sigma^2_{j,j-1}) \right) > 0 \ , \tag{139}$$

Moreover, if the deformation modulus of $\Phi'_{1,0}$ is assumed to be $\Phi'_{0,1}$, the left side of equation (99) exceeds unity, and therefore equation (136) is obviously wrong.

In Nevot and Croce's treatment of the Parratt formalism for the reflectivity calculation including surface and interface roughness,[2] the relations of the Fresnel coefficients between reflection and transmission as Equations (99), (100) and (118) were not shown. Furthermore, the modification of the Fresnel coefficients according to Nevot and Croce has been used for only surface and interface reflection. However, the modification of the transmission coefficients has an important role when the roughness of the surface or interface is high, and the effect of diffuse scattering due to that roughness should not be ignored. The error in Nevot and Croce's treatment[2] originates in the fact that the modified Fresnel coefficients was calculated based on the Parratt formalism which contains the x-ray energy conservation rule at the surface and interface. In the discussion on pp.767-768 of Nevot and Croce's[2], their Fresnel coefficients at the rough interface fulfil x-ray energy flow conservation at the interface, and so diffuse scattering was ignored at the rough interface. In their discussion, the transmission coefficients t_R and t_I were replaced approximately by the reflection coefficients r_R and r_I by the ignoring diffuse scattering term, and according to the principle of conservation energy. The reflection coefficient r_R at the rough interface should be expressed as a function of the reflection coefficient r_I and transmission coefficient t_I. However, the reflection coefficient r_R at the rough interface was expressed only by the reflection coefficient r_I, while the transmission coefficient t_I had already been replaced by the reflection coefficient r_I by the ignoring diffuse scattering term in the relationship based on the principle of the conservation of energy. Thus, the reflection coefficient r_R at the rough interface as equation (11) of p.771 in Nevot and Croce[2] had been expressed with the reflection coefficient r_I only, and this results in the equation was also sure to include the conservation of energy.

The resulting increase in the transmission coefficient completely overpowers any decrease in the value of the reflection coefficient at the rough interface. Thus, because Nevot and Croce's treatment of the Parratt formalism contains an fundamental mistake regardless of the size of the roughness, results using this approach cannot be correct. The size of the modification of the transmission coefficient is one-order smaller than that of reflection coefficient, but, the size of transmission coefficient is one-order larger than the reflection coefficient at angles larger than critical angle. Thus, the errors of transmittance without the modification cannot be ignored.

Of course, there are cases where that Nevot and Croce's treatment can be applied. However, their method can be applied only to the case where there is no density distribution change at all in the direction parallel to the surface on the surface field side, and only when the scattering vector is normal to the surface. A typical example of surface medium to which this model can be applied is one where only the density distribution change in the vertical direction to the surface exists, as caused by diffusion etc. In such a special case, Nevot and Croce's treatment can be applied without any problem. However, because a general multilayer film always has structure in a direction parallel to the surface field side, Nevot and Croce's expression fails even when the roughness is extremely small. The use of only Fresnel reflection coefficients by Nevot and Croce is an fundamental mistake that does not depend on the size of the roughness.

4.4. A New Accurate Formula for the Reflectivity from a Multilayer with a Rough Surface and Rough Interfaces

To proceed, we therefore reconsider the derivation of the average value of the matrix. When the z position of the interface of the 0-th layer and 1-th layer $z_{0,1}$ fluctuates vertically as a function of the lateral position because of the interface roughness, the relations between the amplitudes A_0, A'_0, A_1, and A'_1 are derived by the use of the Fresnel coefficient tensor Φ for refraction and the Fresnel coefficient tensor Ψ for reflection as follows

$$\begin{pmatrix} \exp(-ik_{0,z}z_{0,1}) & 0 \\ 0 & \exp(ik_{1,z}z_{0,1}) \end{pmatrix}\begin{pmatrix} A'_0 \\ A_1 \end{pmatrix} = \begin{pmatrix} \boldsymbol{\Psi}_{0,1} & \boldsymbol{\Phi}_{1,0} \\ \boldsymbol{\Phi}_{0,1} & \boldsymbol{\Psi}_{1,0} \end{pmatrix}\begin{pmatrix} \exp(ik_{0,z}z_{0,1}) & 0 \\ 0 & \exp(-ik_{1,z}z_{0,1}) \end{pmatrix}\begin{pmatrix} A_0 \\ A'_1 \end{pmatrix} \tag{140}$$

$$\begin{pmatrix} A'_0 \\ A_1 \end{pmatrix} = \frac{1}{\exp(-ik_{0,z}z_{0,1})\exp(ik_{1,z}z_{0,1})}\begin{pmatrix} \boldsymbol{\Psi}_{0,1}\exp(i(k_{1,z}+k_{0,z})z_{0,1}) & \boldsymbol{\Phi}_{1,0}\exp(i(k_{1,z}-k_{1,z})z_{0,1}) \\ \boldsymbol{\Phi}_{0,1}\exp(i(-k_{0,z}+k_{0,z})z_{0,1}) & \boldsymbol{\Psi}_{1,0}\exp(i(-k_{0,z}-k_{1,z})z_{0,1}) \end{pmatrix}\begin{pmatrix} A_0 \\ A'_1 \end{pmatrix} \tag{141}$$

$$\begin{pmatrix} A'_0 \\ A_1 \end{pmatrix} = \begin{pmatrix} \boldsymbol{\Psi}_{0,1}\dfrac{\exp(i(k_{1,z}+k_{0,z})z_{0,1})}{\exp(i(-k_{0,z}+k_{1,z})z_{0,1})} & \boldsymbol{\Phi}_{1,0}\exp(i(k_{0,z}-k_{1,z})z_{0,1}) \\ \boldsymbol{\Phi}_{0,1}\exp(i(-k_{1,z}+k_{0,z})z_{0,1}) & \boldsymbol{\Psi}_{1,0}\dfrac{\exp(i(-k_{0,z}-k_{1,z})z_{0,1})}{\exp(i(-k_{0,z}+k_{1,z})z_{0,1})} \end{pmatrix}\begin{pmatrix} A_0 \\ A'_1 \end{pmatrix} \tag{142}$$

Again, we take the average value of this matrix,

$$\begin{pmatrix} A'_0 \\ A_1 \end{pmatrix} = \begin{pmatrix} \boldsymbol{\Psi}_{0,1}\dfrac{\langle\exp(i(k_{1,z}+k_{0,z})z_{0,1})\rangle}{\langle\exp(i(-k_{0,z}+k_{1,z})z_{0,1})\rangle} & \boldsymbol{\Phi}_{1,0}\langle\exp(i(k_{0,z}-k_{1,z})z_{0,1})\rangle \\ \boldsymbol{\Phi}_{0,1}\langle\exp(i(-k_{1,z}+k_{0,z})z_{0,1})\rangle & \boldsymbol{\Psi}_{1,0}\dfrac{\langle\exp(i(-k_{0,z}-k_{1,z})z_{0,1})\rangle}{\langle\exp(i(-k_{0,z}+k_{1,z})z_{0,1})\rangle} \end{pmatrix}\begin{pmatrix} A_0 \\ A'_1 \end{pmatrix} \tag{143}$$

For Gaussian statistics of standard deviation value σ , the Fresnel reflection coefficient Ψ' are as follows

$$\boldsymbol{\Psi}'_{0,1} = \boldsymbol{\Psi}_{0,1}\frac{\langle\exp(i(k_{1,z}+k_{0,z})z_{0,1})\rangle}{\langle\exp(i(-k_{0,z}+k_{1,z})z_{0,1})\rangle} = \boldsymbol{\Psi}_{0,1}\frac{\exp(-\frac{1}{2}(k_{0,z}+k_{1,z})^2\sigma_{0,1}{}^2)}{\exp(-\frac{1}{2}(k_{0,z}-k_{1,z})^2\sigma_{0,1}{}^2)} = \boldsymbol{\Psi}_{0,1}\exp(-2k_{0,z}k_{1,z}\sigma_{0,1}{}^2)\,. \tag{144}$$

$$\boldsymbol{\Psi}'_{1,0} = \boldsymbol{\Psi}_{1,0}\frac{\langle\exp(i(-k_{0,z}-k_{1,z})z_{0,1})\rangle}{\langle\exp(i(-k_{0,z}+k_{1,z})z_{0,1})\rangle} = \boldsymbol{\Psi}_{1,0}\frac{\exp(-\frac{1}{2}(-k_{0,z}-k_{1,z})^2\sigma_{0,1}{}^2)}{\exp(-\frac{1}{2}(k_{1,z}-k_{0,z})^2\sigma_{0,1}{}^2)} = \boldsymbol{\Psi}_{1,0}\exp(-2k_{0,z}k_{1,z}\sigma_{0,1}{}^2)\,. \tag{145}$$

Because x-rays that penetrate an interface reflect from the interface below, and penetrate former interface again without fail, it is necessary to treat the refraction coefficients $\Phi'_{0,1}$ and $\Phi'_{1,0}$ collectively.

$$\begin{aligned}\boldsymbol{\Phi}'_{0,1}\boldsymbol{\Phi}'_{1,0} &= \left\langle \boldsymbol{\Phi}_{0,1}\exp(i(-k_{1,z}+k_{0,z})z_{0,1})\boldsymbol{\Phi}_{1,0}\exp(i(k_{0,z}-k_{1,z})z_{0,1})\right\rangle \\ &= \boldsymbol{\Phi}_{0,1}\boldsymbol{\Phi}_{1,0}\left\langle \exp(i(-k_{1,z}+k_{0,z})z_{0,1})\exp(i(k_{0,z}-k_{1,z})z_{0,1})\right\rangle \\ &= \boldsymbol{\Phi}_{0,1}\boldsymbol{\Phi}_{1,0}\left\langle \exp(i(2k_{0,z}-2k_{1,z})z_{0,1})\right\rangle \\ &= \boldsymbol{\Phi}_{0,1}\boldsymbol{\Phi}_{1,0}\exp(-2(k_{0,z}-k_{1,z})^2\sigma_{0,1}{}^2)\end{aligned} \tag{146}$$

Then the Fresnel coefficients Ψ' and Φ' are reduced as follows

$$\boldsymbol{\Psi}'_{0,1} = \boldsymbol{\Psi}_{0,1}\exp(-2k_{0,z}k_{1,z}\sigma_{0,1}^2)\ , \tag{147}$$

$$\boldsymbol{\Psi}'_{1,0} = \boldsymbol{\Psi}_{1,0}\exp(-2k_{0,z}k_{1,z}\sigma_{0,1}^2)\ , \tag{148}$$

$$\boldsymbol{\Phi}'_{0,1} = \boldsymbol{\Phi}_{0,1}\exp(-(k_{0,z}-k_{1,z})^2\sigma_{0,1}{}^2)\ , \tag{149}$$

$$\boldsymbol{\Phi}'_{1,0} = \boldsymbol{\Phi}_{1,0}\exp(-(k_{0,z}-k_{1,z})^2\sigma_{0,1}{}^2)\ , \tag{150}$$

$$\begin{pmatrix} A'_0 \\ A_1 \end{pmatrix} = \begin{pmatrix} \boldsymbol{\Psi}'_{0,1} & \boldsymbol{\Phi}'_{1,0} \\ \boldsymbol{\Phi}'_{0,1} & \boldsymbol{\Psi}'_{1,0} \end{pmatrix}\begin{pmatrix} A_0 \\ A'_1 \end{pmatrix} \tag{151}$$

Then

$$\begin{pmatrix} A'_0 \\ A_1 \end{pmatrix} = \begin{pmatrix} \boldsymbol{\Psi}_{0,1}\exp(-2k_{0,z}k_{1,z}\sigma_{0,1}{}^2) & \boldsymbol{\Phi}_{1,0}\exp(-(k_{0,z}-k_{1,z})^2\sigma_{0,1}{}^2) \\ \boldsymbol{\Phi}_{0,1}\exp(-(k_{0,z}-k_{1,z})^2\sigma_{0,1}{}^2) & \boldsymbol{\Psi}_{1,0}\exp(-2k_{0,z}k_{1,z}\sigma_{0,1}{}^2) \end{pmatrix}\begin{pmatrix} A_0 \\ A'_1 \end{pmatrix} \tag{152}$$

$$\begin{pmatrix} A'_0 \\ A_1 \end{pmatrix} = \begin{pmatrix} \dfrac{k_{0,z}-k_{1,z}}{k_{0,z}+k_{1,z}}\exp(-2k_{0,z}k_{1,z}\sigma_{0,1}{}^2) & \dfrac{2k_{1,z}}{k_{0,z}+k_{1,z}}\exp(-(k_{0,z}-k_{1,z})^2\sigma_{0,1}{}^2) \\ \dfrac{2k_{0,z}}{k_{0,z}+k_{1,z}}\exp(-(k_{0,z}-k_{1,z})^2\sigma_{0,1}{}^2) & \dfrac{k_{1,z}-k_{0,z}}{k_{0,z}+k_{1,z}}\exp(-2k_{0,z}k_{1,z}\sigma_{0,1}{}^2) \end{pmatrix}\begin{pmatrix} A_0 \\ A'_1 \end{pmatrix} \tag{153}$$

and

$$\boldsymbol{\Phi}'_{0,1}\boldsymbol{\Phi}'_{1,0}-\boldsymbol{\Psi}'_{0,1}\boldsymbol{\Psi}'_{1,0} = \frac{4k_{0,z}k_{1,z}}{(k_{0,z}+k_{1,z})^2}\exp(-2(k_{0,z}-k_{1,z})^2\sigma_{0,1}{}^2)+(\frac{k_{1,z}-k_{0,z}}{k_{0,z}+k_{1,z}})^2\exp(-4k_{0,z}k_{1,z}\sigma_{0,1}{}^2) < 1 \tag{154}$$

The Fresnel refraction coefficients Φ' derived by this method are reduced, and can be used to calculate the reflectivity from rough surface and interface. Therefore, we calculate the reflectivity using these newly-derived Fresnel coefficients in an accurate reflectivity equation of $R_{j-1,j}$ and $R_{j,j+1}$ as follows,

$$\boldsymbol{R}_{j-1,j} = \frac{\boldsymbol{\Psi}'_{j-1,j}+(\boldsymbol{\Phi}'_{j-1,j}\boldsymbol{\Phi}'_{j,j-1}-\boldsymbol{\Psi}'_{j-1,j}\boldsymbol{\Psi}'_{j,j-1})\boldsymbol{R}_{j,j+1}}{1-\boldsymbol{\Psi}'_{j,j-1}\boldsymbol{R}_{j,j+1}}\exp(2ik_{j-1,z}h_{j-1})\ , \tag{155}$$

Based on the above considerations, we again calculated the x-ray reflectivity for the GaAs/Si system, but now considered the effect of attenuation in the refracted x-rays by diffuse scattering resulting from surface roughness. The results are shown as the dashed line in Figure 20 for a surface roughness of 4 nm and flat interface, and the dotted line shows the calculated result in the case that the surface and interface both have a rms roughness of 4 nm.

The ratio of the oscillation amplitude to the size of the reflectivity in the reflectivity curve (dot) in Figure 20 is smaller than that of the reflectivity curve Figure 19. In the reflectivity curve (dashed line), the very large amplitude of the oscillation near an angle of incidence of 0.36° in Figure 19 has disappeared in Figure 20. These results are now physically reasonable. All the strange results seen in Figure 19 have disappeared in Figure 20. It seems natural that the effect of interference does decrease at a rough surface and interface, because the amount of coherent x-rays should reduce due to diffuse scattering.

In concluding, we have investigated the fact that the calculated result of the x-ray reflectivity based on Parratt formalism[1] with the effect of the roughness incorporated by the theory of Nevot-Croce[2] show a strange phenomenon in which the amplitude of the oscillation due to the interference effects increase in the case of the rougher surface.

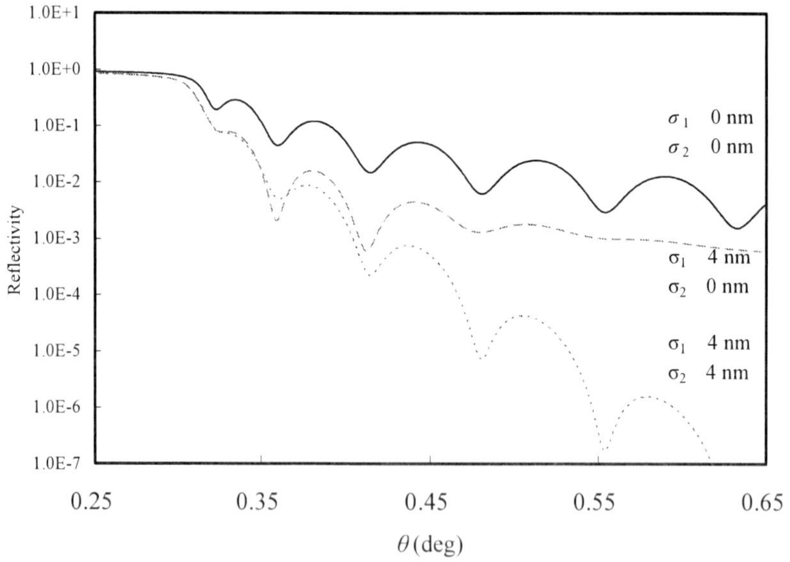

Figure 20. New calculated reflectivities from a GaAs layer with a thickness of 48 nm on a Si substrate. The line is for a flat surface and a flat interface. The dashed curve is for a surface roughness σ_1 of 4 nm and with a flat interface, while the dotted curve is for a surface roughness σ_1 of 4 nm and-interface roughness σ_2 of 4 nm.

We found that the strange result originates in the currently used equation due to a serious mistake where the Fresnel refraction coefficient in the reflectivity equation is increased at a rough interface. We have developed a new formalism that corrects this mistake, producing more accurate estimates of the x-ray reflectivity for systems having surface and interfacial roughness, taking into account the effect of roughness-induced diffuse scattering. The new, accurate formalism is completely described in detail. The x-ray reflectivity R is derived by the use of accurate reflectivity equations for $R_{j-1,j}$ and $R_{j,j+1}$ as following,

$$R = \left|R_{0,1}\right|^2 ,$$

$$R_{j-1,j} = \frac{\Psi'_{j-1,j} + (\Phi'_{j-1,j}\Phi'_{j,j-1} - \Psi'_{j-1,j}\Psi'_{j,j-1})R_{j,j+1}}{1 - \Psi'_{j,j-1}R_{j,j+1}} \exp(2ik_{j-1,z}h_{j-1}) ,$$

$$R_{N,N+1} = 0 ,$$

where the Fresnel coefficients Ψ' for reflection and Φ' for refraction are reduced as follows

$$\Psi'_{j-1,j} = \frac{k_{j-1,z} - k_{j,z}}{k_{j-1,z} + k_{j,z}} \exp\left(-2k_{j-1,z}k_{j,z}\sigma_{j-1,j}^{\ 2}\right), \quad \Psi'_{j,j-1} = -\Psi'_{j-1,j}$$

$$\Phi'_{j-1,j} = \frac{2k_{j-1,z}}{k_{j-1,z} + k_{j,z}} \exp\left(-(k_{j-1,z} - k_{j,z})^2\sigma_{j-1,j}^{\ 2}\right), \quad \Phi'_{j,j-1} = \Phi'_{j-1,j}\frac{k_{j,z}}{k_{j-1,z}}$$

The reflectivity calculated with this new, accurate formalism, gives a physically reasonable result. The use of this equation resolves the strange numerical results that occurred in the previous calculations that neglected diffuse scattering and is expected that buried interface structure can now be analyzed more accurately. Here, the reduced Fresnel coefficients Ψ' and Φ' in the reflectivity equation needs further research and we will continue to refine this theory.

REFERENCES

[1] Parratt LG; *Phys. Rev.* 95, 359 (1954).

[2] Nevot L and Croce P, *Rev. Phys. Appl.* 15 761 (1980).

[3] Marra W C, Eisenberger P and Cho A Y, *J. Appl. Phys.* 50 6927 (1979).

[4] Robinson I K, *Phys. Rev.* B 33 3830 (1986).

[5] Fujii Y, Nakayama T and Yoshida K, *ISIJ International* 44 1549 (2004).

[6] Fujii Y, Komai T, Ikeda K; *Surf.Interface Anal.* 37, 190 (2005).

[7] Holy V, Pietsch U and Baumbach T, *High-Resolution X-Ray Scattering from Thin Films and Multilayers,* Springer, Berlin (1999).

[8] Daillant J and Gibaud A, *X-ray and Neutron Reflectivity, Principles and Applications,* Springer, Berlin (1999).

[9] Shinha SK, Sirota EB, Garoff S, Stanley HB; *Phys. Rev.* B38, 2297 (1988).

[10] Holy V, Kubena J, Ohlidal I, Lischka K and Plotz W, *Phys. Rev.* B 47 15896 (1993).

[11] Vineyard GH; *Phys. Rev.* B50, 4146 (1982).

[12] Fujii Y, Nakayama T; *Surf.Interface Anal.* 38, 1679 (2006).

[13] Fujii Y, Yoshida K, Nakamura T, Yoshida K; *Rev. Sci. Instrum.* 68, 1975 (1997).

[14] Slater C, and Frank NH; *Electromagnetism*, McGraw-Hill, New York (1947).

[15] Predecki P et. al.; *Advances in X-ray Analysis* 36, 237 (1993).

[16] Sakaida Y, Harada S, Tanaka K; *J. Soc. Mat. Sci. Japan* 42-477, 641 (1993).

[17] Fujii Y, Yanase E, Arai K; *Applied Auface Science* 244, 230 (2005).

[18] Yanase E, ZolotarevKV, Nishio K, Kusumi Y, Okado H, Arai K; *J. Synchrotron Rad.* 9, 309 (2002)..

[19] Vidal B and Vincent P 1984 *Applied Optics* 23 1794 (1984).
[20] Boer D K G 1995 *Phys. Rev.* B 51 5297 (1995).
[21] Sakurai K, *Introduction to X-ray Reflectivity,* Kodansha Scientific, Tokyo (2009).
[22] Fujii Y, *Surf. Interface Anal.* 42 1642 (2010).

In: X-Ray Scattering
Editor: Christopher M. Bauwens
ISBN: 978-1-61324-326-8

Chapter 8

SAXS/WAXS CHARACTERIZATION OF SOL-GEL DERIVED NANOMATERIALS

Gang Chen
Ohio University, Athens, Ohio, USA

ABSTRACT

Small- and wide-angle X-ray scattering (SAXS and WAXS) are useful techniques for characterizing structures of disordered materials. An important branch of the disordered materials is sol-gel derived nanomaterials, which have many applications such as in catalysis, sorption, sensing, drug delivery, photovoltaics, and nanofabrication. In this chapter we first review the principles and methodology behind the SAXS and WAXS techniques, and then we focus on applications of these techniques to the characterization of sol-gel derived nanomaterials. Structural information such as nanoscale morphology, fractal dimension, surface area, defects, and short- and intermediate-range order can be obtained. Specific examples are provided with an emphasis on the characterization of the intermediate-range and nanoscale structures of the sol-gel derived nanomaterials. Finally we introduce synchrotron-based SAXS/WXAS techniques and address corresponding new opportunities for investigating nanostructured materials.

1. INTRODUCTION

Even though X-rays were first discovered by Roentgen in 1895, applications of X-rays to the study of materials did not start until 1912 when Laue predicted for the first time that interference of X-rays with a crystal generates a diffraction pattern. [1] Two years later, the Braggs (a farther and son team) developed a theory to determine crystal structure from the diffraction pattern. [2] Since then, X-ray diffraction has been widely used as a tool to study atomic structure of crystals.

X-ray scattering is a process caused by elastic interaction of X-rays with matter. Any ordered or disordered material can be used to generate an X-ray scattering pattern. In contrast, X-ray diffraction patterns can only be observed from materials with periodically arranged structure (e.g., crystals). Therefore, X-ray diffraction is a subset of X-ray scattering. In general, X-ray scattering can be divided into two categories: small- and wide-angle X-ray scattering (SAXS and WAXS), depending on the range of scattering angles. The SAXS refers to scattering angles less than 5° and provides information about inhomogeneity of materials beyond the atomic scale. In contrast, the WAXS has scattering angles larger than 5° and is often used to investigate atomic structure of materials.

In this chapter, we first review briefly the principles behind the SAXS and WAXS techniques. Then we introduce sol-gel derived materials with a focus on nanostructured materials. Applications of SAXS and WAXS to the characterization of sol-gel derived nanomaterials will be discussed in details. Finally, we address recent advancement of SAXS/WAXS techniques based on synchrotron X-rays. A portion of this chapter was published previously in reference 34.

1.1. Theoretical Basis of Small-Angle X-Ray Scattering

X-ray scattering is a powerful tool for investigating structures of a wide range of materials. The scattering process originates from elastic photon-electron interaction during which X-rays are scattered by electrons inside atoms and then distributed at different angles relative to the primary beam. The electron density distribution of a material can be obtained from the scattering angle dependent X-ray intensity. In general, the amplitude of X-rays scattered by a single particle can be expressed as:

$$A(q) = \int_V \rho(r) e^{-iqr} dr \tag{1}$$

where $\rho(r)$ is the electron density of the particle at distance r, q is the momentum transfer with $q = 4\pi.sin\theta/\lambda$, θ is a half of the angle between the primary beam and scattered X-rays, and λ is the wavelength of X-rays. The integration is over the entire volume of the particle.

Let us first consider a scenario where identical particles are dispersed in a deeply diluted solution. The difference in electron density ($\Delta\rho$) between the particle (ρ) and solvent (ρ_0) contributes to the scattering intensity at small angles. The absolute scattering intensity depends not only on $\Delta\rho$ but also on the geometric shape and size of the particles.

For particles with a simple geometric shape, the integral in Equation (1) can be easily solved. For instance, the scattering intensity of spherical particles is given by [3]

$$I(q) = N(\Delta\rho)^2 V^2 \left[3\frac{\sin(qR) - qr\cos(qR)}{(qR)^3} \right]^2 \tag{2}$$

where N is the total number of particles, V is the volume of an individual particle, and R is the radius of the sphere.

If the shape of particles is thin rods with a cylindrical length of L and a radius of R ($L >> R$), then the averaged scattered intensity can be expressed as [4]

$$I(q) = N(\Delta\rho)^2 V^2 \frac{1}{2} \int_0^{\pi} \left(\frac{2}{qL\cos\Theta} \right)^2 \sin^2(\frac{qL}{2}\cos\Theta)\sin\Theta d\Theta \quad (3)$$

Here Θ is the angle between the scattering vector q and the long axis of the rods.

Another example is thin disk particles. Assuming such disks have a radius of R and a thickness of d ($R >> d$), then the scattered intensity is related to the scattering vector (q) through [5]

$$I(q) = N(\Delta\rho)^2 V^2 \frac{2}{q^2 R^2} \left[1 - \frac{J_1(2qR)}{qR} \right] \quad (4)$$

where $J_1(x)$ is the first-order Bessel function.

When the shape of the particles is not known, the scattering functions in Equations (2) - (4) cannot be used. However, it turns out that the scattering function follows a universal expression in the limit of small q: [6]

$$I(q) = N(\Delta\rho)^2 V^2 \exp(-\frac{1}{3} q^2 R_g^2) \quad (5)$$

This equation is called the Guinier law, which allows determination of the radius of gyration (R_g) of particles with unknown sizes and shapes. The Guinier law can be applied only under certain conditions such as dilute and isotropic systems in the limit of small q ($q << 1/R_g$).

If the particles are correlated (e.g., particles in a concentrated solution), then the scattering intensity can be written as

$$I(q) = 4\pi \int_0^{\infty} p(r) \frac{\sin(qr)}{qr} dr \quad (6)$$

Here $p(r)$ is called the distance distribution function. For homogenous particles, $p(r)$ represents the number of distances (with a length of r) within the particles, while for heterogeneous particles this function is proportional to the number of pairs of difference electrons (separated by the length r) within the particles. The distance distribution function drops to 0 at $r \geq D$ with D being the maximum length of the scatters.

The $p(r)$ function is related to the correlation function $\gamma(r)$ by

$$p(r) = V.\gamma(r).r^2 \quad (7)$$

where $\gamma(r)$ is the average of the product of two fluctuations at distance r. Thus the scattering intensity $I(q)$ in Equation (6) can be rewritten as

$$I(q) = V \int_0^{D} 4\pi r^2 \gamma(r) \frac{\sin(qr)}{qr} dr \quad (8)$$

For an ideal two-phase system, the most important prediction from the small-angle scattering theory is that the intensity *I(q)* decreases as ~ q^{-4} at large q. This relationship is named the Porod law, which can be expressed as: [7]

$$\frac{I(q)}{Q} \rightarrow \frac{2\pi}{\phi(1-\varphi)} \frac{S}{V} \frac{1}{q^4} \tag{9}$$

Here Q is called invariant defined as the quantity obtained when the intensity *I(q)* is integrated with respect to q throughout the reciprocal space, ϕ is the volume fraction of one phase, and S and V are the surface area and volume of the phase, respectively. The Porod law can be used to evaluate the specific surface area (V/S), which has dimension of length and is related to the Porod length that is characteristic to the structure. The Porod length l_p is defined as

$$l_p = \frac{4\phi(1-\phi)V}{S} \tag{10}$$

The Porod length is also called the chord length, which is a measure of the average size of the heterogeneities in the two-phase system.

A new function, namely the chord length distribution function (*g(r)*), is related to the correlation function $\gamma(r)$ via [8]

$$g(r) = l_p.\gamma''(r) \tag{11}$$

The *g(r)* function contains very useful structural information. For instance, for a porous material the average pore size, pore-wall thickness, and interpore distance can be derived form *g(r)*). [9,10]

Small-angle scattering can also be used to characterize the fractal dimension of a two-phase system. [11,12] A fractal object retains a self-similarity under scale deformations. That is, if a part of the fractal structure is magnified, the magnified portion will look exactly like the original. For a disordered system, the fractal dimension is equal to the degree of disorder and thus can be described in terms of nonintegral numbers. [13,14] In contrast, a nonfractal object has an integral dimension D (D = 1, 2, 3).

A fractal object can be either mass or surface fractal. In the case of a mass fractal, the mass of the object (M) is related to the radius (r) as well as the mass fractal dimension (D_m) via:

$$M(r) \propto r^{D_m} \text{ or } \rho(r) \propto r^{D_m - 3} \tag{12}$$

where D_m is less than D. The mass fractal dimension is also related to the scattering intensity *I(q)*:

$$I(q) \propto q^{-D_m} \tag{13}$$

This relationship is valid for $1/R << q << 1/a$. Here R is the overall dimension of the objet ($\sim R_g$), and a is the size of the basic building block of the structure and could be as small as the size of an atom or a molecule. A double-logarithmic plot of $I(q)$ vs. q gives a straight line, whose slope is equal to the mass fractal dimension D_m.

Some objects may have a rough surface and thus exhibit fractal properties. Such an object is called a surface fractal. In this case, the scattering intensity is related to the surface fractal dimension (D_s) through

$$I(q) \propto q^{-(6-D_s)} \quad (14)$$

A log-log plot of $I(q)$ against q for the high q range will give a straight line with a slope of $6 - D_s$. For a three dimensional surface fractal, the slope is usually between -4 and -3, which corresponds to a fractal dimensional between 2 and 3.

1.2. Theoretical Basis of Wide-Angle X-Ray Scattering

Wide-angle X-ray scattering is a very useful tool for characterizing atomic structure of materials. A well-known example is X-ray diffraction in which constructive inference of X-rays from periodically arranged atoms generates sharp diffraction peaks. Therefore, X-ray diffraction is mostly used for characterizing ordered materials such as crystals. For a disordered material (e.g., glass), an oscillating diffuse scattering intensity is observed due to lack of regular arrangement of atoms in the material. The diffuse scattering provides structural information about the arrangement of atoms in disordered materials. In this section we focus on the basic principles behind the WAXS method for the characterization of disordered materials. Readers who are interested in the X-ray diffraction technique can find useful references elsewhere. [15]

The scattering method used for obtaining the atomic structure of a partially or fully disordered material is called the pair distribution function (PDF) method. This method is based on the Fourier transformation of the structure factor ($S(q)$)of the material. Considering an array of atoms with all orientation in space, the average X-ray scattering intensity is given by the Debye scattering equation: [16]

$$I(q) = \sum_m \sum_n f_m f_n \frac{\sin q r_{mn}}{q r_{mn}} \quad (15)$$

where q is the scattering vector, m and n are a pair of atoms, r_{mn} is the distance between the atom pair, and f_m and f_n are the atomic form factors of the atoms m and n, respectively. The normalized structure factor is proportional to the scattering intensity function $I(q)$.

To derive $S(q)$ from $I(q)$, one has to apply several corrections. In the case of X-ray scattering, $I(q)$ has to be first normalized according to the flux of the X-rays. Then corrections due to sample absorption, multiple scattering, X-ray polarization, Compton scattering, and atomic form factors need to be applied. Once $S(q)$ is derived, the atomic structure can be expressed in the form of atomic pair distribution function ($G(r)$): [17]

$$G(r) = 4\pi r[\rho(r) - \rho_0] = \frac{2}{\pi}\int_0^\infty q(S(q)-1)\sin(qr)dq \quad (16)$$

Here $\rho(r)$ is the microscopic pair density, and ρ_0 is the average number density. The $G(r)$ function represents deviation of the atomic pair density from the average number density caused by arrangement of atoms. It takes into account the average of every possible atomic pair in the material and thus provides information about the atomic structure.

Crystalline and amorphous materials are distinct in how the atoms are arranged (i.e., the structural ordering). In general, structural ordering of condensed matter can be separated into three length scales: short range, intermediate range and long range. [18] The short-range order has a characteristic length of 2 – 5 Å, which is observed in both crystalline and amorphous materials. The intermediate-range order (IRO) has a length scale of 5 – 20 Å, which is observed only in amorphous materials. The long-range order is beyond 20 Å, and it exists only in crystalline materials.

The intermediate-range order is unique to amorphous materials. It is associated with a special feature of scattering data of amorphous solids, i.e., the first sharp diffraction peak (FSDP). The FSDP is usually located at q = 1 - 2 $Å^{-1}$ and can be measured by WAXS. The FSDP is considered as a pseudo-Bragg peak associated with some type of quasi-periodicity in the real space. The periodicity is given by $2\pi/q_1$, where q_1 is the position of the FSDP, and the correlation length can be expressed as $2\pi/\Delta q_1$ with Δq_1 being the full width at half maximum of the FSDP. The structural origin of the IRO in amorphous solids has been a source of controversy in the past two decades, and various structural models have been proposed. [18-21]

2. SAXS/WAXS Characterization of Sol-Gel Derived Nanomaterials

2.1. Sol-Gel Derived Materials

The sol-gel processing is based on a wet-chemical technique that involves reactions of molecular precursors in solution (sol) to form an integrated network (gel). The reactions usually occur at much lower temperatures than those required for solid-state reactions and thus enable low-temperature processing of materials such as glass and ceramics. Another advantage of the sol-gel processing is generation of nanostructured materials with well-controlled morphology. For instance, colloidal silica with uniform size distribution was developed over a half century ago, [12] and materials with periodically arranged nanosized pores have been synthesized in the past two decades. [22,23]

Among all the sol-gel derived materials, silica is probably the most extensively studied. Typically, the sol-gel derived silica is in the form of colloids or porous solids. A colloid is a suspension of small particles (1nm - 1 μm in diameter) in a solvent. The particles are so small that the gravitational force is negligible when it is compared to the short-range forces such as the van der Waals and the Coulombic forces. Consequently, the colloids exhibit Brownian motion in the solution. Porous materials can be divided into two categories depending on the structure of the pores. If the pores have uniform size and shape and are arranged periodically, then this type is called an ordered porous material. Otherwise, it is called a disordered porous material.

The sol-gel derived silica is usually formed through a two-step process. The first step involves a hydrolysis reaction during which an organosilicate precursor reacts with water to form silicon hydroxide:

$$Si(OR)_4 + 4H_2O \rightarrow Si(OH)_4 + 4R(OH) \quad (17)$$

Here R represents an alkyl (e.g. CH_3, C_2H_5, etc.). The second step is a condensation reaction through which solid silica is formed:

$$Si(OH)_4 + Si(OH)_4 \rightarrow 2SiO_2 + 4H_2O \quad (18)$$

There are a couple of approaches to produce monodispersed silica spheres. A widely used method is the SFB method developed by Stober, Fink and Bohn. [24] In this method, an organosilicate precursor is dissolved in a basic solution that contains water, alcohol and ammonia. By controlling the concentrations of the constituents, the hydrolysis and condensation rates can be controlled and silica spheres with desirable sizes are obtained.

To synthesize porous silica, two major sol-gel based methods have been applied. One is the gelation-drying method. [25] Materials such as aerogel and xerogel are prepared by this approach. In this method, a gel is formed first from a sol, and then the gel is dried either by evaporating the solvent or supercritical drying. In the former case, shrinkage of the gel network typically accompanies with the drying, caused by a large capillary pressure of the solvent trapped in the gel.

As the solvent is completely evaporated, the porous network is generated and the so-formed material is called xerogel. To eliminate the shrinkage during drying, a supercritical drying method has been applied. In this approach, the gel is heated to a high temperature in an enclosed vessel so that the critical point of the liquid-gas phase boundary is passed, and then the solvent-gas phase is exchanged by another gas phase. Since no capillary pressure is involved in this process, the gel network does not shrink, and the porous material formed in this way is called aerogel. The xerogel and aerogel formed by the gelation-drying method have a wide range of pore sizes, and the pores are randomly oriented. Therefore, both materials belong to the disordered porous materials.

The second sol-gel method for producing porous silica is the self-assembly method. [22,23] In this approach, a structural-directing template (e.g., micelles) is dissolved in a sol to form ordered aggregates in the solution. Then the silica precursor fills in the space between the ordered structures and solidifies after the condensation reaction. The template is then removed later through calcination at a high temperature, which leads to the formation of porous silica.

The pores are usually very uniform in size and arranged periodically. Therefore, this type of porous silica is an ordered porous material. For instance, periodic mesoporous silica MCM-41 and SBA-15 are synthesized through this approach. [22,23]

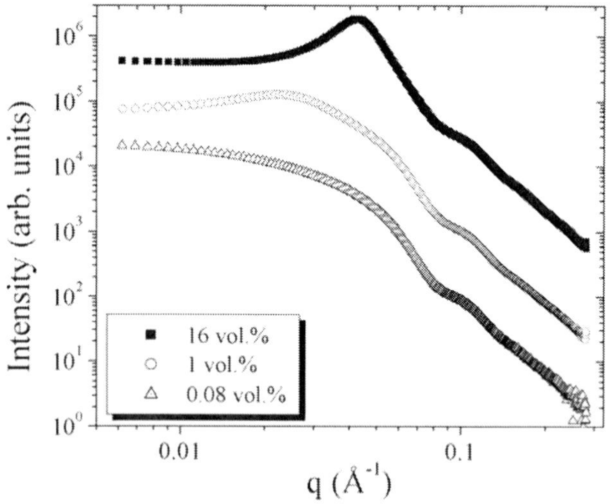

Figure 1. SAXS patterns of silica particles dispersed in water with different volume concentrations.

2.2.1. SAXS Characterization of Sol-Gel Derived Nanomaterials

SAXS has been widely used for characterizing nanostructured materials. In the case of colloidal and mesoporous silica, SAXS can be used to obtain structural information such as particle/pore size distribution, surface area, porosity, and fractal dimension. Figure 1 shows SAXS patterns of colloidal silica (Ludox® SM) with different particle volume fraction (f_V). [26] The scattering patterns contain information about shape and size of the particles (i.e., the form factor) as well as interparticle correlation (i.e., the structure factor). For the deeply diluted solution (f_V = 0.08 vol.%), the particle-particle correlation is negligible. Therefore the scattering pattern is solely due to the particle form factor.

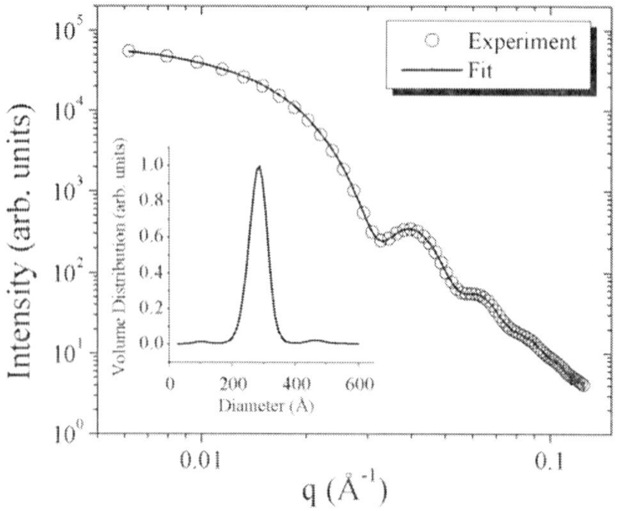

Figure 2. A fit to the SAXS pattern of a diluted silica nanoparticle solution. The derived particle size distribution profile is shown in the inset.

The observed form factor is averaged over all the particles, which may have a size distribution (i.e., polydispersity). To derive the polydispersity of suspended particles from the scattering pattern, one can fit the scattering curve by incorporating a polydispersity function (*D(R)*) into the intensity function. Thus the scattering intensity is given by [27]

$$I(q) = \int_0^\infty D(R).R^6 I_1(qR)dR \qquad (19)$$

where $I_1(qR)$ is the normalized intensity function of a particle with a radius *R*. The *D(R)* function is usually obtained through indirect Fourier transform. Figure 2 shows a fit to the scattering curve of a deeply diluted silica solution, with the inset showing the polydispersity function. The silica particles have an average diameter of 29 nm, and the full width at half maximum of the distribution is measured to be 7 nm.

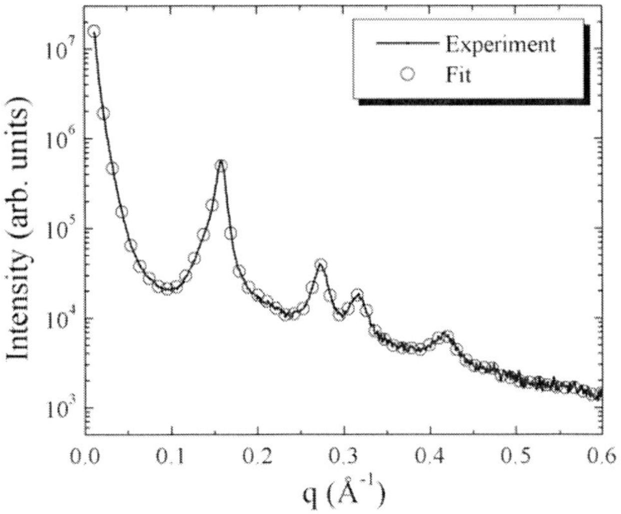

Figure 3. A fit to the SAXS pattern of a periodic mesoporous material to derive the distance distribution function *p(r)*.

For the concentrated colloidal silica (f_V = 16 vol.% in Fig. 1), its scattering pattern shows a hump at ~ 0.03 Å^{-1}, which is superimposed on the form factor of the silica particles. This hump contains information about the structure factor of the colloids. Since the structure factor is due to interparticle correlation, it is useful to introduce a real space equivalent function: the correlation function $\gamma(r)$. The $\gamma(r)$ function is the average of the product of two fluctuations at distance *r* and it is related to the scattering intensity function through Eq. (8).

In the case of an ordered porous material, $\gamma(r)$ is linked to the chord length distribution function *g(r)*, which contains important structural information such as interpore distance, pore diameter and pore-wall thickness. To derive *g(r)*, the first step is to obtain the distance distribution function (*p(r)*) from the scattering intensity profile. In principle, the *p(r)* function can be derived by applying Fourier transform directly on the measured intensity profile. However, this operation is rather difficult due to a termination effect which causes strong Fourier ripples in the *p(r)* function. One way to bypass this obstacle is to apply the indirect

Fourier transform method. The basic approach is to use a set of N cubic B-spline $\phi_n(r)$ to represent $p(r)$: [28]

$$p(r) = \sum_{n=1}^{N} c_n \phi_n(r) \tag{20}$$

where c_n are a series of constant. After applying Fourier transformation to the $\phi_n(r)$ functions, new functions $\psi_n(q)$ are generated. Thus the scattering intensity can be expressed as

$$I(q) = \sum_{n=1}^{N} c_n \psi_n(r) \tag{21}$$

The $p(r)$ function is then derived by fitting the scattering curve with a set of $\phi_n(r)$ functions.

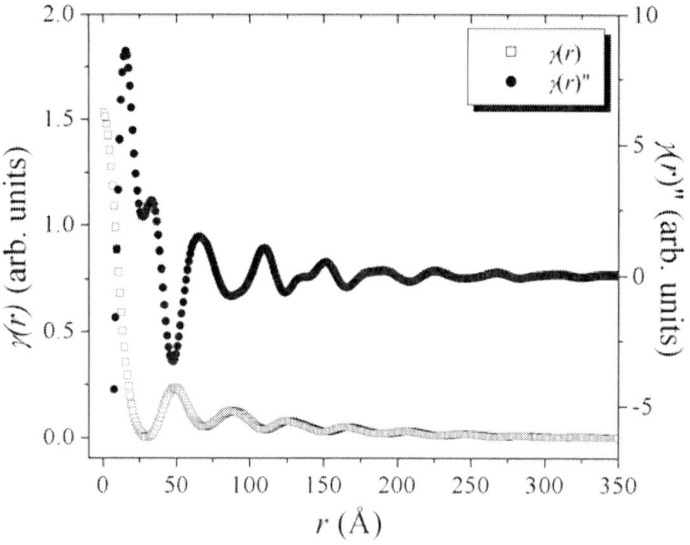

Figure 4. The correlation function $\gamma(r)$ and its second derivative $\gamma(r)''$ obtained from the correlation function $p(r)$.

Figure 3 shows the SAXS pattern of a periodic mesoporous silica sample (MCM-41). The circle symbol curve is a fit to the experimental data (solid curve) using the indirect Fourier transform method, and the $p(r)$ function can be obtained from the fit. The correlation function $(\gamma(r))$ and its second derivative $(\gamma(r)'')$ are calculated from the $p(r)$ function using Eq. (7). Figure 4 shows the $\gamma(r)$ and the $\gamma(r)''$ functions of the MCM-41 sample. The $\gamma(r)''$ function contains information about the porous structure: the second maximum corresponds to the average width of the pore, and the principle minimum after the second maximum is linked to the average interpore distance. SAXS analysis of the MCM-41 sample gives a pore width value of 3.3 ± 0.1 nm and an interpore distance of 4.7 ± 0.1 nm. In comparison, the gas adsorption experiment using N_2 isotherms at 77 K gives a pore width value of 3.2 nm, which is consistent with the SAXS result.

Since the colloidal silica and mesoporous silica are synthesized by different sol-gel methods, one would expect a difference in their surface property (e.g., surface roughness). As discussed previously, sol-gel derived silica can be regarded as fractals, and their surface property is related to the fractal dimension. For a surface fractal, its fractal dimension is between 2 and 3, which can be measured by SAXS. Figure 5 shows a double logarithmic plot of *I(q)* against *q* for colloidal silica (Ludox® TM) and periodic mesoporous silica (MCM-41). A power law relation is observed for both materials. The q range in which the pattern obeys the power law provides roughly the length scale of the fractals of interest. The slops of the straight lines are -4 for the colloidal silica and -3.7 for the periodic mesoporous silica, which were both smaller than -3, suggesting that the fractals are both surface fractals. The fractal dimension can be calculated using Eq. 14 from which a value of 2 is obtained for the colloidal silica and 2.3 for the periodic mesoporous silica. These numbers imply that the silica particles have a smooth surface while the periodic mesoporous silica has a rough pore-wall surface.

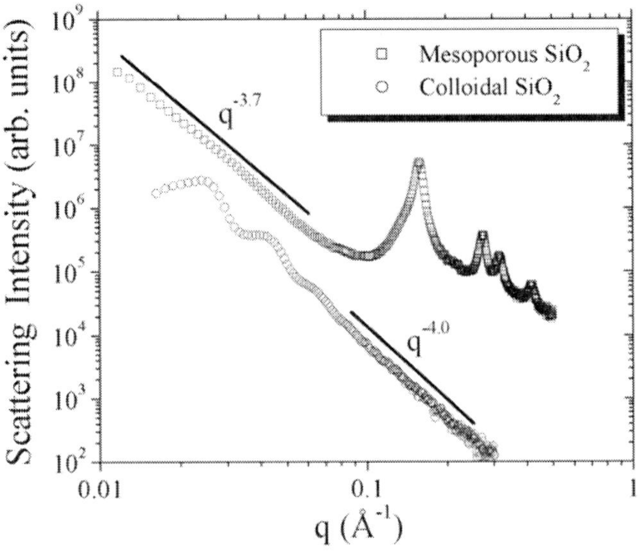

Figure 5. Comparison of the SAXS patterns of periodic mesoporous silica and colloidal silica.

2.3. WAXS Characterization of Sol-Gel Derived Nanomaterials

The sol-gel processing is a solution-based technique, and the materials produced by this approach are usually amorphous. A unique structural feature of amorphous materials is the intermediate-range order, which is regarded as the next highest structural organization beyond the short-range order. [18] The IRO is associated with a feature of diffraction data of amorphous solids, i.e., the first sharp diffraction peak (FSDP). Such a peak is located at values of scattering vector q = 1 – 2 $Å^{-1}$, depending on the material. [18] Although the IRO is observed in many amorphous solids, its origin has been a source of controversy. [18-21] Models based on interstitial voids, [18] pseudo-Bragg planes, [19] cages [20] and cation-cation correlations have been proposed. [21]. In this section, we apply WAXS to the study of prototypical sol-gel derived nanomaterials. We use this example to illustrate that WAXS is a very useful tool to extract structural information encoded in the IRO of such materials.

The materials being investigated are colloidal and periodic mesoporous silica. Their atomic structures are similar to that of a bulk silica glass prepared by the traditional high-temperature processing (i.e., the melt-quenching method). In a perfect silica glass, each oxygen atom bonds to two silicon atoms (such oxygen is called bridging oxygen), and each silicon atom bonds to four bridging oxygen atoms. However, the short-range structure of sol-gel derived silica is not exactly the same as that of the melt-quenched silica glass. It is known that in sol-gel derived silica some of the silicon atoms partially bond to defects such as terminating hydroxyl and/or alkoxy groups. [29, 30] These defects are usually incorporated into the silica network during the sol-gel processing, and they play an important role in the physical and chemical properties of sol-gel derived materials. [31, 32]

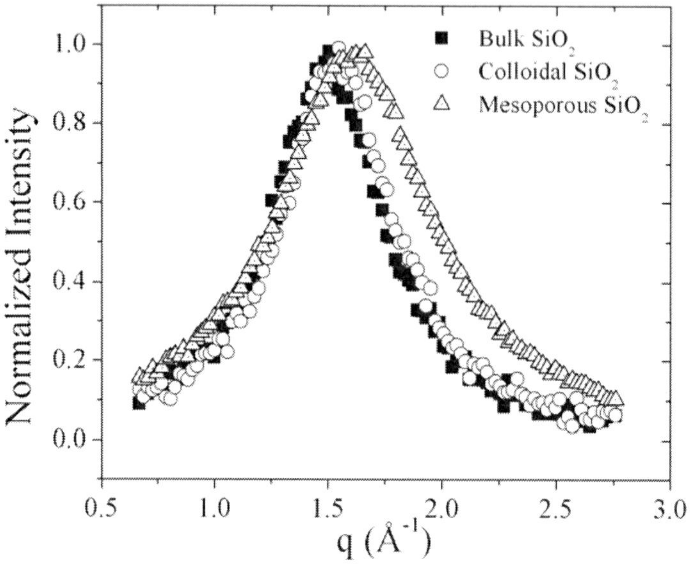

Figure 6. The first sharp diffraction peak of a bulk silica glass, colloidal silica and periodic mesoporous silica.

One would ask a question naturally: how do the defects affect the IRO of sol-gel derived nanomaterials? X-ray diffraction has been used to characterize the structure of crystalline materials for almost a century. From the position and width of the diffraction peaks, one can obtain information regarding the lattice parameters, average crystal sizes, defects, stresses, etc.. [33] Similar idea could also be applied to amorphous solids if the FSDP is selected for analysis. [34] The FSDP is considerably narrower than the other peaks in the structure factor of amorphous solids but much wider than the diffraction peaks of crystals. The broad FSDP makes it not very sensitive to the factors that influent the diffraction peaks of crystals. Therefore, the idea of using FSDP as a defect and strain sensor has long been overlooked. Figure 6 shows the FSDP of a bulk silica glass (square symbol), colloidal silica (circles), and periodic mesoporous silica (triangles). Compared to the bulk glass, the FSDP of the colloidal silica is a little wider and slightly shifts toward a higher *q*. In contrast, the FSDP of the periodic mesoporous silica is much wider and shifts to a much higher q than the other two. This result implies that the IRO of sol-gel derived materials is different from that of the corresponding bulk glasses and it also depends on specific nanoscale morphology. The challenge of applying WAXS to study the IRO of amorphous solids lies in the fact that there is only one FSDP for any material. Therefore, it is impossible to adopt the multiple-peak analysis method used for crystalline materials. One way to bypass the problem is to apply external stimuli to modify the IRO. An example of the stimuli is heat treatment. By tracking the changes of the FSDP induced by thermal annealing, it is possible to study the factors that influent the IRO. We demonstrate applying such an approach to decode structural properties of sol-gel derived nanostructured silica. [34]

The material selected for this study is periodic mesoporous silica (MCM-41), which was synthesized through the sol-gel processing based on established procedures. [35] The MCM-41 powder was first pressed into pellets and then annealed at different temperatures (500 °C to 1100 °C) for 1.5 hours. The effect of thermal annealing on the FSDP of MCM-41 was studied by the WAXS. Figure 7 shows the center position and the width (i.e., full width at half maximum) of the FSDP as a function of annealing temperature. As the annealing temperature was increased from 500°C to 700°C, the FSDP slightly shifted from 1.615 $Å^{-1}$ to 1.605 $Å^{-1}$, respectively. Further increases in the temperature caused the FSDP position to shift toward lower values at a faster pace. When the annealing temperature reached 1100 °C, the peak shifted to 1.509 $Å^{-1}$, which was very close to that of bulk silica glass. [19] Two straight lines were used to fit the FSDP position profiles in the above-mentioned temperature regions. The interception gave a critical temperature of 700 °C ± 50 °C, which corresponded to initiation of significant changes in the FSDP position. Similar to the position profile, the width profile (solid circles) can also be separated into two temperature regions. From 500 °C to 950 °C, the peak width slightly decreased from 0.85 $Å^{-1}$ to 0.79 $Å^{-1}$, respectively. Further increases in the temperature resulted in sharp decreases in the width, which reached 0.58 $Å^{-1}$ at 1100 °C. A critical temperature of 950 °C ± 50 °C can be derived using the method described previously, and it corresponds to initiation of drastic changes in the FSDP width.

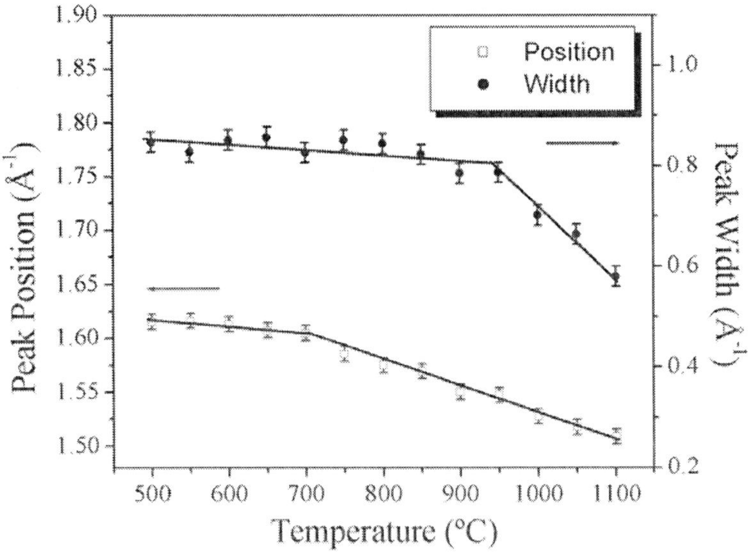

Figure 7. Temperature dependences of the position (hollow square) and width (filled circle) of the first sharp diffraction peak of MCM-41.

Two critical temperatures have been observed for the MCM-41: 700°C and 950 °C. The former corresponds to initiation of drastic changes in the position of the FSDP, while the latter is linked to initiation of significant changes in the width of the FSDP. Generally speaking, X-ray diffraction peaks of single-phase polycrystals can be affected by a few factors. For instance, uniform strain (caused by defects or external stresses) may shift the diffraction peaks, while small grains, defective grains, or non-uniform strain may lead to broadening of the peaks. [33] Similar arguments could also be applied to the FSDP of amorphous solids. In the case of MCM-41, its FSDP shifted toward the bulk silica glass position as the annealing temperature was increased, suggesting annihilation of defects since no external stresses were applied. The defects generate compressive micro-strain on the intermediate-range structural units. Subsequent thermal annealing reduces the defect concentration and thus relief the strain, causing the FSDP to shift toward lower q values, as observed experimentally.

A question remains to be answered: what are the possible defects? In calcined sol-gel derived silica, the dominant defects are terminating hydroxyls (i.e., silanol groups), [36,37] which can react with each other though condensation reactions. It has been well established that vicinal silanol groups start to condense at temperatures 200 °C – 300 °C, while isolated silanol groups condense at much higher temperatures (> 650 °C) due to limited mobility of the isolated hydroxyls. [38-40] Since the MCM-41 was calcined at 500 °C, most of the vicinal silanol groups should be eliminated after calcination. The leftovers are probably the isolated silanol groups. It is interesting to note that the minimum reaction temperature (~ 650 °C) for the isolated silanol groups is in line with the critical temperature (~ 700°) derived from the temperature dependency of the FSDP position. Therefore, the observed shift of the FSDP with temperature is very likely caused by annihilation of defects such as the isolated silanol groups.

The other critical temperature (i.e., 950 °C) is associated with changes in the width of the FSDP. Broadening of the FSDP may be caused by small/defective intermediate structural units and/or non-uniform residual strain. The "size" of an intermediate-range structural unit is related to the correlation length, which is expressed as $2\pi/\Delta q$ with Δq being the width of the FSDP (assuming no other factors causing broadening of the FSDP). To determine which factor plays a dominant role in the annealing process, we conducted a series of experiments on two related amorphous silica: periodic mesoporous silica (SBA-15) and non-mesoporous sol-gel derived silica. The SBA-15 was synthesized using a well-established method, [23] and the sol-gel derived non-mesoporous silica was synthesized with the same method except that no structural-directing surfactants were used. SAXS analysis of the SBA-15 revealed a 2D hexagonal cylindrical pore structure (similar to the MCM-41) with an average interpore distance of 10.4 nm and an average pore diameter of 5.3 nm. In comparison, the SAXS data collected from the non-surfactant silica did not show any mesoporous structure. The specific surface areas of the SBA-15 and the non-mesoporous silica were analyzed using N_2 gas as adsorptive at ~ 77 K (Tristar 3020, Microertics Inc.), and the calculated Brunauer-Emmett-Teller specific surface areas were 511 ± 26 m^2/g and 35 ± 1 m^2/g, respectively. WAXS analysis shows that these two materials have almost the same FSDP position but significantly different peak widths (i.e., 8.5 ± 0.2 $Å^{-1}$ for the SBA-15 and 6.4 ± 0.2 $Å^{-1}$ for the non-mesoporous silica). This discrepancy suggests that the broader FSDP of the SBA-15 is related neither to the correlation length (i.e., the size) nor to the defects inside the intermediate-range structure units, since both materials have the same composition and thermal history as well as identical FSDP positions. The same conclusion could also be applied to the MCM-41 since both the MCM-4 and the SBA-15 have the same type of pore structure and almost identical FSDP width (after calcination at 500 °C).

The only candidate left is non-uniform residual strain, which typical extends over a length scale larger than the size of grains (e.g., the size of intermediate structural units) and is classified as macro-strain. The strain could be generated during the calcination (at 500 °C) when condensation of vicinal silanol groups occurs on the surface of the PMS. It is known that dehydroxylation on the silica surface at temperatures above 200 °C creates strained silicon species. [40,41] Such surface strain could extend over a large area on the surface and is affected by the curvature of the surface [41]. Consequently, the FSDP is broadened when the residual strain acting on the pore walls of the PMS.

The WAXS study of thermally annealed periodic mesoporous silica revealed two critical temperatures. One (c.a. 700 °C) is related to reduction of certain defects (such as the isolated silanol groups) through thermally activated diffusion and subsequential chemical reaction. The other (c.a. 950 °C) is linked to viscous flow, which degrades the mesoporous structures and relieves the residual macro-strain in the pore walls. The residual macro-strain seems to be inherent to the mesoporous structures and thus has implications for the structural properties of mesoporous materials. This study demonstrates that X-ray scattering is a very useful tool to characterize residual strain of different origins in sol-gel derived nanomaterials.

3. Recent Advancement in Synchrotron-Based SAXS/WAXS

Synchrotron radiation originates from acceleration of ultrarelativistic charge particles. In the case of a synchrotron ring, thermoelectrons are accelerated by a linear accelerator and become ultrarelativistic electrons. These electrons are then ejected into the storage ring where magnetic devices such as bending magnets or insertion devices are placed to bend the electron beam. At the site where the electron beam is bent, radiation is emitted in the form of electromagnetic waves (e.g., X-rays). Compared to traditional X-rays sources, synchrotron X-rays have many advantages such as high intensity, broad spectral range, high brightness, small angle divergence, small source size, temporal and spatial coherence, pulsed time structure, and polarization. These superior properties enable new X-ray techniques that are very useful for characterizing sol-gel derived nanomaterials.

One of the synchrotron-based techniques is the grazing incidence small-angle X-ray scattering (GISAXS). It is a combination of X-ray reflectivity with SAXS. X-ray reflectivity is a useful technique for characterizing thin films (< 200 nm thick). Information such as film thickness and surface and interface roughness can be obtained. Therefore, GISAXS is particularly useful for characterizing nanostructured thin films. Platschek et al. applied in situ GISXAS technique to study growth of periodic mesoporous silica confined in a tubular structure. [42] These authors observed various GISAXS patterns that correspond to columnar hexagonal, circular hexagonal and tubular lamellar structures, depending on the structure-directing agents for forming the mesoporous silica. They also observed phase transformation during the growth of the mesoporous silica from which the formation mechanism of the mesostructured silica under confined geometry was proposed. This study demonstrates the usefulness of synchrotron-based GISAXS for the characterization of mesostructured thin films.

Another synchrotron-based technique is the rapid-acquisition pair distribution function (RA-PDF). This method uses high-energy X-rays in combination with an area detector to collect wide-angle scattering patterns from powder samples. The high energy X-rays enable collection of a large range of scattering vector in the reciprocal space and thus can be used to derive high-resolution pair distribution function. The high flux of synchrotron X-rays in conjunction with the area detector also significantly shortens the data acquisition time. Pauly et al. reported application of the RA-PDF to characterize the short-range order of periodic mesoporous silica. [43] These authors compared the pair distribution function of bulk silica glass with that of the mesoporous silica and discovered that the average Si-O bond length of the latter is 0.02 – 0.03 Å longer. They attribute the lengthening of the bond to the porosity of the silica glass. Their study shows that sol-gel derived nanomaterials have complex atomic structure that could affect the physical and chemical properties of these materials.

Conclusions

In this chapter, we reviewed briefly the SAXS and WAXS techniques and address various capabilities of these techniques for use in materials characterization. Specific examples are provided in connection with applications of these techniques to sol-gel derived nanomaterials such as colloidal and periodic mesoporous silica. Derivation of structural information, including particle size and distribution, average pore size and interpore distance, fractal dimension, short- and intermediate-range order, has been demonstrated. Existence of residual strain in the pore walls of periodic mesoporous materials has been inferred from a WAXS study. Publically accessible synchrotron sources provide new opportunities for characterizing nanomaterials, and two of such examples were demonstrated for the studies of sol-gel derived nanomaterials. More new experimental techniques based on small and wide angle X-ray scattering are emerging when the unique properties of synchrotron (e.g., pulsed time structure, coherence, polarization) are considered.

Acknowledgments

A part of the research work was conducted at the Advanced Photon Source (APS) of Argonne National Laboratory. Use of the APS was supported by the U. S. Department of Energy, Office of Science, Office of Basic Energy Sciences, under Contract No. DE-AC02-06CH11357. This work was supported by NSF under Grant No. DMR-0906825.

References

[1] M. Laue, *Ann. Phys.* 41, 989 (1913).
[2] W. H. Bragg, *Science* 40, 795 (1914).
[3] L. Rayleigh, *Proceedings of the Royal Society of London.* Series A 84, 25 (1910).

[4] T. Neugebauer, *Annalen der Physik* 434, 509 (1943).
[5] O. Kratky and G. Porod, *J. Colloid Interface Sci.* 4, 35 (1949).
[6] A. Guinier, *Ann. Phys.* 12, 161 (1939).
[7] G. Porod, *Kolloid Zeitschrift* 124, 83 (1951).
[8] B. Smarsly, M. Groenewolt, and M. Antonietti, edited by N. Stribeck and B. Smarsly (Springer, New York, 2005), p. 105-113.
[9] B. Smarsly, M. Antonietti, and T. Wolff, *J. Chem. Phys.* 116, 2618 (2002).
[10] W. Gille, O. Kabisch, S. Reichl, D. Enke, D. Furst, and F. Janowski, *Microporous Mesoporous Mater.* 54, 145 (2002).
[11] C. Aubert and D. S. Cannell, *Phys. Rev. Lett.* 56, 738 (1986).
[12] H. E. Bergna, in *Collid Chemistry of Silica - an Overview*, 1994, p. 1-47.
[13] B. B. Mandelbrot and C. J. G. Evertsz, *Nature* 348, 143 (1990).
[14] J. Rarity, *Nature* 339, 340 (1989).
[15] B. E. Warren, *X-ray Diffraction* (Addison-Wesley, Reading, MA, 1969).
[16] P. Debye, *Annalen der Physik* 351, 809 (1915).
[17] F. Zernike and J. A. Prins, *Zeitschrift für Physik A Hadrons and Nuclei* 41, 184 (1927).
[18] S. R. Elliott, *Nature* 354, 445 (1991).
[19] P. H. Gaskell and D. J. Wallis, *Phys. Rev. Lett.* 76, 66 (1996).
[20] A. C. Wright, *J. Non-Cryst. Solids* 179, 84 (1994).
[21] H. E. Fischer and et al., *Reports on Progress in Physics* 69, 233 (2006).
[22] C. T. Kresge, M. E. Leonowicz, W. J. Roth, J. C. Vartuli, and J. S. Beck, *Nature* 359, 710 (1992).
[23] D. Y. Zhao, J. L. Feng, Q. S. Huo, N. Melosh, G. H. Fredrickson, B. F. Chmelka, and G. D. Stucky, *Science* 279, 548 (1998).
[24] W. Stober, A. Fink, and E. Bohn, *J. Colloid Interface Sci.* 26, 62 (1968).
[25] C. J. Brinker and G. W. Scherer, *Sol-Gel Science : The Physics and Chemistry of Sol-Gel Processing* (Academic Press, San Diego, 1990), p. 1-18.
[26] G. Chen, W. H. Yu, D. Singh, D. Cookson, and J. Routbort, *Journal of Nanoparticle Research* 10, 1109 (2008).
[27] O. Glatter, *J. Appl. Crystallogr.* 13, 7 (1980).
[28] B. Weyerich, J. Brunner-Popela, and O. Glatter, J. Appl. Crystallogr. 32, 197 (1999).
[29] J. Trebosc, J. W. Wiench, S. Huh, V. S. Y. Lin, and M. Pruski, *J. Am. Chem. Soc.* 127, 3057 (2005).
[30] M. T. J. Keene, R. D. M. Gougeon, R. Denoyel, R. K. Harris, J. Rouquerol, and P. L. Llewellyn, *J. Mater. Chem.* 9, 2843 (1999).
[31] F. Q. Zhang, Y. Yan, H. F. Yang, Y. Meng, C. Z. Yu, B. Tu, and D. Y. Zhao, *J. Phys. Chem.* B 109, 8723 (2005).
[32] M. Broyer, S. Valange, J. P. Bellat, O. Bertrand, G. Weber, and Z. Gabelica, *Langmuir* 18, 5083 (2002).
[33] R. Jenkins and R. L. Snyder, in *Introduction to X-ray Powder Diffractometry* (Wiley, New York, 1996), p. 47-95.
[34] G. Chen and C. Wan, *Appl. Phys. Lett.* 96, 141906 (2010).
[35] M. Kruk, M. Jaroniec, and A. Sayari, *J. Phys. Chem.* B 101, 583 (1997).
[36] L. L. Hench and J. K. West, *Chem. Rev.* 90, 33 (1990).
[37] J. C. Ro and I. J. Chung, *J. Non-Cryst. Solids* 130, 8 (1991).
[38] B. A. Morrow and I. A. Cody, *J. Phys. Chem.* 80, 1998 (1976).

[39] J. B. Peri, *J. Phys. Chem.* 70, 2937 (1966).
[40] I. S. Chuang and G. E. Maciel, *J. Phys. Chem.* B 101, 3052 (1997).
[41] C. J. Brinker and G. W. Scherer, in *Sol-gel Science : the Physics and Chemistry of Sol-gel Processing* (Academic Press, Boston, 1990), p. 617-672.
[42] B. Platschek, R. Kohn, M. Doblinger, and T. Bein, *Langmuir* 24, 5018 (2008).
[43] T. R. Pauly, V. Petkov, Y. Liu, S. J. L. Billinge, and T. J. Pinnavaia, *J. Am. Chem. Soc.* 124, 97 (2001).

INDEX

D

E

F

G

M

N

O

P

Q

R

S

T

U

V

W

X

Y

Z